TWO-PHASE FLOW IN COMPLEX SYSTEMS

TWO-PHASE FLOW IN COMPLEX SYSTEMS

SALOMON LEVY

A Wiley-Interscience Publication
JOHN WILEY & SONS, INC.
New York / Chichester / Weinheim / Brisbane / Singapore / Toronto

This book is printed on acid-free paper. ∞

Copyright © 1999 by John Wiley & Sons. All rights reserved.

Published simultaneously in Canada.

No part of this publication may be reproduced, stored in a retrieval system or transmitted in any form or by any means, electronic, mechanical, photocopying, recording, scanning or otherwise, except as permitted under Sections 107 or 108 of the 1976 United States Copyright Act, without either the prior written permission of the Publisher, or authorization through payment of the appropriate per-copy fee to the Copyright Clearance Center, 222 Rosewood Drive, Danvers, MA 01923, (978) 750-8400, fax (978) 750-4744. Requests to the Publisher for permission should be addressed to the Permissions Department, John Wiley & Sons, Inc., 605 Third Avenue, New York, NY 10158-0012, (212) 850-6011, fax (212) 850-6008, E-Mail: PERMREQ @ WILEY.COM.

This publication is designed to provide accurate and authoritative information in regard to the subject matter covered. It is sold with the understanding that the publisher is not engaged in rendering professional services. If professional advice or other expert assistance is required, the services of a competent professional person should be sought.

Library of Congress Cataloging-in-Publication Data

Levy, Salomon.
    Two-phase flow in complex systems / by Salomon Levy.
       p.   cm.
    ISBN 0-471-32967-3 (alk. paper)
    1. Two-phase flow. 2. Coupled problems (Complex systems)
I. Title.
TA357.5.M84L49   1999   98-53675
532'.051—dc21

Printed in the United States of America.

10 9 8 7 6 5 4 3 2 1

CONTENTS

PREFACE   xiii

1   PROGRAM FOR TESTING AND ANALYSIS OF TWO-PHASE FLOW IN COMPLEX SYSTEMS   1

    1.1   Introduction / 1
    1.2   Large Break in Pressurized Water Reactor / 4
    1.3   Alternative Test and Analysis Programs / 11
    1.4   Elements of an Integrated Testing and Analysis Program / 13
    1.5   Complex System Characteristics / 16
    1.6   Scenarios / 16
    1.7   Phenomena Importance and Ranking / 17
    1.8   Test Matrix and Plan / 23
    1.9   Summary / 25
         References / 27

2   TESTING TWO-PHASE FLOW IN COMPLEX SYSTEMS   28

    2.1   Introduction to Scaling of Tests for Complex Systems / 28
    2.2   Guidelines for Scaling Complex Systems / 29
         2.2.1   Fluid Properties and Initial Conditions / 29
         2.2.2   Geometric Scaling / 30
         2.2.3   Scaling Interactions / 32
         2.2.4   Typical Scaled Integral Facility: Semiscale / 32
    2.3   Hierarchical Two-Tiered Scaling of Complex Systems / 34

vi CONTENTS

      2.3.1   Hierarchy for Volume, Spatial, and Temporal Scales / 36
      2.3.2   Dimensionless Scaling Groupings / 38
      2.3.3   Scaling Hierarchy / 39
  2.4 Illustrations of Top-Down Scaling / 40
      2.4.1   Top-Down Scaling of Spatially Uniform System / 40
      2.4.2   Top-Down Scaling during Reactor Coolant Blowdown Phase / 43
      2.4.3   Top-Down Scaling of Vapor Volume Fraction during Reactor Blowdown / 45
      2.4.4   Top-Down Scaling of Spatially Uniform Containment / 47
      2.4.5   Top-Down Scaling of Fluid Transfer between Volumes / 48
      2.4.6   Top-Down Scaling of Single-Phase Natural Circulation Systems / 51
      2.4.7   Top-Down Scaling of Two-Phase Natural Circulation Systems / 54
            2.4.7.1   Homogeneous Natural Circulation / 54
            2.4.7.2   Steady-State Two-Phase Homogeneous Natural Circulation / 55
            2.4.7.3   Nonhomogeneous Two-Phase Natural Circulation / 56
      2.4.8   Top-Down Scaling of Catastrophe Functions / 58
  2.5 Illustrations of Bottom-Up Scaling / 59
      2.5.1   Bottom-Up Scaling of Peak Fuel Cladding Temperature during a LOCA from Differential Equations / 60
      2.5.2   Bottom-Up Scaling of Peak Fuel Cladding Temperature during a LOCA from Lumped Models / 62
      2.5.3   Bottom-Up Scaling of Containment Phenomena / 68
            2.5.3.1   Description of Passive Nuclear Containment Structures / 68
            2.5.3.2   Nondimensional Groupings for Buoyant Entrainment / 72
            2.5.3.3   Scaling for Circular Buoyant Jet / 73
            2.5.3.4   Scaling for Circular Buoyant Plumes / 74
            2.5.3.5   Scaling for Buoyant Boundary Layers / 75
            2.5.3.6   Scaling Criteria for Buoyant Jet Transition and Ambient Stratification / 76
      2.5.4   Summary: Bottom-Up Scaling / 76
  2.6 Scaling Biases and Distortions / 80
      2.6.1   Heat Loss Bias / 80
      2.6.1   2.6.2   Flow Pattern Bias / 81

2.7 Summary / 83
     References / 84

3 SYSTEM COMPUTER CODES   86

3.1 Introduction / 86
3.2 Conservation Laws in One-Dimensional Two-Phase Flow with
     No Interface Exchanges / 89
     3.2.1 Conservation of Mass / 90
     3.2.2 Conservation of Momentum / 91
     3.2.3 Conservation of Energy / 92
3.3 Discussion of One-Dimensional Two-Phase
     Conservation Laws / 93
     3.3.1 Steady-State Flow / 93
     3.3.2 The Many Densities of Two-Phase Flow / 94
     3.3.3 Nondimensional Groupings / 95
3.4 Homogeneous/Uniform Property Flow / 98
     3.4.1 Homogeneous Flow Conservation Equations / 98
     3.4.2 Closure Laws for Homogeneous Flow / 99
     3.4.3 Homogeneous Viscosity and Thermal Conductivity / 104
            3.4.3.1 Homogeneous Two-Phase Viscosity / 104
            3.4.3.2 Homogeneous Thermal Conductivity / 106
     3.4.4 Summary: Homogeneous/Uniform Property Flow / 106
3.5 Separated or Mixture Two-Phase Flow Models
     with No Interface Exchange / 107
     3.5.1 Single-Phase Models / 108
     3.5.2 Martinelli Separated Model / 109
            3.5.2.1 Lockhart–Martinelli Model / 109
            3.5.2.2 Martinelli–Nelson Model / 113
     3.5.3 Other Separated Flow Models
            with No Interface Exchange / 117
     3.5.4 Drift Flux Model / 118
3.6 Two-Fluid Models Including Interface Exchange / 124
     3.6.1 Two-Fluid Model Conservation Laws with
            Interface Exchange / 125
     3.6.2 Flow Pattern Maps for Computer System Codes / 128
     3.6.3 Closure Laws for Two-Fluid Models with
            Interfacial Exchange / 131
            3.6.3.1 Interface Shear / 132
            3.6.3.2 Wall Shear / 136
            3.6.3.3 Interfacial and Wall Heat Transfer / 138

3.7 Other Aspects of System Computer Codes / 143
    3.7.1 Numerical Solution of System Computer Codes / 143
    3.7.2 Nodalization / 146
    3.7.3 Subchannel Representation / 147
    3.7.4 Validation of System Computer Codes / 153
3.8 Summary / 155
References / 156

4 FLOW PATTERNS   159

4.1 Introduction / 159
4.2 Description of Flow Patterns / 160
4.3 Bubble Flow / 162
    4.3.1 Types of Bubble Flow / 162
    4.3.2 Single-Bubble Rise in Infinite Medium / 164
    4.3.3 Analysis of Single-Bubble Rise in Infinite Medium / 167
    4.3.4 One-Dimensional Frictionless Bubble Flow Analysis / 169
    4.3.5 One-Dimensional Bubble Flow Analysis with Wall Friction / 174
    4.3.6 Bubble Flow Friction Losses / 176
    4.3.7 Bubble Distribution and Interfacial Area in Bubble Flow / 177
    4.3.8 Multidimensional Bubble Flow and Analysis / 182
    4.3.9 Transition from Bubble Flow to Slug Flow / 187
    4.3.10 Summary: Bubble Flow / 190
4.4 Slug Flow / 192
    4.4.1 Velocity of a Single Slug in Vertical Flow / 194
    4.4.2 Vertical Slug Flow Analysis and Tests with No Wall Friction / 195
    4.4.3 Wall Friction in Vertical Slug Flow / 197
    4.4.4 Interfacial Area and Shear in Frictionless Vertical Slug Flow / 198
    4.4.5 Interfacial Area and Shear in Vertical Slug Flow with Wall Friction / 202
    4.4.6 Horizontal Slug Flow Analysis and Tests / 202
    4.4.7 Transition from Vertical Slug to Annular Flow / 207
    4.4.8 Transition from Horizontal Slug to Intermittent or Annular Flow / 209
    4.4.9 Severe or Terrain Slugging / 211
    4.4.10 Summary: Slug Flow / 214

4.5 Horizontal Stratified Flow / 215
    4.5.1 Analysis of Smooth Stratified Flow / 215
    4.5.2 Interfacial Waves and Friction during Stratified Flow / 218
    4.5.3 Summary: Horizontal Stratified Flow / 221
4.6 Annular Flow / 221
    4.6.1 Types of Annular Flow Patterns / 221
    4.6.2 Two-Dimensional Analysis of Ideal Annular Flow / 222
    4.6.3 Modification of Ideal Two-Dimensional Annular Flow Model / 226
    4.6.4 Two-Dimensional Analysis of Annular Flow with a Two-Layer Liquid Film / 227
    4.6.5 Waves and Interfacial Friction Factor in Annular Flow / 230
        4.6.5.1 Ripple Waves / 231
        4.6.5.2 Disturbance Waves / 232
        4.6.5.3 Empirical Correlations for Interfacial Friction Factor for Vertical Annular Flow / 235
        4.6.5.4 Flooding Waves and Their Interfacial Friction Factor / 237
    4.6.6 Drift Flux Model for Annular Flow / 238
    4.6.7 Gas Core with Entrained Liquid in Annular Flow / 239
        4.6.7.1 Liquid Droplet Characteristics / 241
        4.6.7.2 Liquid Droplet Drag Coefficient / 242
        4.6.7.3 Separate Field Momentum Equation for the Liquid Droplets / 244
        4.6.7.4 Entrained Liquid Rate and Entrainment and Deposition Rates / 245
        4.6.7.5 Closure Equations for Annular Flow with Entrainment / 248
    4.6.8 Heat Transfer in Annular Two-Phase Flow / 252
    4.6.9 Transition from Annular to Mist Flow / 253
    4.6.10 Summary: Annular Flow / 254
4.7 Mist/Dispersed Liquid Droplet Flow / 255
4.8 Summary / 258
    References / 258

5 LIMITING MECHANISM OF COUNTERCURRENT FLOODING   264

5.1 Introduction / 264
5.2 Countercurrent Flooding in Simplified Geometries / 266
    5.2.1 Experimental Setups / 266
    5.2.2 Principal Empirical Correlations in Tube Geometries / 268

5.2.3 Flooding Analysis in Tube Geometries / 272
    5.2.3.1 Mechanism of Kinematic Waves / 273
    5.2.3.2 Mechanism of Sudden Change in Wave Motion / 275
    5.2.3.3 Mechanism of Upward Liquid-Film Flow / 276
    5.2.3.4 Mechanism of Droplet Entrainment / 281
    5.2.3.5 Discussion of Flooding Mechanisms / 282
    5.2.3.6 Flooding in Annular Geometries / 283
 5.3 Countercurrent Flooding for Nonequilibrium Thermal Conditions and More Complex Geometries / 287
  5.3.1 Nonequilibrium Thermal Conditions / 287
  5.3.2 Boiling Water Reactor Upper Plenum Reflooding / 287
  5.3.3 Upper Plenum Injection in Pressurized Water Reactors / 289
  5.3.4 Behavior of Core Cooling Injection Ports / 290
  5.3.5 Injection in Downcomer of Pressurized Water Reactors / 291
  5.3.6 Prediction of Countercurrent Flooding in Complex System Computer Codes / 293
 5.4 Summary / 295
  References / 297

6 CRITICAL FLOW   300

 6.1 Introduction / 300
 6.2 Single-Phase Flow / 301
  6.2.1 Propagation of Disturbances in a Compressible Gas / 301
  6.2.2 Critical Flow / 303
  6.2.3 Nonideal Compressible Gas Flow / 305
    6.2.3.1 Critical Flow with Friction / 305
    6.2.3.2 Nonideal Area and Gas Conditions / 306
  6.2.4 Summary: Single-Phase Critical Flow / 310
 6.3 Two-Phase Critical Flow Models / 310
  6.3.1 Homogeneous Equilibrium Critical Flow Models / 311
  6.3.2 Moody Critical Flow Model / 318
  6.3.3 Other Separated Critical Flow Models / 324
 6.4 Nonthermal Equilibrium Critical Flow Models / 326
  6.4.1 Frozen Homogeneous Flow / 327
  6.4.2 Henry–Fauske Nonequilibrium Model / 329
  6.4.3 Water Decompression or Pressure Undershoot Nonequilibrium Model / 331
  6.4.4 Other Nonequilibrium Critical Flow Models / 333
 6.5 Richter Critical Flow Model / 333

CONTENTS   xi

    6.6   Critical Flow Models in System Computer Codes / 337
    6.7   Applications of Critical Flow Models / 340
           6.7.1   Critical Flow through Pipe Cracks / 341
           6.7.2   Critical Flow through a Small Break
                    in a Horizontal Pipe / 347
    6.8   Summary / 351
           References / 353

7   THE FUTURE OF TWO-PHASE FLOW IN COMPLEX SYSTEMS   356

    7.1   The Players / 356
    7.2   The Complex Systems / 358
    7.3   Key Issues in Two-Phase Flow / 360
    7.4   Key Issues in Complex System Computer Codes / 362
    7.5   The Future of Two-Phase Flow in Complex Systems / 364
           References / 368

8   OTHER TWO-PHASE-FLOW COMPLEX SYSTEMS   369

    8.1   Introduction / 369
    8.2   Global Climate Change and Global Warming / 370
    8.3   Global Average Temperatures / 370
    8.4   Global Climate System / 371
    8.5   Sun / 371
           8.5.1   Sun Influence over Climate / 373
           8.5.2   Variations in Sun Output Energy / 375
           8.5.3   Uniform Net Radiation: Single-Temperature Model / 376
    8.6   Atmosphere / 379
           8.6.1   Atmosphere Components / 379
           8.6.2   Simplified Treatment of Dry Air Vertical Temperature / 380
           8.6.3   Simplified Treatment of Vertical Movement
                    of Air with Moisture / 384
           8.6.4   Clouds / 386
           8.6.5   Aerosols / 388
           8.6.6   Atmosphere Circulation / 389
           8.6.7   Atmosphere General Circulation Models / 393
    8.7   Earth / 395
           8.7.1   Oceans / 395
           8.7.2   Snow and Ice / 401
           8.7.3   Land Surface / 402
    8.8   Greenhouse Gases / 406

8.8.1 Carbon Dioxide / 406
8.8.2 Other Greenhouse Gases / 409
8.8.3 Radiation Budget / 411
8.8.4 Radiative Forcing of Greenhouse Gases / 412
8.8.5 Global Warming Projections / 413
8.9 Summary / 416
8.9.1 General Comments / 416
8.9.2 Comments about International Climate Change Program / 416
8.9.3 An Outsider's Comments / 417
References / 419

INDEX   **421**

PREFACE

This book was started in the early 1960s while I was teaching an early bird course at Santa Clara University. At that time, the subject of two-phase flow and heat transfer was of great importance for nuclear power reactors cooled by light water; however, it was not covered in any dedicated textbook, and detailed notes had to be prepared ahead of the lectures. When the course was over, I began to convert the notes into a book but abandoned the project due to a short hiatus of 30 years in management assignments. In the meantime, a variety of two-phase-flow textbooks have been published (References P-1 to P-8) and even a few handbooks [P-9, P-10].

While preparing for retirement, the original notes resurfaced, and to my surprise, I found that my perspectives about two-phase flow had changed because over the years, my focus had shifted toward the testing and analysis of complex systems. Complex systems such as water-cooled nuclear power plants involve a large number of components and phenomena, and the interest is in testing and analyzing their simultaneous behavior and interactions as well as in the study of components and phenomena one at a time. That perspective of overall system performance was lacking in the original notes, and justifiably so. To date, it has continued to receive limited or no coverage in recent two-phase-flow treatises. With some trepidation, I decided to tackle the subject of two-phase flow in complex systems in this book. This approach is different from that of textbooks currently available in that it emphasizes the needs of the overall system rather than the basic conservation equations and the modeling of separate phenomena.

Because the understanding of two-phase-flow problems in complex systems requires extensive testing and analysis, Chapter 1 is devoted to formulating an integrated program of experiments and analytical tools for complex systems. It is based primarily on the organized, systematic, and effective approach developed through the use of code scaling, applicability, and uncertainty (CSAU) evaluation methodology [P-11] in which I was fortunate to be a participant. As was done for that study, in Chapter 1 I encourage the use of a panel of experts to validate the proposed integrated program and to avoid the past pitfalls of overfocusing on phenomena familiar to investigators or not overly relevant to the performance of a complex system. Simply stated, the objective of Chapter 1 is to make the most effective use of available resources to study a complex system.

Since tests are the most important element to an understanding of two-phase flow in complex systems, Chapter 2 deals with the experimental aspects: in particular, with scaling test facilities to ensure that they represent the important elements of the behavior of complex systems. Chapter 2 covers the hierarchical two-tiered scaling approach developed by Zuber [P-12] and subsequent improvements. Due to presently prevailing uncertainties about two-phase-flow physical details, particularly at the gas–liquid interfaces, the value of preserving fluid properties and the time scale in integral experimental facilities is stressed. Also illustrated is how the overall scaling groups are derivable from a top-down or integrated approach. Even in the case of phenomenological details such as those found with rising fuel clad temperature during a loss-of-coolant accident (LOCA), it shows the advantage of an integral or lumped-parameter model. The key scaling parameter during that accident is the transfer of stored energy in the fuel to the coolant—unfortunately, a scaling parameter that was not always recognized or duplicated in many past loss-of-coolant test facilities. It is hoped that the discussion in Chapter 2 will encourage future examination of important scaling issues before test facilities for complex systems are built and operated.

Chapter 3 is concerned with analytical tools, often referred to as system computer codes, which have been employed to predict the behavior of complex systems during transients and accidents. The treatment follows the historical development of such codes, starting from the oversimplifying assumption of equal gas and liquid velocity and temperature (i.e., homogeneous flow), proceeding through separated and mixture models where liquid–gas interfacial interactions are not considered, and ending up with the currently favored two-fluid representation, which relies on flow pattern maps and avoids any constraints about the gas and liquid velocity and temperature. In Chapter 3 we emphasize the trade-off between physically incorrect assumptions or constraints on gas and liquid flow and energy and the increased number of boundary and interface closure laws. Another important conclusion noted in Chapter 3 is the need to empirically specify the fluid properties for the simplest models, in contrast to their direct derivation in the two-fluid models. It is concluded that the development of system computer codes is a remarkable achievement as long as it is recognized that they are only as good as the degree of testing they receive against applicable separate effects, components, and particularly, integral system tests.

In Chapter 4 flow pattern maps and flow pattern models are discussed because the constitutive equations and closure relations are so dependent on how the two phases arrange themselves. Some recent developments are covered to show that despite progress, there are gaps in describing the interfaces and the transfers of momentum and energy at those locations as well as the transitions from one flow pattern to another.

Chapters 5 and 6 deal with a small set of limiting phenomena that are known to have an influence on the safety and economics of complex systems. Chapter 5 covers the countercurrent limiting conditions during which upward flow of one phase may prevent downward flow of the other. Chapter 6 deals with critical or choking flow. These two two-phase-flow topics were selected because their analytical treatment is among the best. In this book they are evaluated against such previously untested geometries as three-dimensional countercurrent flow in the downcomer of a pressurized water reactor and critical flow through piping cracks. The results tend to reinforce the principle that new configurations or conditions in two-phase-flow complex systems must be tested, preferably at close-to-prototypic conditions.

Chapter 7 provides summary comments about the key issues of two-phase flow in complex systems. Also discussed are potential opportunities to improve the understanding, available experimental information, and predictive ability of complex systems. In Chapter 7 it is recognized that the next significant step forward in two-phase flow will come from models and measurements with appropriate instrumentation of the microscopic structures and exchanges at gas–liquid interfaces. Because present system computer codes employ comparatively coarse nodalizations, they would not be able to simulate microscopic conditions until reliable low-numerical-diffusion methods are developed. For that reason, it becomes all the more important to assess uncertainties in computer system code applications and to emphasize tests and model changes where they can be of most benefit.

Chapter 8, a late addition to the book, was prepared to show that the treatment in Chapters 1 to 7, which deals primarily with nuclear power systems, is applicable to other complex systems. The global climate system (in particular, global warming) was selected for that purpose. It is among the most complex systems, if not the most complex. The increased complexity of the global climate system comes from the inherent natural variability of the climate and from the very strong coupling existing among the various subsystems and components and the numerous positive and negative feedbacks they receive from the biosphere. In Chapter 8 an attempt is made to provide a summary of the range of knowledge of the global climate system and of the extensive international efforts to increase our understanding [P-13]. The basic conservation equations are relatively the same; they are nonlinear and are solved by similar numerical methods. The time and space scales are much larger: going from seconds and hours to hours and decades of years and from tens of centimeters to hundreds of kilometers. The global climate system suffers from the same key issues as those raised in Chapter 1 to 7, including lack of a full phenomena identification and ranking table, scaling difficulties, inadequate and too numerous variations of closure equations or parametrizations, potential inaccuracies in the interfacial exchanges, insufficient data and measurements. In many ways, they are the same concerns as those discussed in Chapter 7, which was written and left unchanged on purpose before Chapter 8 was developed. This was my first serious exposure to the global climate system. I found it worthwhile and fascinating to study this system if for no other reason than it supports the emphasis herein on the importance of focusing on the behavior of the overall complex system.

In this book only liquid and gas systems are dealt with; other multiphase circumstances, such as liquid–solid, gas–solid, and liquid–gas–solid mixtures, are covered only briefly in Chapter 8. Many of the concepts developed herein extrapolate to other components, but I decided not to emphasize them because the complications of different or more components would not have added much to the basic message of the book. Similarly, emphasis is placed on momentum transfer at the liquid–gas interfaces. Comparable treatment was not provided for heat transfer at the interfaces because most of the relevant issues can be covered on the basis of momentum exchanges.

Most of the art in the book comes from other sources and, in a few instances, it relies upon curves or data points which come from references other than those listed at the end-of-chapter lists. Readers interested in further details about those curves or data points will need to refer to the original source.

It should be clear from this brief outline that the book does not purport to be a complete literature survey or to include even a noticeable fraction of the literature on two-phase flow and complex systems. It relies, as one would expect, on the author's experience with two-phase flow in complex systems. The perspectives offered are dominated by the desire to deal with overall two-phase-flow complex system behavior. This in no way is meant to detract from the need to study fundamental two-phase-flow phenomena and their coverage in other excellent publications or books that deal with two-phase flow.

I want to take this opportunity to thank my daughter, Linda Smith, and particularly Ms. Joanne Tone for their excellent work, courtesy, and assistance in the preparation of the book. I want to especially express my gratitude to my wife, Eileen, who allowed me to spend so much extra time on my favorite hobbies of two-phase flow and complex systems.

REFERENCES

P-1   Lahey, R. T., Jr., and Moody, F. J., *The Thermal Hydraulics of a Boiling Water Nuclear Reactor,* LaGrange Park, Ill., Monograph Series on Nuclear Science and Technology, American Nuclear Society, 1977.

P-2   Tong, L. S., and Weisman, J., *Thermal Analysis of Pressurized Water Reactors,* 2nd ed., LaGrange Park, Ill., Monograph Series on Nuclear Science and Technology, American Nuclear Society, 1979.

P-3   Tong, L. S., *Boiling Heat Transfer and Two-Phase Flow,* Wiley, New York, 1965.

P-4   Wallis, G. B., *One-Dimensional Two-Phase Flow,* McGraw-Hill, New York, 1969.

P-5   Govier, G. W., and Aziz, K., *The Flow of Complex Mixtures in Pipes,* Robert E. Krieger, Melbourne, Fla., 1977.

P-6   Collier, J. G., *Convective Boiling and Condensation,* McGraw-Hill, New York, 1972.

P-7   Hsu, Y. Y., and Graham, R. W., *Transport Processes in Boiling and Two-Phase Systems,* Hemisphere, New York, 1976.

P-8   Whalley, P. B., *Boiling Condensation, and Gas–Liquid Flow,* Clarendon Press, Oxford, 1990.

P-9   Hestroni, G., *Handbook of Multiphase Systems,* Hemisphere, New York, 1982.

P-10 Cheremisinoff, N. P., and Gupta, R., *Handbook of Fluids in Motion,* Ann Arbor Science, The Butterworth Group, Ann Arbor, Mich., 1983.

P-11 *Application of Code Scaling, Applicability, and Uncertainty Evaluation Methodology to a Large Break, Loss-of-Coolant Accident,* NUREG/CR-5249, EGG-2552, 1989.

P-12 *An Integrated Structure and Scaling Methodology for Severe Accident Technical Issue Resolution,* NUREG/CR-5809, EGG-2659, 1991.

P-13 Intergovernmental Panel on Climate Change (IPCC), *Climate Change 1995: The Science of Climate Change,* Cambridge University Press, New York, 1996.

CHAPTER 1

PROGRAM FOR TESTING AND ANALYSIS OF TWO-PHASE FLOW IN COMPLEX SYSTEMS

1.1 INTRODUCTION

The simultaneous flow of a gas and a liquid occurs in a multitude of chemical, mechanical, nuclear, and space applications. These conditions become especially complex in nuclear power plants, where the degrees of required testing and analysis are extensive and challenging. For that reason, in this first chapter, which deals with the overall behavior of complex two-phase flow systems, we rely heavily on what has been learned from studies about nuclear power plants. However, the results and comments are just as valid for nonnuclear complex circumstances.

The thermal hydraulics prevailing in nuclear power plants has been described in detail previously [1.1.1, 1.1.2]† and are not repeated here. Also, as noted in a historical recounting of the first 50 years of thermal hydraulics in nuclear power [1.1.3], initially the emphasis was placed on the most limiting two-phase phenomena and obtaining test data and creating basic models to describe them. Overall system tests and analyses did not receive enough attention because the technology available was inadequate. Today, the importance of first creating an integrated test and analysis program is recognized. There are several good reasons for that shift.

First, in the case of nuclear power plants, the licensing regulations [10CFR52.47(b)(2)(i)(A)] demand that:

- The performance of each safety feature of the design be demonstrated through analysis, appropriate test programs, experience, or a combination thereof

---
†Triple numbers in brackets refer throughout to end-of-chapter references.

- Interdependent effects among the safety features of the design be found to be acceptable by analysis, appropriate test programs, experience, or a combination thereof
- Sufficient data exist on the safety features of the design to assess the analytical tools used for safety analysis over a sufficient range of normal operating conditions, transient conditions, and specified accident sequences

These requirements, which apply to other complex undertakings, can be satisfied only through a comprehensive interactive test and analysis program at the system level.

Second, large system computer codes are being used to simulate the behavior and to predict the performance of complex systems during normal and transient operations as well as during accidents. The system computer codes are relied upon because it is not practical and is very costly to subject the prototype plant to such conditions. Also, very often, the results from test facilities cannot be applied directly to a complex system. Furthermore, various plant recovery techniques from transient or accidents can be assessed only through system computer codes. A key issue in such simulations is the accuracy of the results predicted and how to provide for their uncertainties to stay within prescribed safety limits.

Third, the system computer codes have to be validated from tests, and the completeness of that supporting set of tests can be established only by looking at the entire system. The tests that need to be performed have been generally subdivided into four kinds of experiments: (1) separate effects tests, (2) component performance tests, (3) integral system tests, and (4) prototype operational tests. The first two types of tests are necessary for fundamental understanding and basic model development, the latter two for checking the overall performance of complex systems and of the system computer code utilized to predict their behavior.

Separate effects tests deal with basic phenomena and processes. They include such experiments as:

- Measurements of important two-phase-flow variables (i.e., gas volume fraction, pressure drop, heat transfer, etc) in simplified and/or other relevant geometries
- Observations of gas–liquid flow patterns and their use in formulating flow regime maps
- Studies of exchange of mass, momentum, and energy at gas–liquid interfaces and the development of corresponding closure equations
- Critical flow through different break sizes and configurations, critical heat flux, boiling transition, countercurrent flow limitation, and other such limiting phenomena

The separate effects tests are carried out to develop empirical correlations and/or to confirm physically based models of the involved phenomena and processes.

Component performance tests are concerned with the behavior of various pieces of equipment (e.g., valves, pumps, separators, fuel assemblies). They are performed on reduced- and in a few cases, full-scale components. Their purpose is to develop empirical correlations for the components or to validate models provided for them in system computer codes.

Integral system tests attempt to simulate the overall behavior of a system being subjected to specific transients or accident events. These tests generally are performed on a reduced scale and are confined to subsystems and components that play an important role in system dynamics and/or subsystem interactions. They provide the best check of system computer codes because they generate data on severe transients and accidents that cannot be carried out in the prototype.

Finally, most prototype designs are subjected to critical startup tests to compare behavior to expectations. Most units also tend to be equipped with on-line instrumentation to record abnormal events beyond the startup period. Even though limited in scope, such operating plant data offer the best validation of computer system codes because they are obtained at full scale and with no compromise in overall system dynamic behavior or subsystems interactions.

A methodology to deal with the combined elements of testing and analysis of complex two-phase flow systems has been developed recently. The first attempt, made in 1989, was called the code scaling, applicability, and uncertainty (CSAU) evaluation methodology [1.1.4]. Its merits were demonstrated during large-break loss-of-coolant accidents (LBLOCAs) in pressurized water reactors (PWRs). The effort showed that it can lead to increased understanding of the two-phase-flow phenomena involved as well as providing a strong rationale for evaluating the uncertainties in the large system computer code utilized to predict them. In 1991 the methodology was expanded to even more extreme conditions when it was applied to formulate an integrated structure and scaling methodology for severe accident technical issue resolution [1.1.5]. A hierarchical two-tiered scaling (H2TS) analysis was proposed in that reference by N. Zuber, who, shortly thereafter, broadened its application to the scaling and analysis of complex systems [1.1.6, 1.1.7]. Additional refinements and variations of such integrated programs have been proposed [1.1.8, 1.1.9] and the approach has been particularly useful in assessing new advanced light-water reactor (ALWR) designs [1.1.10, 1.1.11]. The author has been fortunate to participate in the early development of some of the methodologies and in the latest review of their application to ALWRs. There is no question that they can be of great benefit to the testing and analysis of complex two-phase-flow systems, *particularly if they are employed before the design and details of test facilities and computer system codes are finalized to predict their performance.* For that reason, this book starts with a discussion of integrated programs for testing and analysis of complex two-phase-flow systems. It is a significant departure from all previous two-phase flow treatments, which tend to concentrate on the basic conservation equations and their application to model separate phenomena for which test data are available. In most practical two-phase-flow systems or components, a large number of phenomena occur simultaneously and their importance, range of interest, and interactions can vary not only with the design of the systems and components but also with the specific event being evaluated. By looking at the overall complex system it is possible to:

- Identify all the relevant phenomena and their importance
- Take into account interactions between phenomena and subsystems

4 PROGRAM FOR TESTING AND ANALYSIS OF TWO-PHASE FLOW IN COMPLEX SYSTEMS

- Identify the range and types of necessary tests and how to scale them
- Assess the accuracy of the system computer codes and estimate the uncertainties associated with the system performance predicted

An early focus on overall system behavior has considerable advantages, and this will be illustrated by a closer look at the scenario of a large-break loss-of-coolant accident (LBLOCA) in a pressurized water reactor (PWR). Most of the description of this LBLOCA comes from a compendium of research for LOCA analysis for light-water-cooled nuclear power plants [1.1.12] and from the CSAU report [1.1.4].

1.2 LARGE BREAK IN PRESSURIZED WATER REACTOR

Figure 1.2.1 shows a typical PWR. Just prior to the accident, high-pressure primary water is pumped into the downcomer of the reactor vessel before it rises along the fuel assemblies in the core to remove the heat generated by nuclear reaction. The hot water leaving the vessel enters the U-tubes in the steam generators, where it gives up its heat to secondary water recirculating on the outside of the tubes. The primary loop is closed by returning the water leaving the U-tubes to the reactor vessel. Steam formed on the secondary side in the steam generators flows to the steam turbine generator, where it produces power. After it leaves the low-pressure

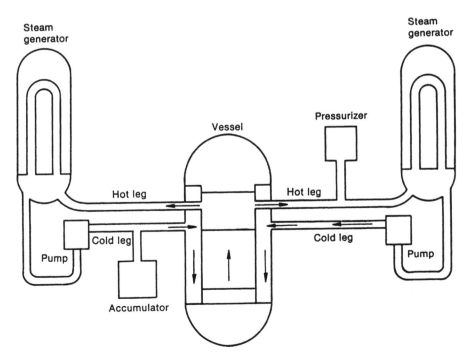

Figure 1.2.1 PWR system immediately before a large cold leg break. (From Ref. 1.1.12.)

stages of the turbine, the secondary steam enters a condenser, where it is condensed and returned to the steam generators in the form of feedwater. Figure 1.2.1 also shows a pressurizer utilized to keep the PWR at a constant high pressure. Also depicted is an accumulator that contains cold water which can be injected by pressurized nitrogen gas into the reactor core when the PWR pressure drops as a result of the large break. All PWRs are equipped with several other emergency cooling systems, which are not illustrated in Figure 1.2.1 but which can add water to the reactor vessel and steam generators during the course of the accident.

Immediately after the break, subcooled water escapes through the large break at a high flow rate and reaches *critical* (choking) *flow* conditions. Due to the differences in flow resistance, the mass outflow on the break side is much larger than from the pump side. The break produces flow stagnation conditions in the core, resulting in a bidirectional flow pattern and *flow reversal* in the core. The reactor vessel depressurizes fast and the water in the upper plenum of the vessel, hot leg, and pressurizer reaches saturation and flashes into steam. Flashing also starts in the core, which reduces power generation and the *critical heat flux* is reached. This involves steam blanketing of the fuel, increasing its temperature rapidly due to poor heat removal at the fuel surface and high energy stored within the fuel. The nuclear reaction shuts down when the pressurizer reaches a preset low-pressure *set point* and the control rods are inserted. The primary coolant pumps trip but continue to rotate in a degraded coast-down mode. Most of the water flow through the intact loop(s) goes out of the break rather than into the core. As the pressure continues to decrease, the water in the cold leg(s) reaches saturation, which leads to steam formation and a sharp decrease in the critical flow through the break.

As the break flow decreases, the cold fluid in the intact loop(s) can be forced into the core by the *pumps* operating in a *two-phase coast-down mode*. This fluid cools the core and the fuel temperature reaches a peak and starts dropping. This first peak and *rewet* occur at relatively high pressure and low steam quality. The liquid in the intact loop(s) eventually flashes and the inlet flow to the core decreases. A second fuel dryout takes place, but this second temperature peak is much reduced, due to little heat remaining stored in the fuel and due to some upward flow of flashing steam and entrained droplets into the core.

This time period of the LBLOCA is referred to as the *blowdown phase*; it starts with the break and terminates when flow from the intact loop(s) accumulator is initiated. A schematic of the flashing transient phenomena at the end of the blowdown phase is provided in Figure 1.2.2. It shows steam and entrained liquid from the downcomer and lower plenum escaping through the broken pipe. However, as cold water flow from the broken and intact loops(s) accumulator occurs, it mixes with steam and produces oscillations due to *direct contact condensation.*

Once the accumulator cold water reaches the downcomer, it can proceed downward due to gravity or be swept out through the break by escaping steam. Early on, some of the accumulator or other injected water can bypass the core, and this is shown idealistically in Figure 1.2.3. A fraction of the penetrating water into the lower plenum is subsequently swept out by steam trying to reach the break. This *countercurrent flow* mechanism in the downcomer will determine the time to refill

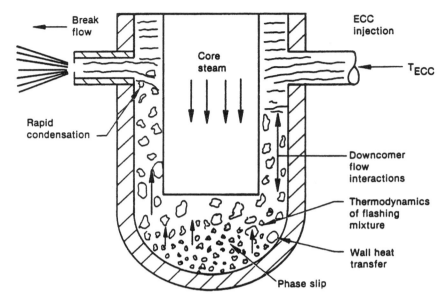

Figure 1.2.2 Schematic of flashing transient phenomena during ECC (Emergency Care Cooling) and blowdown. (From Ref. 1.1.12.)

the lower plenum. It may involve a number of water "dumpings" and "sweep-outs." The amount of water reaching the lower plenum increases with time as the reactor depressurizes and approaches the containment pressure. The steam flow through the core decreases with decreasing depressurization and the fuel in the core producing decay heat starts to heat up again but rather slowly. This phase of the LBLOCA has been called *emergency core cooling (ECC) bypass or refill*. It covers the period of time from the end of blowdown up to when the water reaches the bottom of the core, at which time the *reflood* phase starts.

Due to the high temperatures of the fuel rods, the entire range of thermal regimes is present during reflood: rising single-phase water flow, followed by *nucleate boiling, transition boiling, dispersed droplet flow*, and steam flow. Higher steam velocities and liquid entrainment occur in the central region of the core because it produces more power. The *entrained liquid* has several beneficial cooling effects. As drops are deposited on the hot fuel rods, they reduce its temperature upon impact; also, droplet deentrainment takes place at such flow obstructions as upper tie plate and grid spacers, resulting in local fuel quenching at those locations.

Some of the entrained liquid reaches the upper plenum, where it is *deentrained* and forms a two-phase pool, as illustrated in Figure 1.2.4. Liquid from the upper pool can reenter the outer region of the core because of its reduced power generation and low upward steam velocities. As shown in Figure 1.2.4, a three-dimensional flow pattern develops. In the core, the flow is from low to high power regions, while in the upper plenum, the flow is in the opposite direction.

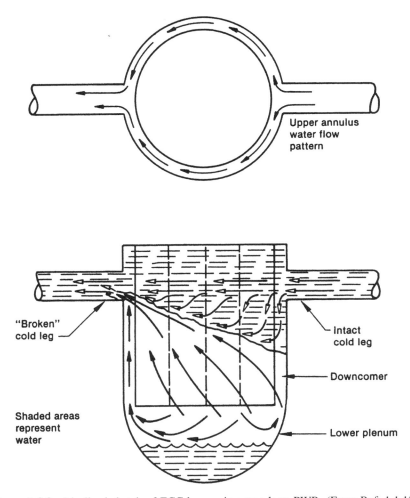

Figure 1.2.3 Idealized sketch of ECC bypass in a two-loop PWR. (From Ref. 1.1.12.)

Liquid from the two-phase pool in the upper plenum may be carried over to the hot leg(s) after some of it has been deentrained on the upper plenum structures. The *carried-over liquid* will evaporate in the steam generators when it reaches them. This causes an increase of pressure in the steam generator and upper plenum. This pressure increase reduces the reflood rate, which in turn lowers the amount of steam generated and liquid entrainment, causing a subsequent increase in flooding rates. This *oscillatory behavior* continues but with decreasing amplitude until the entire core is reflooded. Under some conditions, manometer oscillations can exist between the liquids in the downcomer and core, and they too can influence the reflood rates.

As the water level rises in the core, the fuel is rewetted and its temperature decreases to the water saturation temperature. Prior to the rewet, the fuel will reach a new peak temperature and this peak value rises with reduced reflood rates and

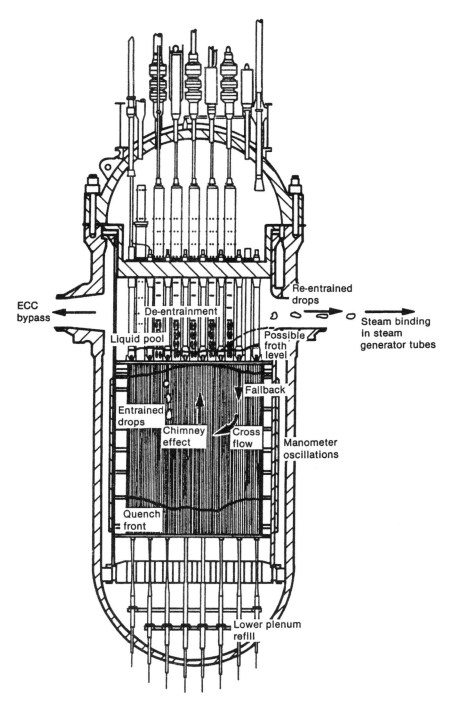

Figure 1.2.4 Refill–reflood phenomena in vessel: deentrainment of liquid on the upper core structure. (From Ref. 1.1.12.)

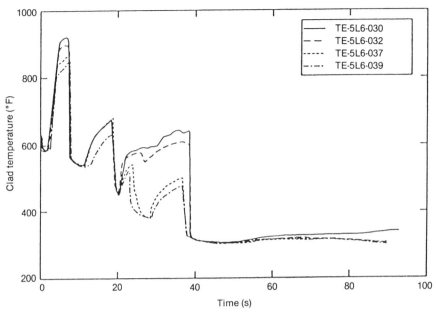

Figure 1.2.5 Measured cladding temperature (TE) during blowdown in LOFT. (From Ref. 1.1.12.)

an increased fuel power generation. In most PWRs, two or more peak clad temperatures are expected during a LBLOCA. Figure 1.2.5 shows actual data obtained in the loss-of-flow test (LOFT) facility L2-2 experiment. LOFT is an integral nuclear system facility which could realistically simulate a PWR LBLOCA. There are three temperature peaks in Figure 1.2.5. The first one, associated with the blowdown phase, is the largest because it involves the initial stored energy in the fuel rods. The second and third peaks, which occur during the refill and reflood phases, are usually much lower because they are caused primarily by decay heat. All three peaks are produced by poor fuel rod-to-coolant heat transfer, and as inferred from the preceding brief description of the LBLOCA, they are not as readily or accurately predictable. The reasons are as follows:

1. A large number of important phenomena are involved. They were italicized in the text and include:
   - Subcooled and saturated critical flows at the break
   - Release of stored and decay heat from the fuel
   - Flashing and voiding in all volumes
   - Critical heat flux (CHF) or departure from nucleate boiling (DNB)
   - Direct contact condensation
   - Two-phase flow through pump

- Countercurrent flooding limitation (CCFL) and emergency core cooling (ECC) bypass in the downcomer
- ECC penetration into the lower plenum
- Quenching (or rewet) of the heated rods by the rising liquid level, the falling water film, or transported water to the core
- Liquid entrainment in the steam at the quench front, upper plenum, and into the steam generators
- Droplet evaporation in the steam generators
- Oscillations in water level, pressure, reflooding rates, and so on.

2. The phenomena vary from one phase of the accident to the next, and they change rapidly with time. Figure 1.2.6 illustrates the flashing and voiding taking place in the lower plenum and core of a typical standard four-loop PWR. The core water content is seen to drop to zero in a few seconds. The entire blowdown phase lasts only 12 seconds and the bottom plenum is refilled with water 33 seconds after the break, at which time the reflood phase starts.

3. Many of the phenomena are three-dimensional, particularly inside the reactor vessel, where the break location naturally sets up asymmetrical flow patterns.

4. The phenomena interact with each other, and some of the interactions produce oscillations in important parameters.

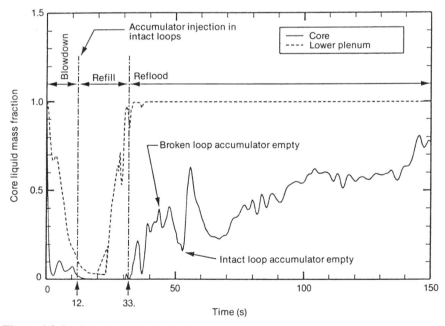

Figure 1.2.6 Calculated liquid fractions in core and lower plenum, showing blowdown, refill, and reflood phases. (From Ref. 1.1.4.)

Under such circumstances it is not surprising that a significant amount of testing and a validated system computer code are now being required to deal with a PWR LBLOCA. The testing and analysis demands or standards have evolved and become much more stringent over the years. A history of that development is helpful because it identifies possible alternative ways of proceeding with integrated test and analysis programs for complex systems.

1.3 ALTERNATIVE TEST AND ANALYSIS PROGRAMS

Test and analysis programs for complex systems can be expected to depend on the prevailing state of knowledge. If there are shortfalls in testing or analysis capability, they could be compensated for by providing additional margins in the predictions of the complex system behavior [1.1.3]. This was the case, for example, for the rules developed over time for loss-of-coolant accidents in nuclear power plants. Initial specific requirements were formulated after extensive public hearings in 1973 and published in January 1974 in Appendix K to the *Code of Federal Regulations*, Title 10, Section 50.46 [1.3.1]. Analysis models used to calculate the thermal-hydraulic performance of emergency core cooling systems (ECCSs) had to conform to the acceptance criteria of Appendix K, which required that:

- The peak fuel (zircaloy) cladding temperature be kept below 2200°F (1477K)
- The maximum cladding oxidation not to exceed 17% of the original cladding thickness
- The hydrogen generation from chemical reaction of the cladding with water or steam stay below 1% of the maximum potential value
- A coolable core geometry be retained
- A means be provided for long-term cooling of the reactor core and its fuel

Appendix K also assured that additional allowance for these criteria would be present by prescribing specific conditions for the analyses. For example, it was required to provide a 20% addition to the American Nuclear Standard standard for decay heat, that is, a 20% increase in the initial fuel stored energy. There were other conservative provisions as well: for example, in the case of pressurized water reactors (PWRs), in the rate of metal–water reaction, degrees of ECCS bypass or water entrainment, and upper plenum deentrainment and fallback of water into the reactor core. Similar conservative parameters were defined for other types of light-water-cooled nuclear reactors and boiling water reactors (BWRs).

The extra margins generated by Appendix K assumptions were estimated as early as 1977 by removing one conservative provision at a time and calculating the corresponding PWR peak cladding temperature. Those results, reproduced in Figure 1.3.1, show that the extra margins were significant. In essence, the regulations prescribed evaluation models that predicted reactor performance only slightly below, the 2200°F from the Appendix K rules. This was clearly more than enough to

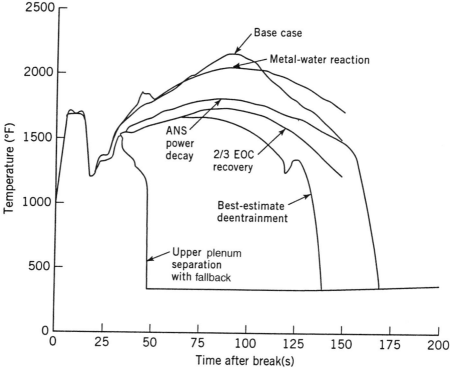

Figure 1.3.1 Effect of selected conservative assumptions on PCT. (From Ref. 1.3.2.)

compensate for any lack of test data or inadequate modeling, particularly when taking into account the regulatory inputs prescribed for the analysis of various transients and accidents. For instance, the LBLOCA had to be assumed an instantaneous, double-ended break of up to the largest pipe connected to the reactor vessel. Complete loss of off-site power and the worst possible single failure of equipment were postulated to occur simultaneously.

Appendix K requirements remained in effect until August 1988, when best estimate (BE) methodology was allowed to be used provided that the uncertainty of the best estimate predictions could be quantified and included in the calculated results for their comparison to the acceptance limits of Appendix K. The Appendix K acceptance criteria were left intact. This relaxation led to the development of the code scaling applicability, and uncertainty (CSAU) evaluation methodology [1.1.4]. The CSAU methodology produced the first integrated systematic evaluation program for the testing and analysis of complex two-phase systems. It could not have been implemented without the extensive testing results and improvements in system computer codes that happened from 1974 to 1988 and which are summarized in Reference [1.1.12].

The efforts to implement the CSAU methodology are clearly much larger than those for the bounding approach that preceded it. However, additional margins were identified by the CSAU program, which could be used to improve the economics and operations of nuclear units. There is also an intangible sense of increased safety which comes from better understanding of prevailing conditions, which may be of even greater value.

Recently, the nuclear industry has decided to introduce advanced light-water reactor (ALWR) designs and to seek their certification by regulatory authorities. The certification is to be applied to yet-to-be-built plants, to be valid for 15 years, and for the plants using that design to be capable of operating for another 60 years. The certification process can be based only on the application of results from prior and ALWR specific testing programs. Also, the proposed safety analysis system computer codes must be validated and shown to predict confidently that the plants meet prescribed safety limits. For that reason, there is little tolerance left in the ability to predict the behavior of the plants being certified, and that low degree of tolerance can be achieved only by employing a thorough and integrated test and analysis program.

1.4 ELEMENTS OF AN INTEGRATED TESTING AND ANALYSIS PROGRAM

The formulation of a good integrated testing and analysis program for a complex two-phase flow system involves a series of steps, including the need to:

1. Establish the *system characteristics*. If the system design is not at the final stage, subsequent design changes and their influence on the integrated testing and analysis plan must be considered.
2. Identify the specific *behavior scenarios* and limits of importance to the safety and economic performance of the system.
3. Select and control (freeze) the *system computer codes* that are to be utilized to predict the scenarios identified. It is presumed that the system computer codes will have been assessed, documented, and validated prior to this application.
4. *Identify the phenomena and interactions involved, and rank them in order of importance for each limiting scenario.*
5. Create a *test assessment matrix* to cover the dominant phenomena and the interactions identified in each scenario. This test matrix should list the available test data and their range of application.
6. Generate a *test plan* to cover the omissions in the available tests or their inadequate coverage of system parameters. The newly required data could come from separate effects, components, integrated systems, and related operational system tests.
7. Subject the new tests and test facilities to *scaling analysis* to assure that they will yield appropriate data.

8. Determine *system computer code accuracy* through both pre- and posttest predictions of the previous and new test results selected. The system computer code comparisons to tests should be performed with computer code nodalization, which is employed to predict full-scale complex system behavior.
9. Assess *uncertainties in system behavior* predictions. The uncertainties can arise from different sources: system computer code, experiments, scaling distortions, and system knowledge uncertainties. Eventually, they need to be combined into an *overall uncertainty*.

These steps are all listed (sometimes in different order or form) in the code scaling, applicability, and uncertainty (CSAU) methodology reproduced in Figure 1.4.1. The CSAU approach and its subsequent refinement provide a logical structure for dealing with complex two-phase flow systems. The specific issues to be raised during preparation of an integrated testing and analysis program are expected to vary with the nature of the complex system being assessed. Typically, the following issues deserve consideration:

- Was an important adverse physical phenomenon overlooked in the testing and analysis program?
- Were the test facilities scaled appropriately or their scaling distortions recognized, and can the system computer code appropriately scale up phenomena observed in scaled-down experiments?
- Was the system computer code overly adjusted or tuned to match certain sets of data or to improve their numerical stability?
- Was sensitivity to system computer code nodalization minimized in predicting the test results and the performance of the prototype complex system?
- Were enough independent or blind predictions carried out to validate the system computer codes ability to predict the tests, in particular the component and integral system tests?
- Was the uncertainty with which the system computer code calculates important parameters reasonably established? How well were the biases accounted for, and how well were the uncertainties combined?

In the sections that follow, additional summary comments are provided about step 1 (system characteristics), step 2 (scenarios), step 4 (phenomena identification and ranking), and steps 5 and 6 (test matrix and plan). The topic in step 7 (scaling and testing of complex systems) is covered in detail in Chapter 2, and steps 3, 8, and 9, which cover system codes and their uncertainties, are discussed in Chapter 3.

1.4 ELEMENTS OF AN INTEGRATED TESTING AND ANALYSIS PROGRAM

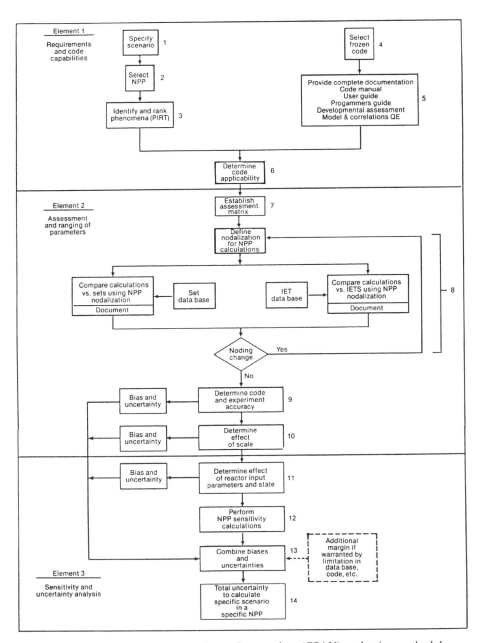

Figure 1.4.1 Code scaling, applicability, and uncertainty (CSAU) evaluation methodology. (From Ref. 1.1.4.)

1.5 COMPLEX SYSTEM CHARACTERISTICS

The complex system being evaluated must have its characteristics well defined, particularly as they relate to two-phase thermal hydraulics. Not only must all the important components be specified geometrically but also how they are coupled together. The system performance parameters must be established (e.g., flow, temperature, pressure, heat generation and removal) both locally and at the inlet and outlet of each component. The expected range of those parameters must be known during normal, transient, and accident conditions so that it can be compared to those present in the available or proposed tests. If gaps exist and are significant, they must be considered in the testing plans.

Another important consideration deals with the system control functions, which have a great influence on the course of the events. The control characteristics must be defined because they determine the start, stop, and transients of various components (pumps, valves, etc.) and how they interact with each other.

The design of most complex systems evolves from earlier implementations, and full advantage should be taken of issues previously evaluated by testing and analysis. For that reason, the focus should be on new and unique features, to ascertain that they do not involve different phenomena or novel interactions. This particular aspect of evolving designs may not have been emphasized enough in CSAU methodology. By contrast, it received adequate attention in the yet-to-be certified passive Advance Light Water Reactors (ALWRs). In fact, a list of unique and new features was compiled, as well as a tabulation of associated phenomena and supporting qualification data. The objective of the evaluation was to identify any new thermal-hydraulic phenomena due to the new features, during normal operation, transients, and accidents. For instance, both passive ALWRs use new depressurization valves and employ natural circulation heat exchangers to remove decay heat. Testing to evaluate their performance was included, and in the case of the heat exchangers, the presence of noncondensable gases and their degrading impact on heat transfer were emphasized.

Similarly, subsystems interaction assumes increased importance in passive ALWRs, where they can interfere or overwhelm the gravity driving forces used in the passive safety systems. The approach used to deal with this issue was to employ the available system computer code to assess the interactions between the different passive systems and between passive and available active systems. Such studies helped determine which interactions were important and had to be incorporated in the planned integral system tests. In a few cases, they even led to changes in the design or planned operating procedures.

1.6 SCENARIOS

Complex systems are subjected to a large number of different scenarios. The events can cover a wide range of conditions, including normal operations, transients, and accidents. Also, the dominant processes and the important phenomena are known to vary from one scenario to the next. Sometimes, the system computer code being employed changes with the scenario being evaluated. Finally, the safety margins to be provided tend to fluctuate with the events; high margins are generally required for high-frequency events and they are lowered as the probability of event occurrence decreases.

In the case of nuclear power plants, the scenarios to be analyzed are specified. So are the parameters that have the most significance. Because the safety of nuclear power plants is related directly to a defense-in-depth strategy, it relies upon maintaining the structural integrity of the three principal barriers to the release of radioactive fission products: the nuclear fuel cladding, the primary system boundary within which the fuel resides, and the containment building that surrounds the primary system. During large-break loss-of-coolant accidents (LBLOCAs) in water-cooled reactors, the water in the reactor vessel escapes from the primary system and the fuel cladding temperature rises. The peak temperature reached determines whether the fuel cladding will fail and release some of the fission products it contains. The focus of tests and the validation of system computer codes to predict LBLOCA events is, therefore, based on that key parameter. For plant transients, the emphasis shifts to avoiding steam blanketing of the fuel or critical heat flux (CHF) or departure from nucleate boiling (DNB). For accidents involving failure to shut down the nuclear reaction, the emphasis moves to the pressure level in the primary system, and so on. Often, to get a meaningful validation of a system computer code, it is necessary to look at several secondary parameters. For example, in a LBLOCA, the reactor system pressure and the water inventory in the reactor deserve to be examined as a function of time as well as the initiation of emergency cooling systems to add water to the reactor. Another important lesson learned from nuclear power plants is the need to understand how events can be influenced by human performance. For example, small loss-of-coolant accidents can take much more time to occur, and there are more chances for human involvement and error. For that reason, human performance deserves careful attention and evaluation during small-break scenarios. This is not the case for large breaks, which happen so fast that the operators do not have a chance to intervene.

It took many years of experience to define the scenarios most important to the safety of nuclear power plants, to specify the important parameters to be tracked and the desired range of safety values during each scenario, and to demand their predictions with a valid and proven system computer code(s). This approach has served the nuclear industry well in terms of safety, and it deserves consideration by other complex chemical, biological, and space applications.

1.7 PHENOMENA IMPORTANCE AND RANKING

For each significant scenario, it is important to:

- Identify all the prevailing phenomena (together with relevant components). It is often advantageous, as in the case of the LBLOCA, to subdivide the scenario

into separate phases when the dominant processes change from one phase to the next.
- Rank the various phenomena identified with respect to their effect on the safety criteria for the scenario. Such ranking is necessary to assure that the important phenomena and components have been subjected to tests and are considered in the system computer code used to predict the behavior of the complex system. The ranking is also needed to guide the selection of phenomena to be utilized in quantifying the uncertainty in the predictions of the system computer code.

This process relies to a considerable extent upon experience and subjective judgment. The prevailing phenomena are found by examining the available test data and system computer code simulations of the specified scenario. It is best done by a group of experts well versed in the tests and computer code analyses carried out to date. This same group of experts working as a team can rank the various phenomena in order of importance. If necessary, sensitivity calculations with the system computer code can be performed for certain selected individual or combined phenomena to validate the judgment of the panel of experts. This was the approach used during CSAU evaluation of the LBLOCA, and its workability is demonstrated by some of the results obtained [1.1.4].

The potentially important phenomena for a LBLOCA in a PWR are listed in Figures 1.7.1, 1.7.2, and 1.7.3 for the blowdown, refill, and reflood phases, respectively. They have been classified by component types to assist in evaluation of the available test data for each component and their modeling in the computer system code. Ranking of the phenomena by the group of experts is provided in Table 1.7.1. The phenomena were ranked from 9, the highest rank, to 1, the lowest. An independent ranking developed by a group of thermal-hydraulic analysts at the Idaho National Engineering Laboratory (INEL) [1.7.1], also shown in Table 1.7.1, tends to support the findings of the team of experts, who concentrated on phenomena ranked 7 and above. It is interesting to note that only three of more than 40 phenomena were judged to be of highest importance during the blowdown phase: fuel-rod stored energy, break flow, and reactor coolant pump degradation. Similarly, during the refill phase only three phenomena were singled out: cold leg and downcomer condensation and multidimensional flow in the downcomer. During reflood an increased number of phenomena received a high rating, including reflood heat transfer, three-dimensional flow and void generation in the core, entrainment–deentrainment in the upper plenum and hot legs, steam binding in the steam generator, two-phase-flow oscillations in the loop, and the presence of noncondensable gases.

All of these highly ranked phenomena are included in the system computer code, TRAC-PF1/MOD1 [1.7.2], used by CSAU to predict LBLOCA in a PWR. Many other phenomena in Figures 1.7.1, 1.7.2, and 1.7.3 are incorporated. However, as pointed out in Chapter 3, the equations describing many of the phenomena in the computer system code are elementary and oversimplified, in order to be able to carry out the analyses within a reasonable amount of computer cost and time. These limitations and the fact that all the phenomena may not be modeled in system

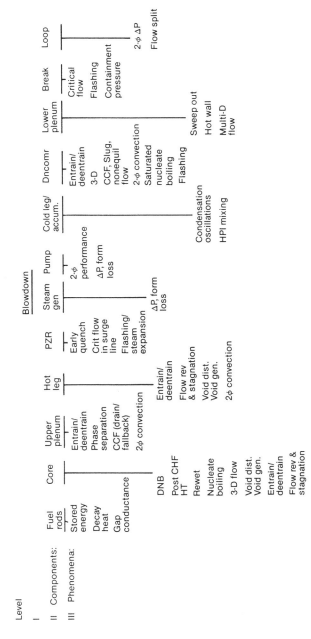

Figure 1.7.1 Component and phenomena hierarchy during LBLOCA blowdown. (From Ref. 1.1.4.)

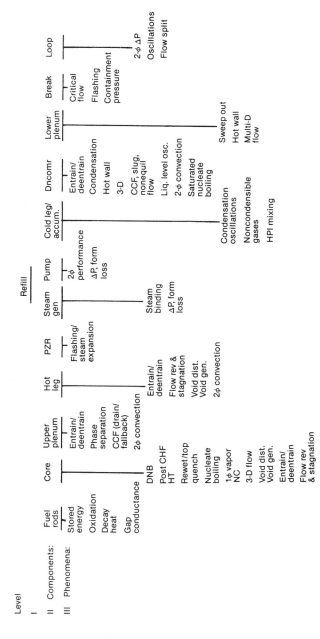

Figure 1.7.2 Component and phenomena hierarchy during LBLOCA refill. (From Ref. 1.1.4.)

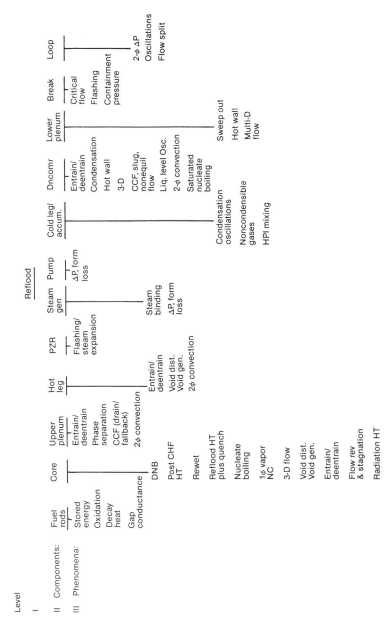

Figure 1.7.3 Component and phenomena hierarchy during LBLOCA reflood. (From Ref. 1.1.4.)

TABLE 1.7.1 Summary of Expert Rankings and AHP Calculated Results

	Blowdown		Refill		Reflood	
	Experts	AHP	Experts	AHP	Experts	AHP
Fuel rod						
Stored energy	9	9	—	2	—	2
Oxidation	—	—	—	1	8	7
Decay heat	—	2	—	1	8	8
Gas conductance	—	3	—	1	8	6
Core						
DNB	—	6	—	2	—	2
Post-CHF	7	5	8	8	—	4
Rewet	8	8	7	6	—	1
Reflood heat transfer	—	—	—	—	9	9
Nucleate boiling	—	4	—	2	—	2
One-phase vapor natural circulation	—	—	—	6	—	4
Three-dimensional flow	—	1	—	3	9	7
Void generation and distribution	—	4	—	6	9	7
Entrainment/deentrainment	—	2	—	3	—	6
Flow reversal/stagnation	—	3	—	1	—	1
Radiation heat transfer	—	—	—	—	—	3
Upper plenum						
Entrainment/deentrainment	—	1	—	1	9	9
Phase separation	—	2	—	1	—	2
CCF drain/fallback	—	1	—	2	—	6
Two-phase convection	—	2	—	1	—	5
Hot leg						
Entrainment/deentrainment	—	1	—	1	9	9
Flow reversal	—	2	—	1	—	—
Void distribution	—	1	—	1	—	4
Two-phase convection	—	2	—	2	—	3
Pressurizer						
Early quench	7	7	—	—	—	—
Critical flow in surge line	—	7	—	—	—	—
Flashing	—	7	—	2	—	2
Steam generator						
Steam binding	—	—	—	2	9	9
Delta-p, form losses	—	2	—	2	—	2
Pump						
Two-phase performance	9	9	—	5	—	—
Delta-p, form losses	—	3	—	3	8	8
Cold leg/accumulator						
Condensation	—	2	9	9	—	5
Noncondensible gases	—	—	—	1	9	9
HPI mixing	—	—	—	3	—	2

TABLE 1.7.1 *(Continued)*

	Blowdown		Refill		Reflood	
	Experts	AHP	Experts	AHP	Experts	AHP
Downcomer						
Entrainment/deentrainment	—	2	8	8	—	2
Condensation	—	—	9	9	—	2
Countercurrent, slug, noneq	—	1	—	8	—	2
Hot wall	—	—	5	4	7	3
Two-phase convection	—	2	—	3	—	2
Saturated nucleate boiling	—	1	—	2	—	2
Three-dimensional effects	—	2	9	7	—	2
Flashing	—	1	—	—	—	—
Liquid level oscillations	—	—	—	3	—	7
Lower plenum						
Sweep out	—	2	7	6	—	5
Hot wall	—	1	—	7	7	6
Multidimensional effects	—	1	—	2	—	7
Break						
Critical flow	9	9	7	7	—	1
Flashing	—	3	—	2	—	1
Containment pressure	—	2	—	4	—	2
Loop						
Two-phase delta-*p*	7	7	—	7	—	6
Oscillations	—	—	7	7	9	9
Flow split	—	7	7	7	—	2

Source: Ref. 1.1.4.

computer codes have led some to question the application of system computer codes to obtain best estimates and uncertainties of the behavior of complex systems. This somewhat "idealistic" perspective, however, cannot detract from the merits of implementing a phenomena identification and ranking program. The process cannot help but provide guidance about additional testing and ways to improve the modeling of phenomena and the system computer codes available.

1.8 TEST MATRIX AND PLAN

Once all the phenomena and interactions during a scenario are identified, a corresponding test matrix needs to be prepared which shows the available tests, their range of application, and their capability of resolving each phenomenon, issue, or interaction. The test matrix should preferably be subdivided into separate effects, components, integral system tests, and operating plant data because they satisfy different objectives. Additional needs for testing or an extension of the available

Cross reference matrix for transients in PWRs

— Phenomena versus test type
● Simulated
○ Partially simulated
— Test facility versus phenomena
● Suitable for code assessment
○ Limited suitability
— Test type versus test facility
● Already performed or planned until 12/84
○ Performed or planned until 12/84, but of limited use

* Volumetric scaling
** For phenomena requiring separate effects test, e.g. pressurizer behavior, refer to small leak cross reference matrix
*** Valve flow behavior will be strongly design-dependent, specific experimental data should be used if possible
**** Problem for scaled test facilities

		Test Type									Test Facility System Tests							
		ATWS	Loss of feedwater, non ATWS	Loss of heat sink, non ATWS	Station blackout	Steam line break	Feed line break	Cooldown prim. feed and bleed	Reactivity disturbance	Overcooling	PWR 1:1*	LOFT 1:50	LSTF 1:48	BETHSY 1:100	PKL-I 1:134	SPES 1:430	LOBI-II 1:712	SEMISCALE 1:1600
Phenomena	Natural circulation in 1-phase flow	●	●	●	●	●	●	●	○	○	●	○	●		○		●	●
	Natural circulation in 2-phase flow	●	●	●	●	—	—	○	○	—	—	○	●		○		●	●
	Core thermohydraulics	●	●	●	●	○	○	○	●	○	○	●	●		—		●	●
	Thermohydraulics on primary side of SG	●	○	○	●	○	○	●	○	●	○	○	●		—		●	○
	Thermohydraulics on secondary side of SG	●	●	●	●	●	●	●	○	●	○	○	●		—		●	○
	Pressurizer thermohydraulics**	●	●	●	●	○	○	○	○	●	○	○	○		—		○	○
	Surgeline hydraulics (CCFL, choking)**	●	●	●	●	○	○	○	○	○	○	○	○		—		○	○
	Valve leak flow***	●	●	●	●	●	●	●	●	●		○	○		○		○	○
	1- and 2-phase pump behavior	●	●	●	●	○	○	○	○	○	○	○	○		—		●	●
	Thermohydraulic-nuclear feedback	●	—	—	—	—	—	—	●	—	●	●	—		—		—	—
	Structural heat and heat losses****	○	○	○	○	○	○	○	○	○	—	○	○		○		○	○
Test Facility System Tests	PWR	—	—	—	—	—	—	—	—	○								
	LOFT	●	●	●	○	—	—	●	●	●								
	LSTF																	
	BETHSY																	
	PKL-1	—	—	—	—	—	—	—	—	—								
	SPES																	
	LOBI-II	○																
	SEMISCALE	—	○	○	○	○	○	○	—	○								

Figure 1.8.1 Cross-reference matrix for large breaks in PWRs. (Reprinted from Ref. 1.8.1 with permission from Elsevier Science.)

tests to cover prevailing gaps in the tabulation are next identified, and finally, a specific additional test plan is formulated for the complex system being designed.

This new testing program and its successful completion may be the most important step in assuring the successful performance and safety of the complex system proposed. *The reason is that testing provides the most reliable means to validate the performance of the complex system.* For a few complex systems, it may be possible to show that their performance can be established only through a comprehensive test program and without resorting to a system computer code. This is most likely for complex systems with limited safety implications and where the prototype unit can be instrumented and subjected to operational tests to generate data for significant scenarios. To be successful, such an operational program may have to be preceded by separate effects tests of the important phenomena and scaled integral tests of the complex system. Even under such circumstances a good test matrix and plan would be valuable in validating the strategy, which relies upon testing exclusively. In most complex systems, the protection of the investment to be made in the prototype unit and the safety implications associated with some scenarios tend to favor the development of a system computer code or a similar means to predict the behavior of the complex system. The primary objective of the test program then shifts to validate the system computer code and the test matrix, and the plan can satisfy that goal only by emphasizing the need to:

- Ascertain that any new facility, particularly any new integral facility, is scaled correctly so that its experimental results are directly applicable to and can test the accuracy of the system computer code.
- Preselect specific tests in the new facilities to check the accuracy of the system computer code rather than adjusting the code to match all the test results.
- Provide available instrumentation to measure the key safety parameters identified by the system computer code.
- Employ other available integral tests, particularly if they are of different scales. Tests performed at different scales provide an independent means to assess the scaling ability of system computer codes.
- Apply engineering judgment to keep the testing and analyses within reasonable costs and schedules. The ranking of phenomena and interactions can guide that judgment. For example, Figure 1.8.1 shows a cross-reference matrix of separate effects and system tests that were judged to be important by the European community in validating computer system codes during a LBLOCA in PWRs [1.8.1]. The emphasis in Figure 1.8.1 is centered on tests carried out at the largest scale because they involve the least distortion.

1.9 SUMMARY

1. Most practical two-phase-flow applications involve complex systems and a multitude of phenomena. Such systems could benefit greatly from an integrated and

systematic program of testing and analysis, particularly if their operations have significant safety and economic implications.

2. A good integrated program of testing and analysis requires a well-defined set of characteristics of the complex system, the safety limits to be maintained during specified scenarios, and the degrees of margin to be provided. This type of information can also help to establish the strategy to be pursued (e.g., whether an analytical tool such as a computer system code is desirable). Most complex systems are best served by an interactive program between testing and analytical modeling at the system level.

3. An identification and ranking of the phenomena involved during all the relevant events is essential to guide the testing required and the features to be incorporated in the system computer code. The ranking can be expected to involve subjective judgment, and it is wise to utilize personnel experienced in the design, operations, phenomena, testing, and analysis of the complex system to formulate or review that judgment.

4. Testing offers the best means to validate the performance of complex systems. The necessary test matrix should be based on a phenomena identification and ranking assessment plan. The matrix needs to cover applicable tests performed previously as well as new tests to check the unique features and phenomena of the complex system proposed and the capability of the system computer code to predict them. If new facilities are included, they need to be subjected to scaling evaluations to achieve similitude between test facility and prototype, at least for the processes of greatest interest. If data from previously available facilities are employed, their scaling distortions must be taken into account.

5. If a system computer code is being used, it should be well documented and validated against test and operational data obtained from earlier versions of the complex system as well as from test results from the newly formulated test matrix. Sensitivity calculations performed with the updated system computer code can be used to establish the uncertainties associated with the predicted behavior of the prototype complex system. Bounding analyses or providing additional margins are alternatives to be considered where inadequate testing or insufficient correlation exists between tests and predictions.

6. Overall, the greatest value of an integrated testing and analysis program may come from the fact that it provides organization, systematization, understanding, and evaluation of the key issues of the complex system being implemented.

7. For those interested in more details about test and analysis programs, the comprehensive program recently published for the simplified boiling water reactor (SBWR) is suggested [1.1.10]. The SBWR design incorporates advanced passive safety features to cool the reactor core and containment of this type of nuclear power plant. Reference 1.1.10 describes the process employed to define the test and analysis needs for the SBWR, the key phenomena and system interactions involved, the new test facilities required, and their scaling adequacy.

REFERENCES

1.1.1 Lahey, R. T., Jr., and Moody, F. J., *The Thermal Hydraulics of a Boiling Water Nuclear Reactor*, La Grange Park, Ill., Monograph Series on Nuclear Science and Technology, American Nuclear Society, 1977.

1.1.2 Tong, L. S., and Weisman, J., *Thermal Analysis of Pressurized Water Reactors*, 2nd ed., La Grange Park, Ill., Monograph Series on Nuclear Science and Technology, American Nuclear Society, 1979.

1.1.3 Levy, S., "The Important Role of Thermal Hydraulics in 50 Years of Nuclear Power Applications," *Nucl. Eng. Des.*, **149**, 1–10, 1994.

1.1.4 *Application of Code Scaling, Applicability, and Uncertainty Evaluation Methodology to a Large Break, Loss-of-Coolant Accident*, NUREG/CR-5249, EGG-2552, 1989.

1.1.5 *An Integrated Structure and Scaling Methodology for Severe Accident Technical Issue Resolution*, NUREG/CR-5809, EGG-2659, 1991.

1.1.6 Zuber, N., *A Hierarchical, Two-Tiered Scaling Analysis*, Appendix D, NUREG/CR-5809, EGG-2659, 1991.

1.1.7 Zuber, N., "A Hierarchical Approach for Analyses and Experiments," in *Multiphase Science and Technology*, CRC Press, Boca Raton, Fla. (to be published).

1.1.8 Wilson, G. E., et al., *Component and Phenomena Based Ranking of Modeling Requirements for the NPR-MHTGR Analysis Code Verification and Validation*, EG&G Report EGG-NPR-9704, June 1991.

1.1.9 Wilson, G. E., and Boyack, B. E., "CSAU: A Tool to Prioritize Advanced Reactor Research," *Proceedings of the 5th Conference on Nuclear Thermal Hydraulics, ANS Winter Meeting*, San Francisco, pp. 377–385, November 26–30, 1989.

1.1.10 *SBWR Test and Analysis Program Description*, GE Report NEDO-32391 Revision C, 1995.

1.1.11 Hochreiter, L. E., and Piplica, E. J., *AP-600 Test and Analysis Plan*, WCAP-14141 (Preliminary Proprietary Report).

1.1.12 *Compendium of ECCS Research for Realistic LOCA Analysis*, NUREG-1230, 1988.

1.3.1 10CFR Part 50, "Acceptance Criteria for Emergency Cooling Systems for Light-Water-Cooled Nuclear Power Plants," *Fed. Reg.*, **39**(3), January 4, 1974.

1.3.2 Steiger, R., *Extended BE/ME Study*, Idaho National Engineering Laboratory Letter STIG-177-77, 1977. Also Boyack, B. E., et al., *Nucl. Eng. Des.*, **119**, 1–15, 1990.

1.7.1 Shaw, R. A., et al., *Development of a Phenomena Identification and Ranking Table (PIRT) for Thermal-Hydraulic Phenomena during a PWR LBLOCA*, NUREG/CR-5074, 1988.

1.7.2 Liles, D. R., et al., *TRAC-PF1/MOD1: An Advanced Best Estimate Computer Program for PWR Thermal-Hydraulic Analysis*, NUREG/CR-3858, 1986.

1.8.1 Wolfert, K., and Brittain, I., "CSNI Validation Matrix for PWR and BWR Thermal-Hydraulic System Codes," *Nucl. Eng. Des.*, **108**, 107–119, 1988.

CHAPTER 2

TESTING TWO-PHASE FLOW IN COMPLEX SYSTEMS

2.1 INTRODUCTION TO SCALING OF TESTS FOR COMPLEX SYSTEMS

The behavior of complex two-phase-flow systems can be predicted only after extensive testing has been performed in separate effects, component, and integral system facilities. The test data are applicable to the prototype system if the test facilities as well as the initial and boundary conditions of the experiments are scaled properly. The scaling methodology employed needs to be defined for each facility (particularly the integral system facility simulating the complex system) and evaluated to ascertain that no test distortions are present that can affect the important physical processes occurring during the scenarios of interest.

In single-phase flow, great strides were made in scaling in the early nineteenth century by employing dimensional analysis and similitude. It was applied by Froude to naval architecture, by Rayleigh to aeronautics, by Reynolds to hydraulics, and by Prandtl, Nusselt, and Stanton to heat transfer. Dimensional analysis revealed that the parameters involved during test work can be arranged into a definite number of independent dimensionless groups and the test results expressed in terms of such groupings. When these coefficients are kept relatively equal in the reduced-scale model and the prototype, the model tests are representative of the behavior of the prototype.

Dimensionless analysis is much more difficult to apply to a two-phase-flow complex system. The only overall model available for the entire system is the system computer code. However, the conservation equations of mass, momentum, and energy employed in that code are not always based on first principles and are averaged spatially. System computer codes are also known to be dependent on a large number (often in excess of 100) of empirical correlations, often referred to as *closure* or *constitutive equations*. Their dimensionless analysis would not be of great help in deciding how to scale required test facilities. For that reason, in Chapter 1 we emphasized how to formulate an integrated test and analysis program for complex systems. The program relied on experience and engineering judgment to identify the important phenomena and their interactions and to assure that the significant ones play relatively the same role in model test facilities and the prototype. However, no quantitative scaling strategy was offered in Chapter 1 to meet that objective. This shortcoming has been tackled in recent years by Zuber [2.1.1], who developed a structured approach to scale complex systems. It addresses the scaling issues in two tiers: a top-down (inductive) system approach, combined with a bottom-up (deductive) process-and-phenomena approach. The objectives of this technique, called *hierarchical two-tiered scaling* (H2TS), were to [2.1.2]:

- Have a method that is systematic, auditable, and traceable.
- Develop a scaling rationale and similarity criteria.
- Assure that important processes have been identified and addressed properly and provide a means for prioritizing and selecting processes to be addressed experimentally.
- Create specifications for test facilities design and operation (test matrix, test initial and boundary conditions) and provide a procedure for conducting comprehensive reviews of facility design, test conditions, and results.
- Assure the prototypicality of experimental data for important processes and quantify biases due to scale distortions or to nonprototypical test conditions.

The practicality of meeting all these objectives was recognized in Reference 2.1.3 by noting that "system tests do not have to provide exact system simulations of the prototype. In fact, it is neither practical nor desirable to attempt to provide such exact simulations." However, the overall (or top-down) and bottom-up behavior of the system test facility cannot be significantly different from that of the prototype. In other words, the system test facility and prototype should exhibit relatively the same degree of importance for the significant phenomena and processes involved and their overall integration. Before discussing how this can be achieved, a few practical guidelines for scaling two-phase-flow complex systems are provided. They are based primarily upon judgment developed over years of experience.

2.2 GUIDELINES FOR SCALING COMPLEX SYSTEMS

2.2.1 Fluid Properties and Initial Conditions

There are significant advantages to using the same fluid properties and initial conditions in the test facility and prototype. Two-phase-flow and heat transfer applications usually involve many properties, such as density (ρ), specific heat (c), viscosity (μ), and thermal conductivity (k). Many of these properties depend on the values of temperature (T) and pressure (p). Under those circumstances, it would be very difficult to preserve nondimensionless groupings without employing the same fluids and initial conditions. This becomes particularly true when changes of phase (i.e., evaporation and condensation) occur or when nonequilibrium liquid–vapor conditions prevail.

In a few cases, the use of a different fluid can be justified. For example, helium gas has been utilized to replace hydrogen gas for test safety reasons. Similarly, air has been found to be an acceptable substitute for pure nitrogen. Still, the experimenter is expected to recognize the limited impact of such fluid changes when correlating the test data. In some circumstances, different fluids have been introduced to allow testing at reduced pressure where visual observations can be made. Refrigerant fluids fall into that category and sometimes have helped improve the physical understanding of two-phase-flow processes. However, seldom have they produced meaningful quantitative correlations when changes of phase and nonequilibrium conditions occur.

Some of the relevant phenomena can happen at different system conditions. For instance, when the reactor core of a nuclear power plant is being reflooded with water after a large-break loss-of-coolant accident (LBLOCA), the pressure in the reactor vessel is low and it would be prohibitively expensive to measure water reflooding heat transfer coefficients in a full-pressure reactor test facility. However, the initial conditions expected in the low-pressure facility must be defined. A system computer code validated up to that phase of the LBLOCA can be employed, as long as the uncertainties in the code predictions for the initial condition of the tests are recognized. Another accepted practice is to employ initial conditions that can be shown to be conservative or to generate conservative test results. Experience has shown that these are reasonable and justifiable compromises between facility costs and exact duplication of initial conditions.

In summary, it is preferable for complex systems involving a change of phase to employ the same fluids, properties, and equivalent initial conditions as the prototype.

2.2.2 Geometric Scaling

There are several potential two-phase-flow guidelines under this topic:

1. Tests should be performed at as large a geometric scale as practical. In nuclear power plants, components that are of a modular nature (i.e., fuel assemblies, steam separators, etc.) tend to be tested at full scale over their entire range of operating conditions. If the component is tested at reduced scale, experience has shown that two-phase-flow tests at about one-third scale may be extrapolable to full scale. However, it is wise to confirm that extrapolation by performing counterpart tests at different scales to support it.

2. It is not necessary or usually advisable to scale down the geometry by the same amount in all three dimensions. Referred to as *linear scaling*, this is particularly not useful when change of phase, heat, or mass transfer is present. Often, it is preferable to retain as close to full-scale dimension in the flow direction as possible if it is important to the phenomena being evaluated. When LBLOCAs were discussed

in Chapter 1, the parameter of most interest was noted to be the peak temperature of the fuel cladding, which was found to be dependent on the heat stored and generated in the fuel and the prevailing heat transfer conditions at that location (i.e., the local streamwise flow conditions in the vertical direction along the fuel). For that reason, preserving the length in the flow or vertical direction is much more important in LBLOCAs than preserving it in the other directions. Scaling for LOCAs in nuclear power plants basically has been a streamtube approach. In other words, the test facility is designed to duplicate a portion of the prototypical reactor flow confined within a streamtube. In the prototype, the streamtube does not have a solid boundary, whereas it would in a reduced-scale facility. The solid walls introduce pressure drop and heat losses in the scaled-down facility which do not exist in the prototype, and it is important that such losses be minimized and that the streamtube being tested be large enough not to affect thermal-hydraulic behavior away from the walls. Most system integral thermal-hydraulic test facilities have tended to preserve the vertical dimension, time, velocity, and heat generation in the flow direction, particularly when flashing, evaporation, or condensation occurs. This has been referred to as volumetric or *power/volume scaling*.

3. For selective phenomenon evaluation it is sometimes desirable to emphasize a dimension transverse to the flow direction. For instance, when interactions of water being sprayed above the reactor core were studied with steam being generated and rising upward, a 30° sector test facility of the top of the reactor core was built [2.2.1]. Prototypical hardware and geometry in the 30° sector were employed in the top region of the core and above it to study the prevailing countercurrent flow of water and steam. Here again there were distortions produced by the walls of the 30° sector and it was necessary to show that they were not significant for the sector size selected. Similarly, if one were interested in the nondimensional grouping controlling the countercurrent flow in the downcomer annulus of a pressurized water reactor (PWR) during a LOCA, the preferred approach would be to preserve the size of the annulus and to use a large enough sector of the full-scale downcomer to be able to disregard the effects of the walls at the two ends of the arc. Unfortunately, as illustrated in Figure 1.2.3, there are also significant asymmetries around the circumference of the downcomer in a PWR during a LOCA, and most tests have opted to employ a reduced scale for the entire vessel and annulus. However, counterpart tests at different scales were necessary, including full-scale experiments in the Upper Plenum Test Facility [2.2.2], to develop acceptable countercurrent flow correlations.

4. As the geometric scale is reduced, it cannot be decreased to the point where it affects the specific phenomena of interest. For instance, in some nuclear power plant containments, the mixing of gases, noncondensable gas distribution, and the temperature and density stratification taking place are important. The scaling of such a test facility must therefore be large enough to evaluate the potential development of thermal plumes and the spreading of temperature and noncondensable maldistributions, while limiting the wall effects produced by the reduced scale.

5. Many components in complex systems are connected by piping, and it is important that the scaled-down versions of such connecting channels not affect

the component transient behavior and the differential pressure between significant components. In practice, the pipe flow areas are oversized for convenience in the scaled facilities, which leads to reduced velocities and form losses in the test facility. The frictional losses tend to be larger, due to the smaller pipe sizes of the test facility, but the total pressure drops dominated by local losses usually end up being somewhat smaller than in the prototype, and local orificing is added to match the total pressure drop. This approach is acceptable as long as the transit times between volumes are small compared to the significant component volume fill times, and the velocities do not become so low as to introduce new phenomena in the model tests.

6. Two-phase flow is known to depend on the flow pattern (i.e., how the gas and liquid arrange themselves). So do the closure and constitutive equations used in system computer code geometries. Scaling requirements are difficult to satisfy across the entire range of flow patterns. For instance, in a recent study of flow-pattern transition boundaries [2.2.3] it was found that distortions are unavoidable. Experience shows again that it is best to focus on the most significant flow pattern (i.e., the pattern that may occur in the prototype) and where and when it has a significant influence on the behavior of the complex system.

2.2.3 Scaling Interactions

Most complex systems employ parallel trains or multiple key components to achieve reliability. Within some components, there are many parallel elements or flow paths (e.g., tubes in a heat exchanger or steam generator). When parallel trains or multiple key components and elements can interact with each other, the interactions must be shown to be insignificant, or the integral test facility must provide for the parallel paths. The same is true of interactions among subsystems of a different nature (e.g., forced and natural circulation subsystems operating simultaneously).

2.2.4 Typical Scaled Integral Facility: Semiscale

Application of the preceding guides can best be illustrated by looking at the semiscale facility located at the Idaho National Engineering Laboratory, which was used to simulate a range of loss-of-coolant accidents (LOCAs) for nuclear power plants. The latest version of that facility, referred to as Semiscale Mod-2C [2.2.4], is depicted in Figure 2.2.1. It is observed that the test arrangement included most of the important components: pressurizer, pumps, loops, steam generators, and reactor vessel. The vessel included an electrical heated core made up of 25 heaters simulating full-length (3.7-m) fuel rods and their power generation and distribution. The tests employed high-temperature, high-pressure light water and reproduced the initial conditions found in full-scale pressurized water reactors (PWRs): pressure, core differential temperature, and cold leg water subcooling. The facility provided for one active (intact) loop and one broken loop, and both loops had full-length steam generator and pump equivalent loop seals. Note that the reactor vessel and broken-loop steam generator were provided with external downcomers to avoid geometric and flow distortions that might have resulted from trying to keep them inside the

2.2 GUIDELINES FOR SCALING COMPLEX SYSTEMS   33

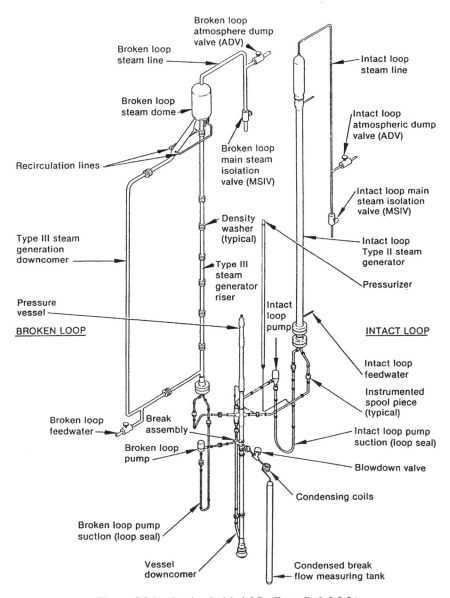

Figure 2.2.1 Semiscale Mod-2C. (From Ref. 2.2.2.)

reactor vessel and the steam generator. Another potential distortion (i.e., excessive heat losses from the tall and skinny semiscale facility) was corrected for by using external heater tapes. A large range of tests was performed from the early 1970s to 1986 in the various versions of the semiscale facility that were built. They included small and large breaks, steam generator tube rupture, natural circulation, power-loss transients, and station blackout events. An important advantage of this reduced-scale (1:1705) facility was the fact that it could be well instrumented and provide information about pressure, peak clad temperature, break flow, liquid level, flows, and volume fractions occupied by the steam as a function of time during a variety of simulated accidents.

In summary, scaling of complex systems needs to be controlled by the significant phenomena and processes being evaluated. For two-phase flow it is best to use as large a scale as practical. In most integral system thermal-hydraulic facilities, a power/volume scale has been the preferred approach. However, the biases in frictional, heat, and other transfer processes produced by power/volume scaling must be taken into account and minimized.

2.3 HIERARCHICAL TWO-TIERED SCALING OF COMPLEX SYSTEMS

The hierarchical two-tiered scaling methodology relies on a *hierarchical structure* to decompose and organize the complex system, starting at the top from the whole system and working downward through subsystems, components, and subcomponents to reach the transfer processes. In that sense, it is very similar to the approach taken in Chapter 1 to deal with the scenario for a large-break loss-of-coolant accident (LBLOCA) in a PWR. There one started with the entire complex system that is the nuclear power plant. It was subdivided next into subsystems, and for a LBLOCA the subsystem of most interest was the primary water reactor system. That subsystem (Table 1.7.1) was subdivided again into several relevant components: reactor pressure vessel, pressurizer, accumulator, steam generator, pump, and loop. The components, in turn, were decomposed (e.g., the reactor pressure vessel into the reactor core, downcomer, upper and lower plenum). Within the reactor core, next came the fuel assemblies and the fuel rods. The corresponding generic hierarchy developed by Zuber for scaling complex systems is shown in Figure 2.3.1. Its similarity to the breakdown in Chapter 1 is strong but not surprising because both approaches employ a top-down process. One significant difference is that Zuber assigned scale measures at each level of the hierarchical structure, as shown in Figure 2.3.1. At the lowest level, the transfer process is characterized by the rate of transfer (temporal scale) and by the transfer area (spatial scale). Because of the presence of other constituents (phases and/or geometries), a third scale measure, the volume fraction, needs to be taken into account. The applicable time (τ), space (L), and volume fraction (α) scales are listed at each structure level of Figure 2.3.1. It should be noted that under the level assigned to phases, the possibility of a gas (g), liquid (f), and solid (s) are recognized. Each geometrical configuration is also described by three field equations, which assure the conservation of mass, energy, and momentum.

The three basic scale measures (mass or volume, length, and time) provided in Figure 2.3.1 vary from level to level. At the low levels, the spatial and temporal scales focus on the local phenomenon or transfer process prevailing at the bottom of the hierarchical structure, and they tend to be small in magnitude. They get averaged or integrated as they are transmitted to the level above. The scales therefore grow with each upward step, and the hierarchical levels shown in Figure 2.3.1 are

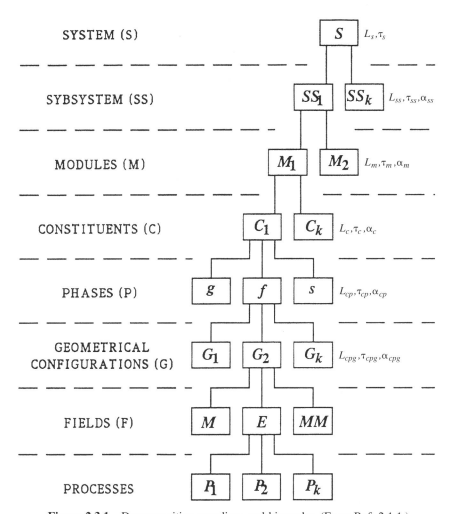

Figure 2.3.1 Decomposition paradigm and hierarchy. (From Ref. 2.1.1.)

isolated from each other because they tend to operate at distinctly different rates and scales. In other words, localized behavior is dealt with at the lower level, while integrated or overall system performance is handled at the higher levels.

The three basic measures that characterize a transfer process can be combined into a characteristic frequency which when multiplied by the temporal scale of the overall system, generates a dimensionless scaling group. Such groupings can be used to establish the relative importance of each transfer process to the overall system and the strength of their interactions. This scaling approach will be demonstrated first generically by relying primarily on material provided in References 2.1.1 and 2.1.2. That discussion is followed by several specific top-down and bottom-up applications to two-phase-flow complex systems.

2.3.1 Hierarchy for Volume, Spatial, and Temporal Scales

Let us first consider a control volume V_{cv}, which is occupied by several interacting constituents. The control volume can be viewed as a window through which one looks at the problem of interest, which can be placed at any hierarchical level, depending on the problem and questions under consideration. The volume V_c occupied by a particular constituent C is a fraction α_c of the control volume V_{cv}, or $V_c = \alpha_c V_{cv}$. This particular constituent C is presumed to involve a liquid and a gas phase. Each phase can have different geometrical shapes (i.e., in the case of the liquid it can be in the form of pools, films, drops, etc). Each geometrical shape also has its own characteristic transfer area which must be recognized. This means that for phase P, the total volume it occupies is V_{cp} and it is a fraction α_{cp} of the volume V_c, or $V_{cp} = \alpha_{cp} V_c$. For a particular geometry G of phase P, the volume it occupies is V_{cpg} and it is a fraction α_{cpg} of the volume V_{cpg}, or $V_{cpg} = \alpha_{cpg} V_{cp}$. This hierarchical volume tree for the control volume V_{cv} is illustrated in Figure 2.3.2, which shows that the volume occupied in the control volume by a specific geometry of a phase of a constituent is a product of three volume fractions, which all have a value less than unity, or $V_{cpg} = \alpha_c \alpha_{cp} \alpha_{cpg} V_{cv}$. The volume V_{cpg} is therefore observed to decrease as one proceeds along the hierarchical levels of Figure 2.3.2.

The spatial scale associated with a specific transfer process is obtained by recognizing that the process occurs from a volume across a specific area, and its value is derived by dividing the transfer area by the volume. For example, for a fluid flowing in a circular pipe of diameter D and length L, the spatial scale for heat transfer would be the heat transfer area divided by the volume, or $\pi DL/(\pi D^2 L/4) = 4/D$. The corresponding spatial scale for pressure loss from the coolant in the

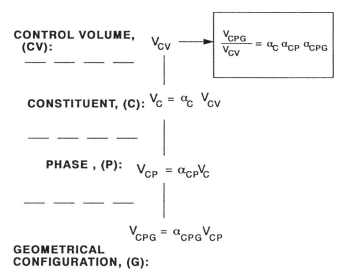

Figure 2.3.2 Hierarchy for volume fractions of geometric configuration G in control volume V_{cv}. (From Ref. 2.1.1.)

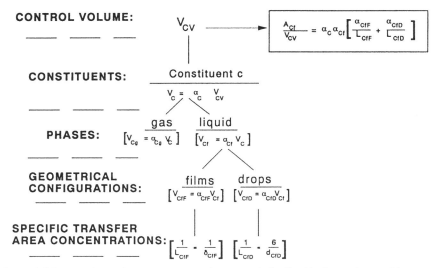

Figure 2.3.3 Total transfer area concentration ($A_{cf}N_{cv}$) for liquid of constituent C in control volume V_{cv} when the liquid is present in a film and drops. (From Ref. 2.1.1.)

pipe is $(\pi D^2/4)/(\pi D^2 L/4) = 1/L$. Similarly, the heat transfer spatial scale is $6/d$ for a sphere of diameter d, and $1/\delta$ for a film of thickness δ. The extension of this concept to a phase of a constituent in a specific geometry is obtained by forming its area/volume ratio, or $A_{cpg}/V_{cpg} = 1/L_{cpg}$; but primary interest is in the spatial scale at the control volume level, or A_{cpg}/V_{cv}, which, according to Figure 2.3.2, is $A_{cpg}/V_{cv} = \alpha_c \alpha_{cp} \alpha_{cpg}(1/L_{cpg})$.

Here again, the importance of the process spatial scale ($1/L_{cpg}$) is seen to decrease when viewed from a total system viewpoint. This same procedure is illustrated in Figure 2.3.3 for a process involving a liquid phase f in two geometric forms: film of thickness δ_{cfF} and spherical drops of diameter d_{cfD}. The proposed methodology is thus capable of distinguishing between different geometric forms.

The temporal scale τ is obtained by recognizing that every transfer process is characterized by a rate and by a transfer area and that the property being transferred can be mass, momentum, or energy. If we let ψ represent the property per unit volume and j_T be the flux of that property from a volume V across a transfer area A_T, the temporal scale τ corresponds to the time it takes the process to transfer the property contained in the volume V across the transfer area A_T, or

$$\tau = \frac{\psi V}{j_T A_T} = \frac{1}{\omega} \qquad (2.3.1)$$

where ω, the specific frequency, represents the number of times per second the amount of the property ψ in volume V is being changed due to the transfer process, $j_T A_T$. For example, for heat transfer at a flux q'' to or from a sphere of density ρ and diameter d with internal energy e per unit mass, the specific frequency is $\omega =$

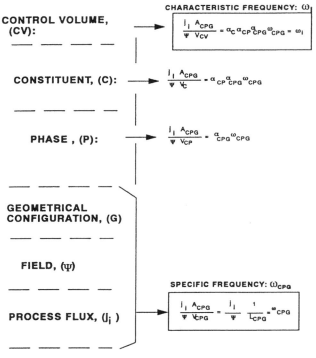

Figure 2.3.4 Hierarchy for the characteristic frequency ω_i (temporal scale) of the transfer process j_i in control volume V_{cv}. (From Ref. 2.1.1.)

$(q''/\rho e)(6/d)$. In the case of enthalpy transfer in a pipe of diameter D and length L, the specific frequency is $\omega = (q''/\rho H)(4/D)$, where H is the fluid enthalpy per unit mass and ρ is the density of the fluid.

The extension of this concept to a transfer process involving several constituents and different phases interacting in control volume V_{cv} is illustrated in Figure 2.3.4. Two specific frequencies are given in Figure 2.3.4: the specific frequency ω_{cpg} at the process level and the specific frequency ω_{cv} at the volume control level, or $\omega_{cv} = \alpha_c \alpha_{cp} \alpha_{cpg} \omega_{cpg}$. The ratio ω_{cpg}/ω_{cv} is nondimensional. That scaling group describes the importance of the specific process at the control volume level. This frequency-time-domain (FTD) method, recommended by Zuber, is quite general and very useful in developing dimensionless groupings.

2.3.2 Dimensionless Scaling Groupings

Multiplying the specific frequency associated with a process by the response time of interest at a desired structure level generates a dimensionless scaling group to assess the importance of the process at that level. To illustrate this approach, let us first look at a single-constituent, single-phase flow condition in a pipe of length L

and diameter D that is being subjected to a heat flux q''. The specific frequency for that process was shown to be

$$\omega = \frac{q''}{\rho H} \frac{4}{D} = \frac{q''A_H}{\rho HV} = \frac{P/V}{\rho H} \quad (2.3.2)$$

where A_H is the heat transfer area of the facility, P the power, and V the associated volume. If the fluid properties and initial conditions are preserved from the test facility to the prototype, and if the specific frequency is to remain the same, Eq. (2.3.2) states that the power/volume ratio needs to be retained from the prototype to the experiment to ensure applicability of test data on the same time scale. This explains why most thermal-hydraulic facilities have employed a power/ volume scaling strategy.

The time of interest t for the process is the transit time through the pipe length L, or $t = L/u$, where u is the fluid velocity and the corresponding nondimensional grouping is

$$\Pi = \frac{q''}{\rho H} \frac{4}{D} \frac{L}{u} = \frac{h}{\rho c_p u} \frac{4L}{D} \quad (2.3.3)$$

In Eq. (2.3.3) one took advantage of the fact that the heat flux q'' is proportional to the heat transfer coefficient h and that the fluid enthalpy is proportional to c_p, with ρ being the fluid density and c_p its specific heat at constant pressure. The temperature differences involved in the top and bottom of Eq. (2.3.3) dimensionalize each other and can be excluded from the proposed relation. The dimensionless grouping derived in Eq. (2.3.3) involves the Stanton number, St, or St $= h/\rho c_p u$ and the geometric group ($4L/D$). These two groupings are known to scale the heat transfer process for single-phase flow in a pipe.

Let us consider another example, the case of cooling or heating of a sphere of diameter d; the specific frequency for that process was shown to be $(q''/\rho e)(6/d)$. At any time t, the applicable nondimensionless grouping is $\Pi = (q''/\rho e)(6/d)t = 6(hd/k)(kt/\rho c_v d^2)$ where q'' was assumed proportional to h and the internal energy ρe to ρc_v (c_v is the specific heat at constant volume). Also, the thermal conductivity k was introduced to form the two well-known dimensionless groupings for this problem of Biot, Bi $= hd/k$ and Fourier, Fo $= kt/\rho c_v d^2 = \kappa t/d^2$, where κ is the thermal diffusivity and is equal to $k/\rho c_v$.

In Reference 2.1.2, Zuber highlighted the significance of this frequency-time-domain (FTD) approach by applying it to numerous transfer processes. He demonstrated not only its usefulness but also its general applicability to scaling. However, as noted in that lecture, the same nondimensional groups could have been derived just as well by writing the appropriate balance equations for the control volume of interest and normalizing them.

2.3.3 Scaling Hierarchy

If a component involves several transfer processes, the dimensionless groups or characteristic time ratios (process/component response ratios) applicable to the vari-

40 TESTING TWO-PHASE FLOW IN COMPLEX SYSTEMS

ous processes can be compared to each other and ranked according to their numerical value. The process with the highest-value dimensionless group has the most relevance to the component, and the test facility should attempt to preserve it particularly in the scaled-down component. This process is repeated at higher levels of the structure of the complex system, and in so doing, a scaling hierarchy can be generated that would rank the various processes in their order of relevance at the top level of the overall system and downward along its lower hierarchical levels. The H2TS methodology is therefore capable of providing an alternative to the experience and engineering judgment ranking approach described in Chapter 1. In fact, the two strategies are not fully independent but rather, complement each other, for several reasons:

- The number of nondimensionless groupings to be preserved is usually large for a complex system, and experience and engineering judgment are still valuable to supplement any final scaling ranking derived from the H2TS methodology.
- There will be distortions for the less important processes in test facilities, emphasizing the phenomena of greatest interest. Their impact is best evaluated from counterpart tests or from system computer code assessments. Here again, experience and engineering judgment are helpful to limit the number of required extra tests and system computer code analyses.

In summary, the hierarchical two-tiered scaling (H2TS) methodology provides a comprehensive and systematic approach to testing and scaling of complex systems. Its effectiveness can benefit from being coupled to knowledge derived from experience and engineering judgment about the complex system being investigated. In the sections that follow, more details are provided about its application to scaling of integrated subsystems and components (top-down approach) and scaling of specific processes that may occur within the subsystems (bottom-up approach). Here again, nuclear reactors and their containments will be employed to illustrate the approach; the concepts are, however, are applicable to other complex two-phase-flow systems.

2.4 ILLUSTRATIONS OF TOP-DOWN SCALING

2.4.1 Top-Down Scaling of Spatially Uniform System

An often used top-down approach to either the entire primary system [also referred to as the *reactor cooling system* (RCS)] or the containment of a nuclear power plant looks at their entire volume V and presumes that it is well mixed. Several constituents may be present in those volumes, and their distribution and properties are assumed to be uniform and in quasi-thermodynamic equilibrium, although changing with

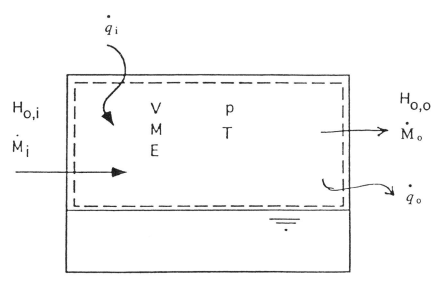

Figure 2.4.1 Control volume (dashed line in the case of a gas space) with incoming and outgoing mass flow and heat rates.

time. Such a spatially uniform system, depicted in Figure 2.4.1, is particularly valid during the blowdown of the RCS or pressurization of the containment by the blowdown from the RCS. It contains a mass M with internal energy E at a pressure p and a temperature T. Its volume V can be time dependent, and mass and energy can flow across its boundary. Changes in the kinetic and potential energy of the mass M are much smaller than changes in its intrinsic internal energy and are neglected, and therefore no momentum conservation equation is employed. The mass conservation equation for Figure 2.4.1 can be written as

$$\frac{dM}{dt} + \sum_o \dot{M}_o - \sum_i \dot{M}_i = 0 \qquad (2.4.1)$$

where $\sum_o \dot{M}_o$ and $\sum_i \dot{M}_i$ are, respectively, the sum of the outgoing and incoming mass flow rates, which often are represented by the symbols $\sum_i W_i$ and $\sum_o W_o$. Their respective total enthalpies are $H_{o,o}$ and H_{oi}. The total enthalpy used herein with a first subscript o includes the kinetic and potential energy. The energy conservation equation is

$$\frac{dE}{dt} = -p\frac{dV}{dt} + \dot{q}_i - \dot{q}_o + \sum_i W_i H_{o,i} - \sum_o W_o H_{o,o} \qquad (2.4.2)$$

where p is the pressure and $\dot{q}_i$ and $\dot{q}_o$ are, respectively, the heat added or leaving the system. The specific internal energy of the system $e = E/M$ obeys a state equation of the form

$$e = \frac{E}{M} = e(p,v,y_j) \tag{2.4.3}$$

where v is the specific volume, $v = V/M$, and y_j represents the mass fraction of the various constituents with $\Sigma_i\, y_j = 1$.

The differential of the specific internal energy is obtained from

$$de = \left(\frac{\partial e}{\partial p}\right)_{v,y_j} dp + \left(\frac{\partial e}{\partial v}\right)_{p,y_j} dv + \sum_j \left(\frac{\partial e}{\partial y_j}\right)_{p,v,y} dy_j \tag{2.4.4}$$

where the subscripts v, p, and y_j mean that they are kept constant in the partial derivatives, while the subscript y means that all y_j are kept fixed except for the one appearing in the partial derivative.

Combining Eqs. (2.4.1), (2.4.2), and (2.4.4) and taking advantage of the definitions of the specific energy e and the specific volume v, the following equation is derived for the pressure rate dp/dt:

$$\frac{V}{v}\left(\frac{\partial e}{\partial p}\right)_{v,y_j}\frac{dp}{dt} = \dot{q}_i - \dot{q}_o + \sum_i W_i\left[H_{o,i} - e + v\left(\frac{\partial e}{\partial v}\right)_{p,y_j}\right]$$

$$- \sum_o W_o\left[H_{o,o} - e + v\left(\frac{\partial e}{\partial v}\right)_{p,y_j}\right] - \left[p + \left(\frac{\partial e}{\partial v}\right)_{p,y_j}\right]\frac{dV}{dt}$$

$$- \frac{V}{v}\sum_j \left(\frac{\partial e}{\partial y}\right)_{p,v,y}\frac{dy_j}{dt} \tag{2.4.5}$$

A similar expression was developed by Moody [2.4.1] for the case of single-component flow, which further simplifies Eq. (2.4.5) to

$$\frac{dp}{dt}\frac{V}{v}\left(\frac{\partial e}{\partial p}\right)_v = \dot{q}_i - \dot{q}_o + \sum_i W_i\left[H_{o,i} - e + v\left(\frac{\partial e}{\partial v}\right)_p\right]$$

$$- \sum_o W_o\left[H_{o,o} - e + v\left(\frac{\partial e}{\partial v}\right)_p\right] - \left[p + \left(\frac{\partial e}{\partial v}\right)_p\right]\frac{dV}{dt} \tag{2.4.6}$$

Equation (2.4.6) as well as the applicable equations of state were nondimensionalized by Moody (2.4.1), who derived a total of 13 model coefficients, with eight of the scaling groups coming from the equation of state. Such a large number of nondimensional groups, for the equations of state, even for single-component flow, provides the strongest argument for preserving the fluid properties in scaling complex systems. The preceding top-down controlling equations can be simplified even more by looking at specific characteristics of the system and of the event under consideration.

2.4.2 Top-Down Scaling during Reactor Coolant Blowdown Phase

As noted previously, it is advantageous to deal with specific transients and accidents and preferably with limited portions of such events (e.g., the blowdown phase of a LBLOCA in a PWR). During that very short period of time, there is no change in the volume of the reactor coolant system (i.e., $dV/dt = 0$); there is no opportunity to add new coolant, or $\Sigma_i W_i = 0$, and the only flow escaping the reactor vessel is the mass flow rate being discharged from the vessel, which during the early phases is the critical mass flow, or W_c.

If the mass conservation equation (2.4.1) is nondimensionalized by using the following elements:

- For volume: V_r, reactor or primary system volume
- For density: ρ_r, initial density in primary system and $v_r = 1/\rho_r$
- For mass flow rate: W_c, initial critical flow rate from vessel
- For time: τ_r, reference time for reactor system

a nondimensional time grouping emerges:

$$\Pi_t = \frac{V_r \rho_r}{W_c \tau_r} \qquad (2.4.7)$$

The choice of the reference time scale is arbitrary, but its value should be chosen to make the magnitude of the nondimensional time derivatives on the order of unity. This can be achieved by using

$$\Pi_t = 1 \quad \text{or} \quad \tau_r = \frac{V_r \rho_r}{W_c} \qquad (2.4.8)$$

The reference time, τ_r, obtained from Eq. (2.4.8) has a physical meaning because it represents the time to blow down the primary system. *An important aspect of the nondimensionalized process used in this book and especially recommended for complex systems is to employ reference parameters with physical and practical meaning.*

If the energy equation (2.4.6) also is nondimensionalized by using

- For heat addition: $\dot{q}_{i,r}$, initial decay heat
- For pressure: Δp_r, pressure difference between reactor vessel and containment
- For total enthalpy: ΔH_r or preferably H_{fg}, heat of vaporization in the case of a flashing component

two nondimensional groups emerge:

$$\Pi_1 = \frac{\dot{q}_{i,r}\tau_r}{\Delta p_r V_r} \quad \text{and} \quad \Pi_2 = \frac{H_{fg}}{\Delta p_r/\rho_r} \tag{2.4.9}$$

For a large break, the reactor vessel depressurization is seen to be globally controlled by the critical flow leaving the vessel and the net heat addition to the reactor coolant system. The net heat entering the primary system generally can be neglected during the blowdown process by comparison to the energy leaving it through the break. Under those circumstances, one is left with a single important nondimensional grouping, Π_2, which is referred to as the *enthalpy–pressure number* because it relates the enthalpy and pressure scales. This significant grouping is scaled exactly when the fluid properties and initial conditions are preserved from the prototype to the model facility.

Furthermore, with no heat addition or removal, Eq. (2.4.6) is simplified to

$$\frac{dp}{dt} = -\frac{v/V}{(\partial e/\partial p)_v} G_c A_B \left[H_{o,o} - e + v \left(\frac{\partial e}{\partial v}\right)_p \right]$$

where G_c is the critical mass flow rate per unit area and A_B is the break area. The critical mass flow rate G_c is known to be a function of the pressure p and the total enthalpy $H_{o,o}$ or $G_c = G_c(p, H_{o,o})$. With the same fluid properties and initial conditions, the critical flow and the pressure rate dp/dt will be preserved from prototype to the scaled model if the ratio $A_B/V = 1/L_B$ is maintained.

The preceding example for blowdown from the reactor vessel during a LBLOCA shows that a top-down uniform system approach is capable of identifying the most significant enthalpy–pressure nondimensional group and of specifying the spatial scale, A_B/V, and the time scale, $\tau_r = V_r\rho_r/W_c$, of most importance to that scenario. Nondimensional analysis was employed in this top-down example, but similar results could have been obtained from the frequency-time domain (FTD) recommended by Zuber (2.1.2). For example, one of the specific frequencies, ω, could have been derived from the ratio of heat addition or removal, $\dot{q}_{i,r}$, to the heat capacity of the receiving volume, $V_r \rho_r \Delta H_r$, so that $\omega = \dot{q}_{i,r}/V_r\rho_r \Delta H_r$. Multiplying this frequency ω by the system time scale yields

$$\Pi_3 = \frac{\dot{q}_{i,r}\tau_r}{V_r\rho_r \Delta H_r} = \frac{\dot{q}_{i,r}\tau_r}{\Delta p_r} \frac{\Delta p_r}{V_r \Delta H_r \rho_r} = \frac{\dot{q}_{i,r}}{W_c \Delta H_r} \qquad (2.4.10)$$

The nondimensional grouping Π_3 appearing in Eq. (2.4.10) involves the groups Π_1 and Π_2 obtained from previous nondimensional analysis, since $\Pi_3 = \Pi_1/\Pi_2$. This grouping Π_3 has been referred to as the *phase change number*, particularly when ΔH_r is replaced by the heat of vaporization H_{fg}.

If we define the system scale R as the ratio of prototype to test facility power input, and if prototypical fluids are used in the model, the phase change Π_3 group would be preserved only if the ratio prototype/test facility critical flow W_c is R. To preserve the time scale τ_r of Eq. (2.4.7), the ratio prototype/test facility volume would have to be R and a power/volume scaling strategy would result for the blowdown phase. The nondimensional group Π_1, requires the ratio of pressure drops to be 1. Under those circumstances, the ratio prototype/test facility flow area would be R to match the ratio for W_c, and the fluid velocity through the break would be the same for the prototype and test facility. If the same principles are applied to the reactor core, the ratio prototype/test facility flow area would be R and the fuel-rod heat flux and length and the velocity along it would be the same and would produce the same fuel temperature. The top-down methodology therefore requires a power/volume scaling strategy in the reactor core and overall facility. It matches the approach used successfully in the semiscale facility discussed in Section 2.2.4 and in other integral test facilities for nuclear reactors.

Despite its effectiveness, top-down scaling has several limitations that need to be recognized. They come primarily from the assumption of spatially uniform conditions. For example, the large break is usually located in the cold piping leg of a PWR, to maximize the escaping flow rate from the reactor vessel because the water is the most subcooled at that position. This subcooling, which increases the critical flow rate, cannot be considered in the proposed top-down approach because the entire system is presumed to be well mixed. Similarly, the location and geometry of the break will be shown to have an influence on the critical flow rate. Such and other similar scaling refinements can be taken into account only by complementing the top-down process with relevant bottom-up treatment. In addition, it is important to evaluate the distortions (e.g., heat losses) produced by the top-down scaling strategy to ascertain that they do not affect the results generated by the test facility.

2.4.3 Top-Down Scaling of Vapor Volume Fraction during Reactor Blowdown

During the blowdown phase, it may be necessary to know the volume or mass fraction of the vapor and liquid phases in the reactor system. For example, this information is needed to specify the liquid level or the flow pattern in the reactor coolant system. If one writes a mass balance equation for each phase, there results

$$\frac{dM_l}{dt} = -\sum_o W_{l,o} - \dot{M}_{fg} \quad \text{and} \quad \frac{dM_g}{dt} = -\sum_o W_{g,o} + \dot{M}_{fg} \quad (2.4.11)$$

where $\dot{M}_{fg}$ is the evaporation rate and the subscripts l and g are used to represent the liquid and vapor phases, respectively. The term $\dot{M}_{fg}$ can be obtained by multiplying the liquid mass balance by the liquid enthalpy H_l and the vapor equation by the gas enthalpy H_g in the blowdown vessel and adding them, or

$$H_{fg}\dot{M}_{fg} = H_l \frac{dM_l}{dt} + H_g \frac{dM_g}{dt} + \sum_o W_{l,o} H_l + \sum_o W_{g,o} H_g \quad (2.4.12)$$

If one rewrites the energy equation (2.4.2) in terms of the fluid enthalpy by recognizing that

$$E = Me = M(H - pv) = MH - pV = M_l H_l + M_g H_g - pV$$

and combines Eq. (2.4.2) with Eq. (2.4.12), there results

$$H_{fg}\dot{M}_{fg} = \dot{q}_i + \frac{dp}{dt}\left(V - M_l \frac{dH_l}{dp} - M_g \frac{dH_g}{dp}\right) - \sum_o W_o(H_{o,o} - H) \quad (2.4.13)$$

Many investigators have tended to neglect the last term in Eq. (2.4.13) because the properties of the fluids outside the break are the same for a spatially uniform system as those in the reactor vessel. Another justification for dropping the last term is the assumption made at the start of Section 2.4.1 that changes in kinetic and potential energy are much smaller than changes in internal energy. If the resulting simplified expression for the evaporation rate $\dot{M}_{fg}$ is next introduced into the vapor mass balance equation (2.4.11), the following expression is obtained for the rate of change of the volume fraction α:

$$\rho_g \frac{d\alpha}{dt} = -\frac{1}{V}\sum_o W_{g,o} + \frac{\dot{q}_i}{H_{fg}V}$$

$$+ \left(\frac{1}{H_{fg}}\right)\frac{dp}{dt}\left[1 - (1-\alpha)\rho_l \frac{dH_l}{dp} - \alpha\rho_g \frac{dH_g}{dp}\right] - \alpha \frac{d\rho_g}{dp}\frac{dp}{dt} \quad (2.4.14)$$

where the vapor volume fraction α is defined from $M_g = V\rho_g \alpha$ with $M_l = V\rho_l(1-\alpha)$. Equation (2.4.14) does not provide any new nondimensional group except for reinforcing the need to substitute H_{fg} for ΔH_r. It also shows that if the gas volume fraction α is to be preserved from prototype to model, it is advantageous to utilize the same fluid properties because Eq. (2.4.14) depends on several additional liquid and vapor properties and their derivatives with respect to pressure.

2.4.4 Top-Down Scaling of Spatially Uniform Containment

Let us consider the early portion of a blowdown of the reactor system into a large dry containment full of air. Again, all the constituents present in the containment are presumed to be well mixed and at the temperature and pressure prevailing in the containment. During this portion of the blowdown, there is no or minimal leakage from the containment, so $\Sigma_o W_o = 0$ in Eqs. (2.4.1) and (2.4.2). The heat losses from the containment are negligible, so $\dot{q}_0 = 0$. The steam water mixture W_i entering the containment from the reactor vessel is at critical flow, and it has a total enthalpy $H_{o,i}$. Initially, it contains no other constituents, so the airmass in the containment, M_j, remains constant.

If we nondimensionalize the overall mass balance equation, Eq. (2.4.1), employing the containment volume V_c, the initial containment air density ρ_{c0}, and the blowdown critical flow W_c, it would lead to a containment reference time

$$\tau_c = \frac{V_c \rho_c}{W_c} \qquad (2.4.15)$$

This new reference time τ_c represents the time to replace the original mass of air in the containment by a corresponding steam–water mixture. It is readily related to the previous reactor reference time by

$$\frac{\tau_c}{\tau_r} = \frac{V_c \rho_c}{V_r \rho_r} = \frac{V_c}{P} \frac{P}{V_r} \frac{\rho_c}{\rho_r} \qquad (2.4.16)$$

where P represents the power generation. Equation (2.4.16) states that a power/volume scaling needs to be employed for the containment if the coupling between the time behavior of the containment and reactor system is important and needs to be maintained. Further, the ratio ρ_c/ρ_r must be preserved.

The corresponding containment energy equation using enthalpy is

$$\frac{d(MH)}{dt} = V\frac{dp}{dt} + \dot{q}_i + \sum_i W_i H_{o,i} \qquad (2.4.17)$$

If we nondimensionalize Eq. (2.4.17), the same two nondimensional groups, Π_1 and Π_2, would emerge except for the introduction of τ_c and the role of properties associated with the original gaseous components of the containment volume and any that might be generated subsequently in the reactor (e.g., hydrogen). Some of the water entering the containment will evaporate, and this evaporation rate and the water and steam volume fraction can be determined by writing mass balances for the liquid and vapor reaching the containment as was done in Section 2.4.3. The resulting expression will be quite similar to Eq. (2.4.14) except for an additional term to take into account the change in enthalpy of the gases originally present in the containment.

It is worthwhile to note that for the dry type of containment discussed above, the pressure history is of much less interest than the peak containment pressure, which can be calculated readily from an integrated heat balance using the stored energy within the reactor cooling system and the initial energy of gases in the containment. (The regulations also require that the containment be designed to handle a specified fraction of the fuel cladding chemically reacting with water–steam.) The dry containment design pressure therefore can be bounded without performing scaled experiments. The need for tests arises later during the event when the containment performance can be influenced by the rate, temperature, and burning of hydrogen generated from the fuel cladding and water chemical reaction. At that time the maldistribution of gases within the containment compartments and the stratification can become important, and such conditions are best handled by scaling the relevant bottom-up processes. The same is true for the pressure suppression containment used for BWRs. In this type of containment, the peak pressure is determined by the quenching capability of vents delivering water–steam released from a LOCA in the reactor to a suppression pool. There are containment loads associated with the quenching process, and these phenomena again require bottom-up scaling. In other words, there are limitations to the top-down scaling of containment long-term performance and the bottom-up scaling of some phenomena becomes relevant, as discussed later.

2.4.5 Top-Down Scaling of Fluid Transfer between Volumes

Most complex systems consist of volumes connected by piping through which mass and energy are transferred. These exchanges lead to changes in the conditions of the various volumes. For instance, the flows between volumes are driven by the pressure difference between them and proper global (top-down) scaling of these flows and pressure behavior is important in configuring the model facility. Let us consider first the case of adiabatic one-dimensional single-phase flow between two volumes at pressures p_1 and p_2, as depicted in Figure 2.4.2. The piping connecting the volumes is made up of a number of segments identified by the subscript n having a flow velocity U_n, a flow area A_n, and length L_n. The density of the fluid, ρ, is assumed to be constant because the pressure and temperature difference between the two volumes is small. Conservation of mass requires that

$$A_n U_n = A_r U_r \qquad (2.4.18)$$

2.4 ILLUSTRATIONS OF TOP-DOWN SCALING 49

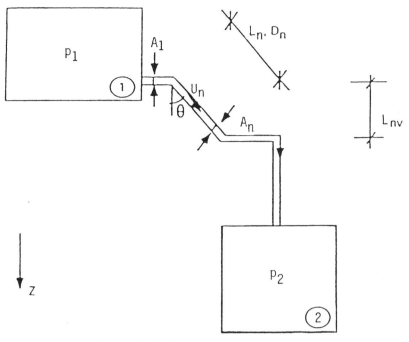

Figure 2.4.2 Pipe connecting two volumes.

where A_r and U_r are, respectively, a reference cross-sectional area and velocity.

An integral conservation momentum equation can be written between the two volumes such that

$$p_2 - p_1 = -\rho \sum_n \frac{A_r}{A_n} L_n \frac{dU_r}{dt} - \rho \left(\frac{A_r^2}{A_2^2} - \frac{A_r^2}{A_1^2} \right) U_r^2 + \sum_n \rho g \, L_{nv}$$

$$- \rho \frac{U_r^2}{2} \sum_n \left(\frac{4 f_n L_n}{D_n} + K_n \right) \frac{A_r^2}{A_n^2} \quad (2.4.19)$$

where L_{nv} represents the vertical projection of the various piping segments which are not horizontal, D_n is the hydraulic diameter, f_n the friction factor, and K_n the form losses associated with each piping segment. The expression for the pressure difference can be simplified to

$$p_2 - p_1 = -\rho L_i \frac{dU_r}{dt} + \rho_g L_g - F' \frac{\rho U_r^2}{2} \quad (2.4.20)$$

where L_g is the sum of the vertical projections L_{nv}, or $L_g = \Sigma_n L_{nv}$. L_i is the equivalent inertia length of the piping, and F' accounts for the total frictional, form, and acceleration losses, or

$$F' = \sum_n \left(\frac{4f_n L_n}{D_n} + K_n\right) \frac{A_r^2}{A_n^2} + 2\left(\frac{A_r^2}{A_2^2} - \frac{A_r^2}{A_1^2}\right) \qquad (2.4.21)$$

If Eq. (2.4.20) is nondimensionalized using the initial values of velocity $U_{r,i}$, density ρ_i, pressure drop Δp_i, and an inertial reference time $\tau_i = L_i/U_{r,i}$ for reference variables, three dimensionless groupings appear and they are

$$\Pi_4 = \rho_i \frac{(U_{r,i})^2}{\Delta p_i} \qquad \Pi_5 = \rho_i g \frac{L_g}{\Delta p_i} \qquad \Pi_6 = F'\rho_i \frac{(U_{r,i})^2}{\Delta p_i} \qquad (2.4.22)$$

The grouping Π_6 equals $F'\Pi_4$, so that it is sufficient to preserve Π_4 and F'. An alternative to the inertial time, $L_i/U_{r,i}$, is the transient fluid time τ_{tr} in the piping such that

$$\tau_{tr} = \frac{\sum_n (A_n/A_r) L_n}{U_{r,i}} = \frac{L_v}{U_{r,i}} \qquad (2.4.23)$$

and the ratio of this transient time to the inertial time is equal to

$$\frac{\tau_{tr}}{\tau_i} = \frac{\sum_n (A_n/A_r) L_n}{\sum_n (A_r/A_n) L_n} = \frac{L_v}{L_i} \qquad (2.4.24)$$

which is another dimensionless geometric grouping that may need to be preserved.

When the connecting piping is part of a complex system, the piping time scales between subsystems are generally much smaller than the time scales for the overall system (τ_s) or for the communicating volumes (τ_v) and

$$\tau_i \ll \tau_v \gg \tau_{tr} \quad \text{with} \quad \tau_v \ll \tau_s$$

This means that the fluid velocities in the connecting piping do not have to be scaled exactly and advantage is taken of that flexibility in most test facilities by adding orifices to the piping primarily to preserve the groupings Π_6 and Π_5 and the pressure drops between volumes. Also, it should be noted that the proposed model for top-down scaling of fluid transfer between volumes is applicable to a blowdown event where critical flow conditions no longer exist.

More recently, it has been employed by Gamble et al. [2.4.2] to evaluate the global momentum behavior of a specific complex system, the simplified boiling water reactor (SBWR). In this global formulation:

- The significant flow paths were identified as well as the involved subsystems, components, and connecting piping. A total of five volumes/components and 11 flow paths were specified for the SBWR.
- Because the participating volumes/components and flow paths vary with the event being considered and sometimes change during portions of the events, the number of volumes/components were adjusted to match the conditions of interest.
- Momentum conservation equations were written for each flow loop or line in the system and the equations were combined whenever possible to eliminate intermediate pressures. Also, mass continuity at junction points was employed to eliminate flow variables when possible.
- The remaining set of flow paths and equations were nondimensionalized and arranged in matrix form to represent the global system.
- The magnitude of the nondimensional groups appearing in the equations determined the relative importance of various nondimensional parameters and helped to identify the adequacy of the test facilities and the presence, if any, of scaling distortions.

The methodology developed in Reference 2.4.2, which was suggested by Wulff [2.4.3] originally, provides an elegant momentum alternative to the hierarchical treatment employed in H2TS.

If two-phase flow occurs in the connecting piping, well-mixed or homogeneous conditions are presumed to exist in order to be consistent with the uniform conditions assumed for the entire system. The preceding derivation then would apply except for replacing the single-phase fluid density with the homogeneous density ρ_H and for defining a homogeneous viscosity μ_H to calculate frictional losses (see Chapter 3). Additional scaling groups will surface under nonhomogeneous two-phase-flow conditions and they are best covered in the next example of top-down scaling of natural circulation systems.

2.4.6 Top-Down Scaling of Single-Phase Natural Circulation Systems

Scaling criteria for light-water reactors under single- and two-phase natural circulation were first proposed by Ishii and Kataoka [2.4.4]. Subsequent evaluations were offered by Kiang [2.4.5] and Kocamustafaogullari and Ishii [2.4.6]. Let us consider first single-phase natural circulation with no heat losses from the loop to the environment. If we integrate the one-dimensional flow conditions around the loop, Eqs. (2.4.9) and (2.4.10) apply except for $p_2 - p_1 = 0$. There results

$$A_n U_n = A_r U_r \tag{2.4.18}$$

$$\rho \frac{dU_r}{dt} \sum_n \frac{A_r}{A_n} L_n = \sum_n \beta g \rho L_{nv} \Delta T_{nv} - \rho \frac{U_r^2}{2} \sum_n \left(\frac{4 f_n L_n}{D_n} + K_n \right) \frac{A_r^2}{A_n^2} \tag{2.4.25}$$

where β is the coefficient of expansion for the fluid and ΔT_{nv} the temperature difference in the loop between the vertical projections of the cold and hot leg.

The energy equation for the heated section where the fluid temperature is T_h along the flow direction z, the fluid velocity U_h, and the specific heat C_p, and q_h''' is the heat generation per unit volume is

$$\rho C_p \left(\frac{\partial T_h}{\partial t} + U_h \frac{\partial T_h}{\partial z} \right) = q_h''' \tag{2.4.26}$$

As done in Section 2.4.2, Eq. (2.4.25) can be simplified to

$$\rho L_i \frac{dU_r}{dt} = \beta g \rho L_g \Delta T - F \frac{\rho U_r^2}{2} \tag{2.4.27}$$

where L_i is the equivalent inertia length of the piping and L_g is the equivalent hydrostatic length such that $L_g \Delta T = \sum_n L_{nv} \Delta T_{nv}$. The symbol F represents the friction and form pressure losses and is nondimensional:

$$F = \sum_n \left(\frac{4 f_n L_n}{D_n} + K_n \right) \frac{A_r^2}{A_n^2} \tag{2.4.28}$$

If we preserve the same fluid from the model to the prototype and if we nondimensionalize Eqs. (2.4.18), (2.4.25), and (2.4.27) by using the initial velocity $U_{r,i}$ and the initial temperature difference ΔT_r during steady-state natural circulation, the density ρ, and a reference inertial time scale $\tau_i = L_i / U_{r,i}$, three dimensionless groups emerge:

$$\Pi_7 = \frac{g \beta \Delta T_r L_g}{(U_{r,i})^2} \qquad \Pi_8 = F \qquad \Pi_9 = \frac{q''' L_i}{\rho C_p \Delta T_r U_{r,i}} \tag{2.4.29}$$

The grouping number Π_7 is the Richardson number and Π_9 has been referred to as the heat source number.

If the heat generation is not uniform over the heated length L_h, it is more appropriate to use an integral form of Π_9, or

$$\Pi_9 = \frac{L_i \int_0^{L_h} q''' \, dz}{L_h \rho C_p \, \Delta T_r \, U_{r,i}} \tag{2.4.30}$$

In this form Π_9 represents the ratio of total heat source addition to the corresponding axial fluid energy change. In addition, it includes another geometric ratio, L_i/L_h. So a total of four integral lengths (L_g, L_i, L_v, L_h) were identified by the top-down scaling process and they are all important. The length L_g multiplied by the density difference prevailing over it establishes the driving head for the natural circulation loop, and the corresponding and controlling Richardson number results by forming the ratio of driving to velocity head. The inertial length L_i, divided by the reference velocity $U_{r,i}$, produces the inertial time scale τ_i, which has a key role in defining the rate of transient flow decay or increase due to power changes. According to Eq. (2.4.23), the length L_v establishes the transit time, τ_{tr}, around the loop and influences the periodicity of loop flow oscillations, should they occur. The heated length L_h becomes important in those cases where there is a feedback (as in a nuclear reactor) between the density and power generation in the heat-producing portion of the loop. The associated power oscillations are related to the heating time scale $\tau_h = L_h/U_{r,i}$.

If a power/volume scaling strategy is used from prototype to model, the flow areas will be scaled by the scaling ratio R. The heat generation per unit volume q''', the reference velocity $U_{r,i}$, the lengths L_g, L_i, L_v, and L_h, and the time scales τ_i, τ_{tr}, and τ_h will all be preserved. With the same fluid properties, the Richardson number Π_7 is the same for model and prototype and so is the non-dimensional grouping Π_9. Finally, the initial reference velocity $U_{r,i}$ at steady-state conditions is given by

$$U_{r,i} = \left(\frac{2\beta g L_g \, \Delta T_r}{F}\right)^{1/2} \tag{2.4.31}$$

which requires the pressure loss factor F to be the same for the model and prototype.

It should be noted that with the proposed top-down approach, the experimenters are left with a fair amount of flexibility in matching the lengths L_g, L_i, L_v, and L_h and the pressure loss factor F because they are integral properties around the natural circulation loop. This means that scaled facilities can utilize orifices and different pipe sizes and lengths as long as the integral values of L_g, L_i, L_v, and L_h, F, and ΔT_r are relatively preserved.

Instead of the stream tube approach or power/volume scaling proposed above, Ishii and Kataoka [2.4.4] considered the possibility of reducing the reference length scale from model to prototype. If we define that length ratio as L_R and the corresponding reference velocity ratio as U_R, the Richardson number will be satisfied only if $U_R = \sqrt{L_R}$ and the time ratio t_R would equal $t_R = \sqrt{L_R}$. The heat generation per unit volume ratio q'''_R becomes $q'''_R = U_R^3/L_R^2 = 1/\sqrt{L_R}$ as long as the fluid properties are preserved. This reduced-length model has been used in some instances where the vertical lengths in the model become excessive.

2.4.7 Top-Down Scaling of Two-Phase Natural Circulation Systems

Ishii and Kataoka also examined natural circulation in a loop with two-phase (liquid–vapor) flow with vapor formation. The driving head becomes dominated by the presence of vapor and the fraction of area α it occupies. The pressure drop also increases sharply due to the presence of vapor, and the natural circulation performance becomes quite dependent on the formulation employed to predict the volume fraction and pressure drop for two-phase flow.

2.4.7.1 Homogeneous Natural Circulation The simplest approach again would be to use homogeneous flow conditions with the liquid and vapor having the same velocity. Because of the possible impact of vapor on the density, Eq. (2.4.18) is best rewritten as

$$\rho_n A_n U_n = A_r U_r \rho_l \tag{2.4.32}$$

where the liquid density ρ_l is used for reference purposes. Equation (2.4.25) becomes

$$\rho_l \frac{dU_r}{dt} \sum_n \frac{A_r}{A_n} L_n = \sum_n g\, \Delta\rho_{nv}\, L_{nv} - \rho_l \frac{U_r^2}{2} \sum_n \frac{\rho_l}{\rho_n}\left(\frac{4 f_n L_n}{D_n} + K_n\right)\frac{A_r^2}{A_n^2} \tag{2.4.33}$$

where $\Delta\rho_{nv}$ is the density difference between the vertical projections of the cold and hot legs. Here again, a hydrostatic length L_g can be introduced such that $L_g\,\Delta\rho = \Sigma_n\,\Delta\rho_{nv}\,L_{nv}$ and the corresponding Π_7 group becomes $\Pi_7 = L_g\,\Delta\rho/(U_{r,i})^2$.

In the vertical projections where single-phase conditions prevail, one could continue to employ the form $\Sigma_n\,\beta g\rho L_{nv}\,\Delta T_{nv}$ utilized in Eq. (2.4.25). When some of the fluid is evaporated or a two-phase mixture starts to occur, the heated length L_{sh} where the fluid loses its subcooling $\Delta H_{sub} = H_{sat} - H_i$ must be identified. H_{sat} is the saturated fluid enthalpy and H_i its enthalpy at the inlet of the heated section. For that portion of the heater the applicable nondimensional grouping Π_q becomes

$$\Pi'_q = \frac{L_i \int_0^{L_{sh}} q''' dz}{L_{sh}\rho_l\,\Delta H_{sub}\,U_r} \tag{2.4.34}$$

where L_{sh} is the subcooled heated length.

The corresponding grouping for the evaporating length is

$$\Pi''_q = \frac{L_i \int_{L_{sh}}^{L_h} q''' dz}{(L_h - L_{sh})\rho_l H_{fg} U_r} \tag{2.4.35}$$

where H_{fg} is the heat of vaporization and L_h is the total heated length. Note that the grouping Π_q'' contains the previously identified phase change number Π_3. As written, Eq. (2.4.35) permits the calculation of the fraction of evaporated liquid, χ, as a function of position in the heated portion of the loop. The corresponding difference in density $\Delta\rho$ is obtained from

$$\Delta\rho = \rho_l - [\rho_l(1-\alpha) - \rho_g\alpha] \qquad (2.4.36)$$

where α is the volume fraction occupied by the vapor and ρ_g its density. If we nondimensionalize Eq. (2.4.36), one new grouping appears, $\Pi_{10} = \rho_g/\rho_l$. For homogeneous flow where the liquid and vapor velocities are assumed equal, the vapor volume fraction, α is related to the mass fraction χ (see Chapter 3) according to

$$\alpha = \frac{\chi}{\chi + (\rho_g/\rho_l)(1-\chi)} \qquad (2.4.37)$$

It can be inferred from the above that two-phase homogeneous natural circulation is quite similar to single-phase behavior except for the modifications made to the Richardson number Π_7, and the introduction of a density ratio and two heat source groupings: Π_q', a subcooling number, and Π_q'', a phase change number. The inertial length L_i and the transit length L_v are the same; however, the hydrostatic head length is dominated by the presence of vapor, and the heated length L_h needs to be broken down into a subcooling length L_{sh} and a boiling length ($L_h - L_{sh}$). The reference parameters should again be given physical meanings. For example, the reference velocity U_r should be selected as the steady-state liquid value, and the density difference $\Delta\rho_r$ could be chosen as the steady-state value at the exit of the heated section.

2.4.7.2 Steady-State Two-Phase Homogeneous Natural Circulation

Reyes [2.4.7] has developed a cubic equation to predict the fluid velocity U_r in a steady-state, two-phase fluid thermosyphon for both subcooled and saturated flow conditions. For the simplified case of a saturated natural circulation loop, he found that

$$U_r^3 + C_3 U_r^2 + C_2 U_r + C_1 = 0 \qquad (2.4.38)$$

where

$$C_3 = \frac{P}{A_c} \frac{\Delta\rho}{\rho_l\rho_g H_{fg}} \frac{1+F_{TP}}{F_T} \qquad C_2 = \left(\frac{P}{A_c}\right)^2 \left(\frac{\Delta\rho}{\rho_l\rho_g H_{fg}}\right)^2 \frac{F_{TP}}{F_T} \qquad (2.4.39)$$

$$C_1 = -\left(2g\frac{PL_h}{A_c}\right)\frac{\Delta\rho}{\rho_l\rho_g H_{fg}}\frac{1}{F_T}$$

where P is the power generated, $\Delta\rho = \rho_l - \rho_g$, F_{TP} the two-phase pressure drop, and F_T the total pressure drop, including the single-phase pressure drop.

In deriving Eqs. (2.4.38) and (2.4.39), Reyes used a simplified term for the buoyancy force,

$$(\rho_l - \rho_h)gL_h = \rho_l\left[1 - \frac{\rho_l}{1 + \chi_e(\Delta\rho/\rho_g)}\right]gL_h \qquad (2.4.40)$$

where χ_e is the heater exit quality and L_h is the distance between the thermal centers of the heat source and the heat sink.

If we divide Eq. (2.4.38) by U_r^3, three nondimensional groupings result:

$$\frac{C_3}{U_r} = \frac{P}{A_c U_r} \frac{\Delta\rho}{\rho_l \rho_g H_{fg}} \frac{1 + F_{TP}}{F_T}$$

$$\frac{C_2}{U_r^2} = \left(\frac{C_3}{U_r}\right)^2 \frac{F_{TP} F_T}{(1 + F_{TP})^2} \qquad (2.4.41)$$

$$\frac{C_1}{U_r^3} = \frac{C_3}{U_r} \frac{2gL_h/U_r^2}{1 + F_{TP}}$$

These three groupings,

$$\frac{P}{A_c U_r} \frac{\Delta\rho}{\rho_l\rho_g H_{fg}} \frac{1 + F_{TP}}{F_T} \qquad \frac{F_{TP} F_T}{(1 + F_{TP})^2} \qquad \frac{2gL_h/U_r^2}{1 + F_{TP}}$$

reproduce the scaling laws of Ishii and Kataoka [2.4.4] by requiring that the ratio of two-phase pressure drop, F_{TP}, total pressure drop F_T, and of the grouping $\Delta\rho/\rho_l\rho_g H_{fg}$ in the prototype and model be the same. Also, they show that the ratio of prototype to model velocity, U_r, must be proportional to the square root of the length ratio and that the ratio power/core volumetric flow should be constant. If it is preferred to have the same velocity in the model and prototype, the thermal lengths and the ratio of power to flow area must be the same, or a power/volume scaling strategy must be employed.

2.4.7.3 Nonhomogeneous Two-Phase Natural Circulation

For most liquid–vapor flows, the vapor velocity is expected to exceed that of the liquid. This means that the vapor volume fraction calculated from Eq. (2.4.37) is too high and so is the driving head. This is somewhat compensated by the homogeneous pressure drop relationship in Eq. (2.4.33), which is high. For those reasons, formulations other than for homogeneous flow conditions have been employed by several authors. For example, Ishii and Kataoka [2.4.4] employed a drift flux model (see Chapter 3) where the drift velocity U_{gj}, which is defined as the void weighted average velocity of the vapor phase with respect to the velocity of the center of volume of the mixture, is given by

$$U_{gj} = 0.2\left(1 - \sqrt{\frac{\rho_g}{\rho_l}}\right)j + 1.4\left(\sigma g \frac{\rho_l - \rho_g}{\rho_l^2}\right)^{1/4} \quad (2.4.42)$$

In Eq. (2.4.42), U_{gj} is the gas drift velocity, j the total volumetric flux or total liquid and vapor volumetric flows divided by the total flow area, and σ the surface tension. Equation (2.4.42) introduces a new nondimensional grouping to be preserved, $\Pi_{11} = U_{gj}/U_r$. Equation (2.4.42) can be rewritten to specify a relation between the vapor void fraction α, its mass fraction χ, and the total mass flow rate per unit area G, such that

$$\alpha = \frac{\chi}{\left[1.2 - 0.2\sqrt{\frac{\rho_g}{\rho_l}}\right]\left[\frac{\rho_g(1-\chi)}{\rho_l} + \chi\right] + 1.4\left[\sigma g \frac{(\rho_l - \rho_g)}{\rho_l^2}\right]^{1/4}\frac{\rho_g}{G}} \quad (2.4.43)$$

Ishii and Kataoka presumed that Eqs. (2.4.42) and (2.4.43) were valid irrespective of flow pattern. Furthermore, they employed a homogeneous pressure drop relation even though the vapor and liquid velocities are not assumed to be equal in Eq. (2.4.42). It can be seen from this top-down treatment of a two-phase natural circulation loop that additional groupings associated with closure equations (i.e., for volume fraction and pressure drop) as well as property groupings of the liquid and vapor phases (i.e., ρ_g/ρ_l and $\{\sigma g \left[(\rho_l - \rho_g)/\rho_l^2\right]\}^{1/4}$) must be taken into account. In addition, distortions produced by differences in flow pattern from model to prototype or inconsistencies in the formulation of closure equations must be considered. These shortcomings, however, do not detract from the ability of the top-down methodology to identify the dominant nondimensional groupings that must be preserved for scaling purposes.

However, as for the previous top-down examples, a note of caution is in order. For instance, the possibility of natural circulation loops to go unstable is of great interest, and such local parameters as liquid subcooling and the two-phase pressure drop along the heater are all known to influence the start of instabilities. They need to be considered during separate studies of loop instability. In recent years, many stability-related thermal-hydraulic phenomena have been described by catastrophe functions that can define the boundaries of instability. Here again, the top-down scaling of such catastrophe functions can then be employed [2.4.8] to predict natural circulation instabilities. This will be illustrated in the next and last application of the top-down methodology to natural circulation loops.

2.4.8 Top-Down Scaling of Catastrophe Functions

In recent years, catastrophe functions have been used to describe many important thermal-hydraulic stability phenomena, including countercurrent flow-limiting conditions [2.4.9] or discontinuous behavior of a two-phase natural circulation loop [2.4.10]. The most elementary form of catastrophe functions is a polynomial function of a simple state variable ψ, or

$$\phi = \psi^k + \sum_{n=1}^{k-1} C_{k-n}\psi^{k-n} \tag{2.4.44}$$

where n are integers from 1 to $k-1$ and C_{k-n} are the control parameters. For example, as discussed in Section 2.4.7.2, Reyes [2.4.10] has developed a cubic equation to predict the fluid velocity under steady-state, two-phase, homogeneous natural circulation conditions for a saturated loop:

$$\phi = U_r^3 + C_3 U_r^2 + C_2 U_r + C_1 = 0 \tag{2.4.45}$$

where U_r is the velocity of the single-phase liquid at the core inlet and the control parameters for no subcooling in the loop are given by Eq. (2.4.39).

Catastrophe theory relies upon setting the derivatives of the catastrophe function ϕ with respect to U_r equal to zero and solving them to develop a relationship or catastrophe manifold between the control parameters. These relations can reveal bifurcations or transition jumps that define boundaries at which small changes in the control parameter may result in significant changes to the state of system. In the case of Eq. (2.4.45), if $U_{r,c}$ is used to define the bifurcation or critical velocity of the catastrophe function and its derivatives are set to zero, there results

$$U_{r,c}^3 + C_3 U_{r,c}^2 + C_2 U_{r,c} + C_1 = 0$$
$$3U_{r,c}^2 + 2C_3 U_{r,c} + C_2 = 0 \tag{2.4.46}$$
$$6U_{r,c} + 2C_3 = 0$$

If we solve Eqs. (2.4.46) for the control parameters, we get

$$C_3 = -3U_{r,c} \qquad C_2 = U_{r,c}^2 \qquad C_1 = -U_{r,c}^3$$

which yields the following only nontrivial and real relation between (C_2, C_1), or

$$\left(\frac{C_2}{3}\right)^{1/2} = (-C_1)^{1/3} \tag{2.4.47}$$

Utilizing the control parameter expressions from Eqs. (2.4.39) for a homogeneous and saturated natural circulation loop, there results

$$\left[\left(\frac{P}{\rho_g A_c H_{fg}}\right)^2 \left(\frac{\Delta\rho}{\rho_l}\right)^2 \frac{1}{gL_h}\right]^{1/3} - 2^{1/3}\sqrt{3}\left(\frac{F_T^{1/3}}{F_{TP}}\right)^{1/2} = 0 \qquad (2.4.48)$$

If we replace the term $P/\rho_g A_c H_{fg}$ by the gas volumetric flux j_g and we recognize that typically $\rho_l \gg \rho_g$, so that $\Delta\rho/\rho_l \approx 1$, Eq. (2.4.48) can be simplified to

$$\left(\frac{j_g^2}{gL_h}\right)^{1/3} - 2.18\left(\frac{F_T^{1/3}}{F_{TP}}\right)^{1/2} = 0 \qquad (2.4.49)$$

The term j_g^2/gL_h is a two-phase Froude number, and Eq. (2.4.49) establishes the vapor flux value where the loop can go unstable and its dependence on the thermal center distance L_h, the two-phase pressure drop factor F_{TP}, and the total pressure drop factor F_T.

If the control parameters derived by Reyes [2.4.7] for homogeneous natural circulation with ΔH_{sub} subcooling are employed, the following criterion is obtained for the onset of unstable natural circulation flow with subcooling [2.4.10]:

$$\left[\frac{P^2}{2gL_h(A_c\rho_g H_{fg})^2}\right]^{1/3}\left[F_T - (F_T + F_{TP})\frac{\rho_l}{\rho_g}\frac{\Delta H_{sub}}{H_{fg}} + F_{TP}\left(\frac{\rho_l}{\rho_g}\right)^2\left(\frac{\Delta H_{sub}}{H_{fg}}\right)^2\right]^{1/6}$$

$$= \sqrt{3}\left[F_{TP} + \frac{\rho_l}{\rho_g}\frac{\Delta H_{sub}}{H_{fg}} 2gL_h \frac{(A_c\rho_g H_{fg})^2}{P^2}\right]^{1/2} \qquad (2.4.50)$$

Equation (2.4.50) establishes the unstable value of the parameter $P^2/2gL_h(A_c\rho_g H_{fg})^2$ in terms of the pressure drop factors F_T and F_{TP} and the property grouping $(\rho_l/\rho_g)(\Delta H_{sub}/H_{fg})$.

This last example of top-down scaling is another excellent demonstration of the merits of that methodology. It not only can identify the important nondimensional global groupings but can be extended to complicated situations and system conditions not considered in the past.

2.5 ILLUSTRATIONS OF BOTTOM-UP SCALING

Top-down scaling cannot claim and does not assure that the detailed behavior of the prototype is simulated correctly in the model. To deal with specific detailed phenomena, top-down scaling needs to be complemented by bottom-up scaling assessments. The methodology for bottom-up scaling relies on defining the significant prevailing processes and employing the frequency-time domain or the applicable conservation equations to generate the nondimensional scaling groups. Even though this approach is not new, it has not always been used correctly or consistently to improve the design of experiments or the correlation of their data. This will be illustrated in the bottom-up scaling test data obtained in various facilities for the first peak fuel temperature during a loss-of-coolant accident (LOCA) in a pressurized water reactor. That detailed example will be followed by several that emphasize containment phenomena because of the recent and increased interest in that field.

2.5.1 Bottom-Up Scaling of Peak Fuel Cladding Temperature during a LOCA from Differential Equations

In a nuclear reactor, the fuel rod consists of stacked UO_2 pellets of radius a which are surrounded by a thin metallic cladding (e.g., zircaloy) of radius b. The pellets and cladding are presumed to be subject to a gas gap conductance h_g. Heat is removed from the cladding by coolant at temperature T_∞, and the heat transfer coefficient at that interface is h. Immediately after a LOCA in a nuclear reactor, reactor core flow decreases and coolant is expelled by flashing from the core. A large drop in cladding-to-coolant heat transfer occurs and the fuel clad temperature rises rapidly, due primarily to the redistribution of the stored energy in the fuel. Shortly thereafter the reactor shuts down and heat generation in the fuel drops to a fraction of its original value, which to a smaller extent can add to the cladding temperature rise. The heat conduction differential equations at a radius r within the pellet and cladding are provided in Reference 2.5.1. They are

$$k_f \left(\frac{\partial^2 T_f}{\partial r^2} + \frac{1}{r} \frac{\partial T_f}{\partial r} \right) + q''' = (\rho C_p)_f \frac{\partial T_f}{\partial t} \quad 0 < r \le a$$

$$k_c \left(\frac{\partial^2 T_c}{\partial r^2} + \frac{1}{r} \frac{\partial T_c}{\partial r} \right) = (\rho C_p)_c \frac{\partial T_c}{\partial t} \quad a \le r \le b$$

(2.5.1)

where T is the temperature, q''' the uniform heat generation per unit volume, t the time, ρ the density, C_p the specific heat, and k the thermal conductivity. The subscripts f and c are used to represent the fuel and cladding. Note that for simplification reasons, the thermal conductivities k_f and k_c were assumed to be constant and axial conduction was neglected. In addition, the following boundary conditions must be satisfied

$$k_f \frac{\partial T_f}{\partial r} = h_g(T_f - T_c) = k_c \frac{\partial T_c}{\partial r} \quad \text{at } r = a$$

$$k_c \frac{\partial T_c}{\partial r} = h(T_c - T_\infty) \quad \text{at } r = b$$

(2.5.2)

2.5 ILLUSTRATIONS OF BOTTOM-UP SCALING

One can solve Eqs. (2.53) and (2.54) to obtain the initial steady-state temperature distributions within the pellet and cladding. Of particular interest are the initial average fuel and cladding temperatures $\overline{T}_{f,i}$ and $\overline{T}_{c,i}$, which according to Reference 2.5.1) are given by

$$\overline{T}_{f,i} = T_\infty + \frac{q'(0)}{2\pi k_f}\left(\frac{1}{4} + \frac{k_f}{hb} + \frac{k_f}{h_g a} + \frac{k_f}{k_c}\ln\frac{b}{a}\right)$$

$$\overline{T}_{c,i} = T_\infty + \frac{q'(0)}{2\pi k_f}\left[\frac{k_f}{hb} + \frac{k_f}{k_c}\left(\frac{1}{2} - \frac{a^2}{b^2 - a^2}\ln\frac{b}{a}\right)\right] \quad (2.5.3)$$

$$\overline{T}_{f,i} - \overline{T}_{c,i} = \frac{q'(0)}{2\pi k_f}\left(\frac{1}{4} + \frac{k_f}{h_g a} - \frac{k_f}{k_c}\frac{1}{2} + \frac{b^2}{b^2 - a^2}\frac{k_f}{k_c}\ln\frac{b}{a}\right)$$

In Eq. (2.5.3), $q'(0)$ is the linear heat generation rate at time zero or $q'(0) = q'''(0)\pi a^2$. At time $t = 0$ it is also assumed that the heat transfer coefficient h drops to a film boiling value h_{fb} and that T_∞ is replaced by the prevailing saturation temperature T_s. Both h_{fb} and T_s vary with time as the reactor pressure decreases and the cladding temperature rises. At time $t = t_d$ the linear heat generation decreases to the decay fraction β of $q'(0)$, with β being a slowly decaying function of time.

In Reference 2.5.1, Eqs. (2.5.1) and (2.5.2) were solved in closed form after they were nondimensionalized to generate the following eight nondimensional groups:

$$T^+ = \frac{k_c T}{q'''(0)}a^2 \qquad t^+ = \frac{k_c t}{a^2}\frac{1}{(\rho C_p)_c} \qquad \frac{(\rho C_p)_f}{(\rho C_p)_c}$$

$$\frac{k_f}{k_c} \qquad \frac{a}{b} \qquad \frac{h_g b}{k_c} \qquad \frac{hb}{k_c} \qquad q'''^+ = \frac{q'''}{q'''(0)} \quad (2.5.4)$$

In Reference 2.5.2, the number of required nondimensional groups were reduced to four:

$$\frac{q'(0)(b-a)^2}{k_f a^2}\frac{(\rho C_p)_c}{(\rho C_p)_f}\frac{1}{\text{PCT} - T_{c,i}} \qquad \frac{h_g(b-a)}{k_f} \qquad \frac{h(b-a)}{k_f} \qquad \frac{(\rho C_p)_c}{(\rho C_p)_f}\frac{b-a}{b} \quad (2.5.5)$$

where PCT is the peak clad temperature and $T_{c,i}$ is the initial clad temperature. It is apparent from the above that there are varying ways to identify the appropriate scaling groups and to reduce their number. However, because there is a distinct advantage to minimize the nondimensional groupings, an integral or lumped-parameter technique rather than a detailed representation of the process or phenomenon will be used here. It evaluates the time behavior of fuel and cladding at their average conditions.

2.5.2 Bottom-Up Scaling of Peak Fuel Cladding Temperature during a LOCA from Lumped Models

If we use average temperatures to represent the fuel and cladding, Eqs. (2.5.1) can be rewritten as

$$q' = C_f \frac{d\overline{T}_f}{dt} + \frac{\overline{T}_f - \overline{T}_c}{R_f}$$
$$\frac{\overline{T}_f - \overline{T}_c}{R_f} = C_c \frac{d\overline{T}_c}{dt} + \frac{\overline{T}_c - T_\infty}{R_c}$$
(2.5.6)

where q' is the linear nuclear generation rate, which varies with time, C_f the heat capacitance of the fuel pellet, or $C_f = \pi a^2 (\rho C_p)_f$, and C_c that of the cladding, or $C_c = 2\pi b(b - a)(\rho C_p)_c$. R_f is the combined heat transfer resistance of the pellet, gap, and cladding, or

$$R_f = \left(\frac{1}{8\pi k_f} + \frac{1}{2\pi a h_g} + \frac{b - a}{2\pi a k_c} \right)$$
(2.5.7)

and R_c the resistance of the coolant film, or $R_c = 1/2\pi bh$. Equations (2.5.6) have an inherent advantage over the differential equations, (2.5.1), because they already incorporate the boundary conditions, Eqs. (2.5.2). Note that the cladding was presumed to be thin enough to use $2\pi b(b - a)$ for its cross-sectional area and to employ a linear temperature distribution across the cladding.

If we reference all temperatures to the initial cladding temperature $\overline{T}_{c,i}$ and nondimensionalize the first of Eqs. (2.5.6) employing a reference time τ_r, and the initial conditions $q'(0)$ and $\overline{T}_{f,i} - \overline{T}_{c,i}$, two nondimensional groupings emerge:

$$\frac{q'(0) R_f}{T_{f,i} - T_{c,i}} \quad \text{and} \quad \frac{C_f R_f}{\tau_r}$$
(2.5.8)

Once again, the choice of the reference time τ_r is arbitrary, but its value should be chosen to make the magnitude of the nondimensional derivatives of the order of unity, so that

$$\tau_r = C_f R_f$$
(2.5.9)

The reference time of Eq. (2.5.9) has a physical meaning because it represents the time constant for a change of fuel average temperature to be felt at the cladding outside surface. Similarly, the first grouping has a physical meaning since it describes the condition satisfied by the initial steady-state temperature difference between fuel and cladding or from the first of Eqs. (2.5.6) with $dT_f/dt = 0$:

$$\overline{T}_{f,i} - \overline{T}_{c,i} = q'(0) R_f$$
(2.5.10)

If we assume that steam blanketing or film boiling occurs on the outside of the fuel rod at time $t \geq 0$, the coolant temperature will approach the saturation temperature, or $T_\infty \to T_s$. The resistance under film boiling conditions will become $R_c \to R_{cb} = 1/2\pi b h_{fb}$ where h_{fb} represents the film boiling coefficient. The second of Eqs. (2.5.6) can then be written as

$$\frac{\overline{T}_f - \overline{T}_c}{R_f} = C_c \frac{d\overline{T}_c}{dt} + \frac{\overline{T}_c - \overline{T}_{c,i}}{R_{cb}} + \frac{\overline{T}_{c,i} - T_s}{R_{cb}} \quad \text{at } t > 0 \quad (2.5.11)$$

If we again reference all temperature to the initial cladding temperature and nondimensionalize Eq. (2.5.11), three nondimensional groupings emerge:

$$\frac{C_c R_f}{\tau_r} = \frac{C_c}{C_f} \quad \frac{R_f}{R_{cb}} \quad \frac{\overline{T}_{c,i} - T_s}{\overline{T}_{f,i} - \overline{T}_{c,i}} \quad (2.5.12)$$

The first grouping in Eq. (2.61) establishes the ratio of heat capacity of the cladding with respect to that of the fuel, or

$$\frac{C_c}{C_f} = \frac{b^2}{a^2}\left(1 - \frac{a}{b}\right)\frac{(\rho C_p)_c}{(\rho C_p)_f}$$

The second grouping compares the cladding capability to receive heat from the fuel to transferring it to the coolant, and the last grouping is satisfied automatically if an overall power/volume scaling approach is utilized because it would preserve the pressure behavior, local hydraulics, and saturation temperature during the LOCA.

In summary, in addition to matching the linear heat generation q' as a function of time, the fuel/cladding radius ratio b/a and flow area characteristics, the product $C_f R_f$, and the ratios C_f/C_c and R_f/R_{cb} and initial condition $q'(0)$ $[R_f/(\overline{T}_{f,i} - \overline{T}_{c,i})]$ must be reproduced in model and prototype to attain equivalent peak clad temperature behavior. The lumped-parameter methodology is thus seen to have the advantage of dealing with the overall fuel-rod resistance R_f instead of the various separate resistances of the fuel, gap, and cladding.

The scaling groups can be simplified further for special circumstances. For example, we could combine the last two terms on the right-hand side of Eq. (2.5.11) to get

$$\frac{\overline{T}_c - T_s}{R_{cb}} = q'_{fb} = 2\pi b h_{fb}(\overline{T}_c - T_s) \quad (2.5.13)$$

where q'_{fb} is the linear film boiling heat removal rate, which varies with the cladding temperature $\overline{T}_c$ and saturation temperature T_s. For simplification purposes, one could presume that the saturation temperature T_s does not change and that the linear film boiling heat removal rate is at its minimum value q'_{mfb}, and that they remain at those values during the short heat-up time of the cladding.

The last two nondimensional groupings in Eq. (2.5.12) can then be combined to obtain one nondimensional group:

$$\frac{q'_{mfb} R_f}{\overline{T}_{f,i} - \overline{T}_{c,i}} = \frac{q'_{mfb}}{q'(0)} \tag{2.5.14}$$

The value of q'_{mfb} and its corresponding $(\overline{T}_c - \overline{T}_s)_{mfb}$ can be obtained from Berenson [2.5.3]:

$$q''_{mfb} = \frac{q'_{mfb}}{2\pi b} = 0.09 \, \rho_g H_{fg} \left(\frac{\rho_l - \rho_g}{\rho_l + \rho_g}\right)^{1/2} \left(\frac{g\sigma}{\rho_l - \rho_g}\right)^{1/4}$$

$$(\overline{T}_c - \overline{T}_s)_{mfb} = 0.127 \frac{\rho_g H_{fg}}{k_g} \frac{\mu_g^{1/3}}{(\rho_l + \rho_g)^{2/3}} \frac{\sigma^{1/2}}{[g(\rho_l + \rho_g)]^{1/6}} \tag{2.5.15}$$

where ρ_l and ρ_g are the liquid and vapor densities, H_{fg} the heat of vaporization, k_g the vapor thermal conductivity, and μ_g the vapor viscosity. All of these properties must be evaluated at the prevailing and presumed relatively constant saturation temperature T_s. Under those circumstances, the grouping $q'_{mfb}/q'(0)$ will be satisfied automatically if a power/volume scaling approach is employed.

It is possible to simplify the lumped model further by combining the two Eqs. (2.5.6) and utilizing the expression (2.5.13); a first-order differential equation can then be derived for $\overline{T}_f - \overline{T}_c$ at time $t \geq 0$, or

$$\frac{d(\overline{T}_f - \overline{T}_c)}{dt} + (\overline{T}_f - \overline{T}_c)\left(\frac{1}{C_c} + \frac{1}{C_f}\right)\frac{1}{R_f} - \frac{\beta q'(0)}{C_f} - \frac{q'_{fb}}{C_c} = 0 \tag{2.5.16}$$

In Eq. (2.5.16) it is assumed that the linear decay heat generation in the fuel rod can be represented by $\beta q'(0)$, where β is a specified function of time. The solution to Eq. (2.5.16) for the special case where β is a constant and the minimum linear film boiling heat flux q'_{mfb} is used:

$$\overline{T}_f - \overline{T}_c = (\overline{T}_{f,i} - \overline{T}_{c,i}) e^{-[(C_f + C_c)/R_f C_c C_f]t} + \frac{R_f C_c C_f}{C_f + C_c}\left[\frac{\beta q'(0)}{C_f} + \frac{q'_{mfb}}{C_c}\right] \tag{2.5.17}$$

If we introduce Eq. (2.5.17) into the second of Eqs. (2.5.6) and integrate, there results

$$\overline{T}_c - \overline{T}_{c,i} = \frac{C_f}{C_f + C_c}(\overline{T}_{f,i} - \overline{T}_{c,i})(1 - e^{-[(C_f + C_c)/R_f C_c C_f]t})$$

$$+ \frac{q'(0)}{C_f + C_c}\int_0^t \beta \, dt - \frac{1}{C_c + C_f}\int_0^t q'_{mfb} \, dt \tag{2.5.18}$$

The first term on the right-hand side of Eq. (2.5.18) establishes how heat stored in the fuel raises the cladding temperature; the second term determines how additional decay heat generation in the fuel would increase the cladding temperature, and the third term accounts for heat removal by film boiling from the cladding and its eventual temperature decline [as long as $q'_{mfb} \geq \beta q'(0)$].

Equation (2.5.18) provides the best picture of what needs to be preserved from prototype to model:

- The initial linear generation rate and its decay fraction after the break.
- The heat removal by the coolant from the cladding. This requires that the thermal-hydraulic characteristics at the peak temperature location are reproduced from model to prototype (i.e., film boiling and/or fuel surface rewetting).
- The fuel-rod integral time constant $R_f C_f$ and the capacity of the fuel and cladding C_f and C_c.

Equation (2.5.18) permits the calculation of peak clad temperatures in several simplified cases. Let us first consider the case of a fuel rod with no heat generation or heat removal after the break. The peak cladding temperature (PCT) would reach an equilibrium value

$$\text{PCT} - \overline{T}_{c,i} = \frac{C_f}{C_c + C_f}(\overline{T}_{f,i} - \overline{T}_{c,i}) = q'(0)\, R_f \frac{C_f}{C_c + C_f} \quad (2.5.19)$$

For a typical reactor fuel rod[†] one would get $\text{PCT} - \overline{T}_{c,i} \approx 84.7 q'(0)$, where $q'(0)$ is in kW/ft and $\text{PCT} - \overline{T}_{c,i}$ is in °F. This relation reproduces the line correlation derived by Catton et al. [2.5.4] except that a constant of 60 instead of 85 is drawn in Figure 2.5.1 to fit the tests at various scales. The value calculated for the exponent in Eq. (2.5.16) is $-1.34t$, and it would take about 3 seconds to reach equilibrium conditions.

Let us next consider the case where the reactor is shut down as the break occurs, the decay heat fraction β is constant and at 15% of the original heat generation, and the minimum film boiling linear rate is constant and calculated at a saturation temperature of 584°F (290°C) or $q'_{mfb} = 5.85$ Btu/sec-ft (6.17 kW/ft). The peak cladding temperature in Eq. (2.5.18) was found to be 1180°F (910 K) for an initial linear fuel heat generation rate of 10 kW/ft and 1605°F (1172 K) for a linear heat generation rate of 15 kW/ft. These two values, plotted in Figure 2.5.1, are in excellent agreement with the line drawn through the data by Catton et al. [2.5.4]. This

---

[†]A fuel rod of 0.35 in. (0.89 cm) was used. The cladding was zircaloy with an outside diameter of 0.415 in. (1.06 cm) and an inside diameter of 0.358 in. (0.91 cm). The zirconium properties were taken at 600°F (589 K): $k_c = 7.37$ Btu/hr-ft-F (12.8 W/mK²); $\rho_c = 410$ lb/ft³ (657 kg/m³); $C_{p,c} = 0.079$ Btu/lb-°F (0.33 kW/kg · K). The fuel properties were taken at 980°F (800 K); $k_f = 2.37$ Btu/hr-ft-F (4.1 W/mK); $\rho_f = 685.5$ lb/ft³ (10,980 kg/m³); $C_{p,f} = 0.072$ Btu/lb-F (0.299 kW/kg · K). The water surface tension at 584°F (563 K) was taken as 0.01671 N/m (0.00115 lb/ft). The pellet-to-clad gap conductance was assumed to be 1000 Btu/hr ft² °F (5.67 kW/m²K).

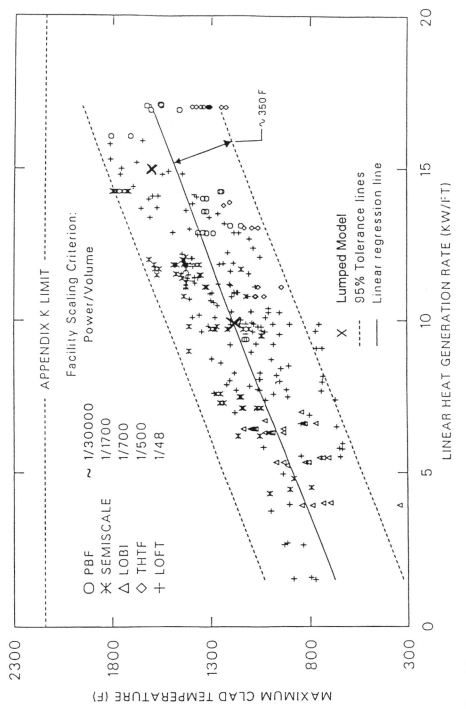

Figure 2.5.1 Maximum measured clad temperature from several scaled experimental facilities (301 data points) during blowdown. (Reprinted from Ref. 2.5.4 with permission from Elsevier Science.)

agreement, however, may be fortuitous because heat loss by radiation from the cladding and the increase in film boiling and radiation heat transfer with rising surface temperature were neglected. Such effects may have been compensated by the decreasing pressure and the added storage of heat in the fuel because its thermal conductivity decreases with rising fuel temperature and linear heat generation.

Let us remember that the primary objective of this bottom-up example was not to get an exact prediction of the fuel clad behavior but rather to define the most relevant scaling groups. The first one according to this simplified model is the nondimensional temperature,

$$(\text{PCT} - \overline{T}_{c,i}) \frac{1 + C_c/C_f}{R_f q'(0)}$$

The second one is the nondimensional time,

$$t_p \frac{1 + C_c/C_f}{R_f C_f}$$

where t_p is the measured time to reach the peak clad temperature. Plotting these two nondimensional groups in Figure 2.5.1 would have helped reduce the scatter of the data or identify tests where scaling of the fuel-rod behavior was inadequate. Another nondimensional group of interest is the film boiling ratio of Eq. (2.5.14), which needs to be preserved in the various test facilities. One way to assess whether it was satisfied would have been to plot the measured time to reach peak clad temperature in the tests versus the prediction from Eq. (2.5.18), or

$$t_p = -\frac{R_f C_f C_c}{C_f + C_c} \ln \frac{C_c}{C_f + C_c} \left[\frac{q'_{mfb}}{q'(0)} - \beta \right] \qquad (2.5.20)$$

This comparison would help to show not only how approximate is the concept of using the minimum heat flux q'_{mfb} but also to yield a comparison of the decrease in heat transfer in the various facilities shortly after the break.

In summary, this example of bottom-up scaling of phenomena demonstrates that one needs to:

- Take into account all significant contributors to the phenomena involved: in this case the initial conditions, which establish the stored energy in the fuel, all the resistances to heat transfer from the fuel to the clad and the fuel time constant of heat transfer, as well as the hydraulic and heat transfer removal conditions prevailing at the peak cladding temperature location.
- Utilize, if practical, an integral (lumped) model to the phenomena which minimizes the number of separate scaling groups to be considered.

- Employ nondimensional groupings to plot the available test data from various facilities. This would help identify inadequate test scaling and/or approximations introduced by simplifications in the conservation equations.

All the data points in Figure 2.5.1 omit water rewet of the fuel rods and therefore apply only to film boiling conditions. Additional hydraulic scaling considerations not considered herein would have to be included if rewet is to be reproduced in the test facility.

2.5.3 Bottom-Up Scaling of Containment Phenomena

In this section we consider the phenomenon of mixing in large stratified volumes both for gases (steam–nitrogen and steam–air–hydrogen) and for water in pools (of the condensing and heat sink type). Mass, energy, and species transport occurs in a variety of stratified applications. It is found not only in nuclear reactor containments, but also in many building fires; heating, ventilation, and air cooling (HVAC) systems; and in several ecological systems. Both air and water bodies, where dispersal of pollutants from smokestacks, toxic chemical releases, and thermal and sewage discharges, are of great interest. Inside such stable stratified layers, mixing processes are usually driven by buoyant jets, buoyant plumes, and wall natural convection boundary layers, and it is important to correctly scale their behavior from the prototype to the test facility. Before describing the applicable models, a short description of nuclear reactor containments and their mixing issues will be provided.

2.5.3.1 Description of Passive Nuclear Containment Structures Two new passive containment designs have been proposed for the advanced PWR (AP-600), being developed by Westinghouse, and the simplified boiling water reactor (SBWR), being developed by General Electric. Both concepts are expected to work without previously employed support systems such as ac power supplies, component cooling or service water forced circulation, and HVAC systems. Instead, passive systems employing gravity and natural circulation flow, pressurized gas bottles, large water pools, check valves, and power from dc batteries are utilized. Both reactor types rely on automatic depressurization systems to bring the reactor pressure down and to permit gravity flooding of the reactor core. Steam released by depressurization in both reactors is condensed in large pools of water which become thermally stratified as the quenching process takes place.

In the AP-600 the early stages of depressurization discharge steam through quenchers to the in-containment refueling water storage tank (IRWST), where it is condensed. The last set of depressurization valves is located on lines connected to the hot legs of the reactor water reactor cooling system and discharge directly into the containment. Also, steam escaping through a break or subsequently produced by decay heat generated in the reactor core is released directly into the containment structure, which is cooled by the passive cooling system depicted in Figure 2.5.2. Containment cooling uses an elevated water storage tank to spray water on the outside of the steel shell of the containment, which in turn condenses steam inside it. The surrounding concrete shield building is arranged to promote the natural

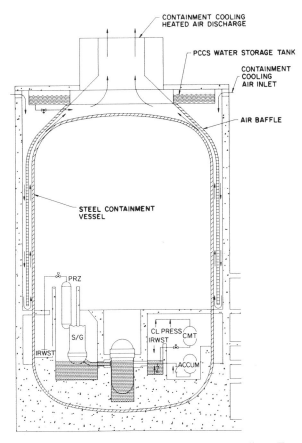

Figure 2.5.2 AP-600 passive safety injection long-term cooling. (From Ref. 1.1.3.)

circulation of air around the containment, which aids in the cooling function and sweeps evaporating water away from the exterior of the shell. Figure 2.5.2 also illustrates the flooded containment and reactor core recovery after an accident by employing water in the core makeup tanks, the accumulators, and the IRWST. The issues here are early stratification of water in the IRWST and long-term stratification of the water in the flooded containment. Another important effect is that of the temperature and gas concentration gradients in the dome of the containment when air, steam, and hydrogen [required to be presumed released by Nuclear Regulatory Commission (NRC) regulations] can stratify to raise potential safety issues unless adequate mixing can be shown to prevail due to jets and wall boundary layers.

In the case of the SBWR, the depressurization system employs safety relief valves, which are located on the main steamline and piped to the suppression pool shown in Figure 2.5.3. It is complemented by special depressurization valves (DPVs) located on stub lines extending from the upper part of the reactor vessel, which discharge directly into the drywell of the containment. Steam from the DPVs and steam escaping through a break raise the drywell pressure, clearing the vents between the drywell and

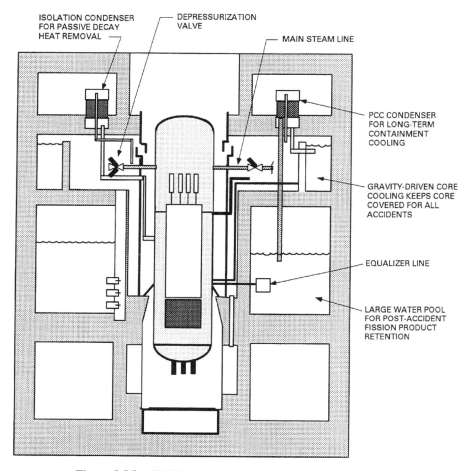

Figure 2.5.3 SBWR passive safety systems. (From Ref. 1.9.1)

suppression pool. First, noncondensable gases (nitrogen) and then steam flow through the vents into the suppression pool, where the noncondensables rise above the pool while the steam is condensed within it. Decay heat produced in the reactor also generates steam which is released into the drywell and condensed in the heat exchangers of the passive containment cooling system (PCCS). These heat exchangers are located in large pools of water that act as heat sinks and are located above the reactor. The heat exchangers are open to the drywell and the steam entering them is condensed and returned to the reactor vessel through gravity drain lines while the entrained noncondensable gases are vented back to the upper part of the suppression pool. The issues here are thermal stratification of the various pools, particularly of the suppression pool, due to blowdown through the main containment vents and the energy added by the PCCS condenser vents at more shallow depths. Another important effect is the stratification and concentration of gases in the wetwell and drywell gas spaces following blowdown or vacuum breaker opening between wetwell and drywell.

2.5 ILLUSTRATIONS OF BOTTOM-UP SCALING   71

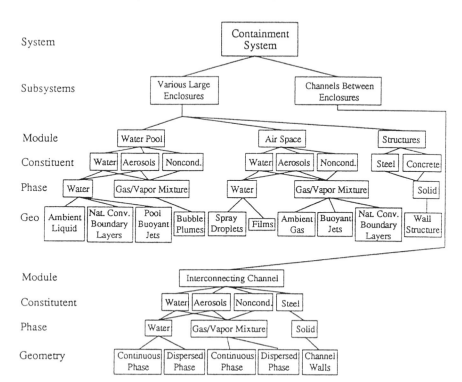

Figure 2.5.4 Hierarchical scaling for large interconnected LWR enclosures. (Reprinted from Ref. 2.5.6 with permission from Elsevier Science.)

In terms of performance, the stratification and degree of subsequent mixing in large gas and water volumes in containment structures are important because they can affect the peak containment pressure. For example, thermal stratification of a water pool increases the pool surface temperature and raises the associated vapor partial pressure. Similarly, gas concentration stratification in the upper containment areas could influence the condensation capability of the PCCS heat exchangers or could enhance the opportunity for hydrogen combustion and again raise the peak containment pressures.

In the sections that follow, we rely on the work of Peterson et al. [2.5.5, 2.5.6] to discuss mixing in large stratified volumes. Peterson employed the Zuber hierarchical methodology, and his suggested structure for scaling containment systems is reproduced in Figure 2.5.4. Light-water reactor (LWR) containment systems are made up of large enclosures and interconnecting channels. The large enclosures are subdivided into modules of water pool, air space, and structures. The primary constituents are water, aerosols, and noncondensables for the water pool and air space and steel and concrete for the structures. Each phase, in turn, can exist in several geometric forms, such as a continuous liquid phase, droplets, films or boundary layers, and pool buoyant jets. In this section we consider the scaling of the latter specific geometrical configurations: buoyant jets, buoyant plumes, and wall boundary layers. For all three configurations, mixing takes place by entraining fluid from the large containment module volumes, and the degree of mixing will be the highest for buoyant jets, less for buoyant plumes, and smallest for boundary layers because their entrainment decreases or because they are constrained by the wall surfaces along which they form.

2.5.3.2 Nondimensional Groupings for Buoyant Entrainment

Let us consider a one-dimensional buoyant jet, plume, or boundary layer of cross-sectional area A_b entraining fluid along a perimeter P_b from a large stratified field of volume V_{sf}. The volumetric flow within the buoyant jet, plume, or boundary layer is J_b at the vertical position z, and the corresponding volumetric flow in the stratified field is J_{sf}. If the entrainment velocity at that location is U_e, we can write

$$P_b U_e = \frac{dJ_b}{dz} = -\frac{dJ_{sf}}{dz} \qquad (2.5.21)$$

If the buoyant jet, plume, or boundary layer has a source strength J_{bo} and is active over a distance H_{sf} in the stratified fluid, the specific frequencies ω_{sf} and ω_b for the stratified volume and buoyant geometry are, respectively,

$$\omega_{sf} = \frac{\int_0^{H_{sf}} P_b U_e \, dz}{V_{sf}} \qquad \omega_b = \frac{\int_0^{H_{sf}} P_b U_e \, dz}{V_b} \qquad (2.5.22)$$

where V_{sf} is the stratified volume and V_b the buoyant geometry volume, or $V_b = \int_0^{H_{sf}} A_b \, dz$.

We can also employ the source strength J_{bo} to calculate scaling times τ_{sf} and τ_b to fill the volumes V_{sf} and V_b:

$$\tau_{sf} = \frac{V_{sf}}{J_{bo}} \qquad \tau_b = \frac{V_b}{J_{bo}} \qquad (2.5.23)$$

The relevant nondimensional groupings for buoyant entrainment are, therefore,

$$\Pi_{sf} = \Pi_b = \omega_{sf} \tau_{sf} = \omega_b \tau_b = \frac{\int_0^{H_{sf}} P_b U_e \, dz}{J_{bo}} = \frac{\int_0^{H_{sf}} (dJ_b/dz) \, dz}{J_{bo}} \qquad (2.5.24)$$

2.5 ILLUSTRATIONS OF BOTTOM-UP SCALING

It is also important to note that

$$\frac{\tau_{sf}}{\tau_b} = \frac{V_{sf}}{V_b} \gg 1$$

and the ratio of residence times in the containment to that in the buoyant volume is much greater than 1 because the stratified containment volume is much larger than that of the buoyant jets, plumes, or boundary layers. This means that the entrainment behavior can be predicted from quasistate correlations rather than from the detailed transient turbulent transport within the entrainment configuration.

In the case of several buoyant entrainments, Eq. (2.5.24) can be expanded to

$$\Pi_{sf} = \frac{\Sigma^n \int_0^{H_{sfb}} P_b U_e \, dz}{\Sigma^n J_{bo}} \tag{2.5.25}$$

where n is the number of buoyant entrainments mixing the containment volume, J_{bo} the source strength of each buoyant configuration, and H_{sfb} the vertical distance over which it acts. J_{bo} and H_{sfb} can vary, depending on the source strength and location of the buoyant source.

2.5.3.3 Scaling for Circular Buoyant Jet This special case is illustrated in Figure 2.5.5, which shows a jet of initial diameter d_{bjo} and strength J_{bjo} entraining fluid in a large stratified volume of height H_{sf}. According to List [2.5.7], for a

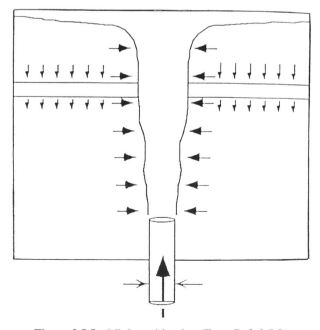

Figure 2.5.5 Mixing with a jet. (From Ref. 2.5.5.)

circular buoyant jet of diameter d_{bj} and volumetric flow J_{bj}, the entrainment rate at position z is

$$\frac{dJ_{bj}}{dz} = \pi d_{bj} U_e = \alpha_T \sqrt{8\pi M} = \alpha_T \sqrt{8\pi M_0} \quad (2.5.26)$$

where α_T is Taylor's jet entrainment constant, typically taken at 0.05, and M is the jet momentum that is presumed to be conserved along the jet path, or

$$M = \frac{J_{bj}^2}{\pi d_{bj}^2/4} = \frac{J_{bjo}^2}{\pi d_{bjo}^2/4} = M_o \quad (2.5.27)$$

The entrainment rate is therefore

$$\pi d_{bj} U_e = 4\alpha_T \sqrt{2} \frac{J_{bjo}}{d_{bjo}} \quad (2.5.28)$$

According to Eq. (2.5.28), the entrainment flow per unit area is independent of the distance from the jet entrance and is about 28% of the jet flow per unit area. The controlling nondimensional entrainment grouping is derived from Eq. (2.5.24) and is for a height H_{sf},

$$\Pi_{sf} = \Pi_{bj} = 4\alpha_T \sqrt{2} \frac{H_{sf}}{d_{bjo}} \quad (2.5.29)$$

Equation (2.5.29) states that similar mixing can be expected between full- and reduced-scale buoyant jets if the height/diameter ratio is preserved.

2.5.3.4 Scaling for Circular Buoyant Plumes

Here we are interested in a gravity-driven plume produced by the addition of a buoyant fluid into another stratified fluid. The buoyancy flux B is now related to the density difference between the ambient fluid ρ_{sf}, the injected fluid ρ_{bpo}, and the plume volumetric source strength J_{bpo}, or

$$B = g \frac{\rho_{sf} - \rho_{bpo}}{\rho_{sf}} J_{bpo} \quad (2.5.30)$$

According to List [2.5.7], for round turbulent buoyant plumes in unstratified volumes,

$$\frac{dJ_{bp}}{dz} = \pi d_{bp} U_e = \frac{5}{3} k_p B^{1/3} z^{2/3} \quad (2.5.31)$$

where J_{bp} is the volumetric flow at location z, d_{bp} the plume diameter, and k_p a constant with a value of 0.35. If we apply List correlations to stratified volumes and introduce the Richardson number, $\mathrm{Ri}_{d_{bpo}}$, based on the initial buoyant plume conditions, or

$$\mathrm{Ri}_{d_{bpo}} = \frac{gd_{bpo}(\rho_{sf} - \rho_{bpo})/\rho_{sf}}{U_{bpo}^2} = \left(\frac{\pi}{4}\right)^2 \frac{gd_{bpo}^5(\rho_{sf} - \rho_{bpo})/\rho_{sf}}{J_{bpo}^2} \quad (2.5.32)$$

there results from Eq. (2.5.31)

$$\Pi_{bj} = \Pi_{sf} = \frac{5}{3}\left(\frac{4}{\pi}\right)^{2/3} k_p \cdot \mathrm{Ri}_{d_{bpo}}^{1/3} \left(\frac{H_{sf}}{d_{bpo}}\right)^{5/3} \quad (2.5.33)$$

which states that similar mixing can be expected between full- and reduced-scale buoyant plumes when the plume height/diameter ratio and its Richardson number based on the initial plume conditions are preserved.

2.5.3.5 Scaling for Buoyant Boundary Layers In this case, the buoyant boundary layer is due to gravity-driven natural convection along a flat surface. For turbulent flow, the method of solution was proposed originally by Eckert and Jackson [2.5.8]. The technique assumes power profiles for the velocity and temperature within the boundary and employs integral methods to establish the entrainment flow and heat transfer rate. Levy [2.5.9] extended that solution from vertical to inclined and horizontal flat surfaces. The primary interest in containment scaling is in the integrated heat transport q_w within the boundary layer. For a plate inclined at an angle θ, the total heat transport over a length H_{sf} is

$$q_w = 0.0246k(T_w - T_{sf})P_{bl} \cdot \mathrm{Gr}_{H_{sf}}^{2/5} \cdot \mathrm{Pr}^{7/15}(1 + 0.494\,\mathrm{Pr}^{2/3})^{-2/5}(\sin\theta)^{2/5} \quad (2.5.34)$$

when k is the fluid thermal conductivity, P_{bl} the heat transfer perimeter, T_{sf} the containment temperature, T_w the surface temperature, $\mathrm{Gr}_{H_{sf}}$ the Grashof number based on H_{sf}, or

$$\mathrm{Gr}_{H_{sf}} = \frac{\beta g(T_w - T_1)H_{sf}^3}{\nu^2}$$

Pr the Prandtl number, β the fluid thermal expansion coefficient, and ν its kinematic viscosity. The corresponding entrained fluid or boundary layer flow over the length H_{sf} is

$$\int_0^{H_{sf}} U_e P_{be}\, dz = 0.098\mathrm{Gr}_{H_{sf}}^{2/5} \nu P_{bl}(1 + 0.494\mathrm{Pr}^{2/3})^{-2/5} \cdot \mathrm{Pr}^{-8/15}(\sin\theta)^{2/5} \quad (2.5.35)$$

For a horizontal plate facing upward, the corresponding heat transport is

$$q_w = 0.067k\,(T_w - T_{sf})P_{bl} \cdot \mathrm{Gr}_{H_{sf}}^{4/11} \cdot \mathrm{Pr}^{9/33}(1 + 0.441\mathrm{Pr}^{2/3})^{-4/11} \quad (2.5.36)$$

Equations (2.5.34), (2.5.35), and (2.5.36) show that the prototype and model heat transport and boundary layer flow will be similar if the Grashof and Prandtl numbers are preserved.

2.5.3.6 Scaling Criteria for Buoyant Jet Transition and Ambient Stratification

Buoyant jet behavior can be expected to transition to a buoyant plume away from the discharge point. For example, Gebhart et al. [2.5.10] give the length scale H_{trans} for the transition from a forced jet to a buoyant plume as

$$\frac{H_{\text{trans}}}{d_{bjo}} = \mathrm{Ri}_{d_{po}}^{1.4}\left(\frac{\rho_{bpo}}{\rho_{sf}}\right)^{1/2} \quad (2.5.37)$$

Again, if the Richardson number and fluid properties are preserved, the transition from jet to plume behavior scales properly.

Others [2.5.11] have suggested that a single pulse forced jet will tend to form a mushroom cap at its leading edge after a distance $H_{\text{cap}} = 0.4 U_{bjo} t_{\text{cap}}$ and that the cap will move at a constant velocity $0.4 U_{bjo}$ beyond that height. The cap transition time t_{cap} can be found from $t_{\text{cap}} = \int_0^{H_{\text{cap}}} (dz/U_{bj})$, and by using Eq. (2.5.25), we find that $H_{\text{cap}} = 10.6 d_{bjo}$, which compares favorably to other hydrodynamic transition lengths.

Experimental data also exist for the breakdown of large stratified fields. Peterson [2.5.6] shows that for round free buoyant jets the breakdown depends on the same two groupings H_{sf}/d_{bjo} and $\mathrm{Ri}_{d_{po}}$ as shown in Figure 2.5.6. He also points out that stratification will almost always prevail for buoyant plumes and wall boundary layers.

2.5.4 Summary: Bottom-Up Scaling

Even though the preceding discussion of bottom-up scaling dealt with a small sample of potential phenomena, one can generalize several of the findings:

1. Bottom-up scaling may identify nondimensional groupings different from those obtained from top-down scaling. Seldom will the two sets of nondimensional groupings complement each other as they did in the case of peak fuel temperature discussed in Section 2.5.1. In that specific instance, the top-down power/volume approach ensured that the fluid velocity and hydraulic conditions were similar in the prototype and the reduced scale facility at the location of interest. Focus could then be placed on the fuel behavior and its characteristics to get relevant fuel bottom-up scaling groupings. In general, the two (bottom-up and top-down) sets of nondimensional groupings cannot be satisfied simultaneously, and the scaling strategy must recognize that potential conflict.

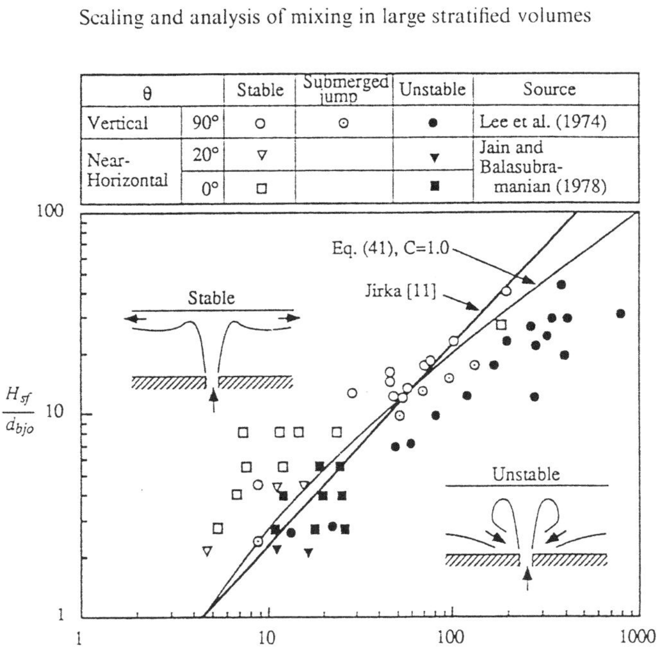

Scaling and analysis of mixing in large stratified volumes

θ		Stable	Submerged jump	Unstable	Source
Vertical	90°	○	⊙	●	Lee et al. (1974)
Near-Horizontal	20°	▽		▼	Jain and Balasubra-manian (1978)
	0°	□		■	

$\dfrac{H_{sf}}{d_{bjo}}$

Eq. (41), C=1.0
Jirka [11]
Stable
Unstable

Figure 2.5.6 Stability criterion for round, free, buoyant jets. (Reprinted from Ref. 2.5.6 with permission from Elsevier Science.)

2. Different phenomena may require alternative scaling approaches. For example, during the early phase of the blowdown in the SBWR of Figure 2.5.3, the primary interest is in the clearing of the main containment vents between the drywell and the suppression chamber. The initial escape of gases through the vents and the subsequent condensation of steam released from the break can produce significant loads on the containment structure. Full-scale vent tests have been found to be the only way to obtain meaningful load data, and the tests have been carried out with a full-scale sector of the containment vents. Another important consideration would be the long-time scaling of the structural response of the prototype and scaled containment facility. Beyond the blowdown phase, the SBWR containment pressure is controlled by the capability of the PCCS heat exchangers to remove decay heat and to permit noncondensable gasses to reach the suppression pool. Natural circulation prevails and a power/volume strategy employing a full-height test facility and PCCS heat transfer tubes has been the preferable choice.

3. Several bottom-up phenomena can occur at the same time. Pilch et al. [2.5.11] have shown (Figure 2.5.7) possible modes of hydrogen combustion in a PWR dry containment under severe accident conditions which involve fuel melting and its escape from the reactor. If the reactor pressure vessel fails while the reactor is at high pressure, the flashing water blowdown will entrain molten core debris, which can accelerate the containment pressure increase. Also, hot hydrogen produced in

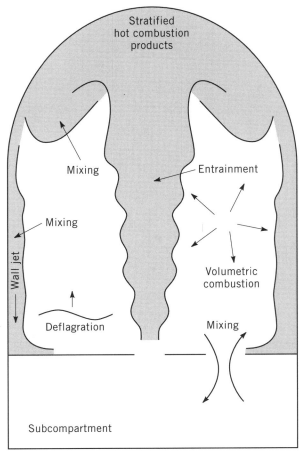

Figure 2.5.7 Hydrogen combustion during severe accident events with molten fuel release from the reactor vessel. (From Ref. 2.5.11.)

the subcompartment vents into the upper dome of the containment, where it entrains gases (particularly, oxygen) prevailing within the containment and burns as a jet diffusion flame. The jet will also strike the dome and produce further mixing and burning of the hot combustion products with the dome atmosphere. Furthermore, a cold boundary layer could be formed along the circumferential walls of the containment, mix with the containment atmosphere, and possibly burn. When several bottom-up phenomena happen simultaneously, as in Figure 2.5.7, it is essential to prioritize them and to put scaling emphasis on the most relevant ones. Because the reactor blowdown is a significant contributor to peak containment pressure, interest is in the amount of hydrogen that might burn on that time scale, and the bottom-up phenomenon of most interest is the fastest-developing one (i.e., the burning hydrogen jet and its entrainment of containment atmosphere over a time scale of a few seconds). The buoyant plume nondimensional groupings of Section 2.5.2.4 are therefore the most important ones in scaling hydrogen burning in a dry containment. In Reference 2.5.11, Pilch calculated time scales for the various phenomena depicted in Figure 2.5.7, and he confirmed the dominance of the buoyant plume behavior.

4. In some cases it is possible to adjust the test configuration to eliminate or reduce the conflict between top-down and bottom-up nondimensional groups. This can be done, for example, for the return of noncondensable gases from the PCCS heat exchangers to the suppression pool of the SBWR containment. If steam is entrained with gases into the suppression pool at a rate J_{bpo} and temperature T_0, the energy brought into the pool above the pool temperature T_p will remain in the plume. The average plume temperature T_z at an elevation z above its release point can be obtained from

$$J_{bpo}(T_0 - T_p) = J_{bp}(T_z - T_p) \quad (2.5.38)$$

If we integrate Eq. (2.5.31) with respect to z, there results

$$J_{bp} = J_{bpo} + k_p B^{1/3} z^{5/13} \quad (2.5.39)$$

so that

$$\frac{T_z - T_p}{T_0 - T_p} = \frac{J_{bpo}}{J_{bpo} + k_p B^{1/3} z^{5/3}} \approx \frac{J_{bpo}}{k_p B^{1/3} z^{5/3}} \quad (2.5.40)$$

With a power/volume scaling approach to the SBWR containment, the ratio of pool suppression area and return flow to it is R from the prototype to the reduced scaled facility, while the vertical dimension ratio is 1. If the fluid properties are preserved, the ratio of local pool temperature difference from prototype to model becomes

$$(T_z - T_p)_R \approx (J_{bpo})_R^{2/3} = R^{2/3} \quad (2.5.41)$$

Equation (2.5.41) states that the temperature of the plume reaching the pool surface will be smaller in the model than in the prototype when the vent submergence is retained. However, it is most likely that the vent from the PCCS heat exchangers into the suppression pool will not utilize an open-ended submerged vertical pipe but rather, a pipe terminating in a horizontal pipe or quencher with multiple small holes to improve the condensation process. If the quencher contains N holes, the rate of flow through each nozzle is J_{bp}/N and Eq. (2.5.41) becomes

$$(T_z - T_p)_R \approx \frac{R^{2/3}}{N_R^{2/3}} \quad (2.5.42)$$

where N_R is the ratio of holes from prototype to model. By selecting $N_R = R$, the temperature ratio $(T_z - T_p)_R$ approaches 1 and it resolves the conflict or bias generated by Eq. (2.5.41). In most other conflicts between top-down and bottom-up scaling, a bias will continue to prevail, and it needs to be quantified as illustrated in the next section. However, the physical importance of each existing bias to the overall behavior of the complex system needs to be assessed physically. For example, in the case of the suppression pool temperature bias discussed above, the hot water reaching the top of the pool will spread horizontally. This horizontal spreading will occur with the velocity of a gravity wave and will take place within a few seconds. Over the long-time response of interest to the containment pressure, all the water in the pool above the PCCS vent release point will therefore tend to reach the same temperature in the prototype and model since the heat delivered to that portion of the pool and its amount of water will both have been scaled by the ratio R. The bias produced by Eq. (2.5.41) is important, therefore, only on a very short time scale and can be neglected in terms of long-term containment pressure considerations.

2.6 SCALING BIASES AND DISTORTIONS

Scaling biases can originate from a variety of sources, including:

- Inherent shortcomings of the test facility such as increased heat losses due to power/volume scaling.
- Failure to duplicate important characteristics of the prototype. It was noted previously that employing nonprototypic fluids under nonequilibrium or change of phase conditions may produce biases that are not readily assessable.
- Different input conditions or input conditions prescribed from system computer codes. In the latter case, a conflict of interest could exist since the purpose of the test facility is to validate the complex system computer code.
- Inability to satisfy all the important nondimensional groupings because of conflicting scaling requirements.
- Inability to scale the various phases of a complex system transient or accident because the controlling phenomena and their relative importance change during the various phases of the event(s).

All such scaling biases, regardless of their source, must be evaluated and quantified to ascertain that their impact on the test facility results can be tolerated. Two specific examples are considered here for illustrative purposes: heat loss and flow pattern biases.

2.6.1 Heat Loss Bias

Let us consider a cylindrically configured facility losing heat at its circumference. Let us also assume that a power/volume scaling approach has been adopted. Under those circumstances, the ratio of the power and cross-sectional area of the prototype to that of the model is R. The corresponding ratios of their outside perimeter and

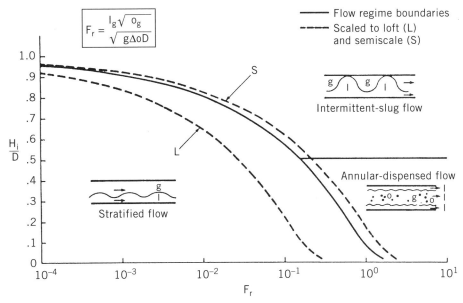

Figure 2.6.1 Dukler–Taitel flow regime map. (From Ref. 2.6.1.)

therefore heat loss surface area are $\sqrt{R}$. This means that the fraction of input heat being lost to the environment will be much greater for the model than for the prototype. Several approaches have been employed to reduce this impact. The simplest is to increase the insulation of the model. Another commonly used method is to add heaters on the outside perimeter of the model, adjusting them to get the correct heat loss. Finally, the heat input to the model has been increased in a few cases to compensate for the increased heat loses. This last technique, however, can influence the thermal-hydraulic behavior within the model and should be used judiciously.

2.6.2 Flow Pattern Bias

During a small-break LOCA in a horizontal pipe, the flow pattern in that pipe can influence the mass flow rate through the break. In particular, when the test facility obeys power/volume scaling, horizontal flow pattern transitions in the model prototype can be different and need to be assessed. Zuber considered this problem in Reference 2.6.1 and relied on the flow pattern map developed by Taitel and Dukler [2.6.2], which is reproduced in Figure 2.6.1. According to Figure 2.6.1, flow regime boundaries depend on the dimensionless liquid depth H_l/D and on the Froude number for the vapor:

$$\mathrm{Fr} = \frac{j_g \sqrt{\rho_g}}{\sqrt{g(\rho_l - \rho_g)D}} \quad (2.6.1)$$

TABLE 2.6.1 Scale Distortion for Horizontal Flow Pattern Transition

	Diameter D	Scale Ratio R	Scale Distortion Δ
Prototype PWR	29 in. (73.7 cm)	1	1
LOFT facility	11 in. (28 cm)	64	0.176
Semiscale facility	1.34 in. (3.4 cm)	1500	1.46

where H_l is the height of the liquid in a horizontal pipe, D its diameter, j_g the superficial gas velocity, ρ_l and ρ_g the liquid and gas densities, and g the gravitational constant.

Figure 2.6.1 shows that the transitions from stratified to slug or annular dispersed will be the same in the model and prototype if their values of Froude number and nondimensional liquid depth agree. With power/volume scaling the ratio of power in the prototype to that in the model is R and so are the ratio of pipe flow areas and gas flows, or

$$\frac{P_p}{P_m} = \left(\frac{D_p}{D_m}\right)^2 = R = \frac{j_{gp}}{j_{gm}}\left(\frac{D_p}{D_m}\right)^2 \tag{2.6.2}$$

where P is the power produced, D the pipe diameter, and the subscripts p and m represent the prototype and model. If the same fluids are used in the model and prototype, the ratio of their Froude numbers becomes

$$\frac{Fr_p}{Fr_m} = R\left(\frac{D_p}{D_m}\right)^{-5/2} \tag{2.6.3}$$

which produces a scale distortion

$$\Delta = \frac{1}{R}\left(\frac{D_{p_p}}{D_m}\right)^{5/2}$$

This scale distortion was calculated in Reference (2.6.1) and the results are tabulated in Table 2.6.1. Referring to Figure 2.6.1, these scale distortions values mean that when the prototype is on the flow pattern transition line, the LOFT facility would be below it and still in the stratified region, while the semiscale facility would be above it and in the slug or annular dispersed flow region. It is also observed that the semiscale facility comes much closer to replicating the prototype behavior. In fact, in Reference 2.6.1 it is shown that for Fr = 0.157, the prototype and semiscale facility would have relatively the same void fraction of 0.49 to 0.5 and that the nondimensional liquid depths would be only 10% apart (0.5 for the prototype and 0.56 for the semiscale facility).

The preceding two examples of scaling distortion are typically unavoidable. They were selected not so much for that reason but rather, to illustrate that distortions must be recognized and quantified during the scaling process.

2.7 SUMMARY

1. A comprehensive and prioritized test and analysis program for the complex system under evaluation should precede the scaling efforts because it identifies the elements of system behavior, component performance, and phenomena understanding, which demand additional experiments and improved modeling.

2. Scaling of test facilities and their capability to simulate the complex system should be performed before the test facilities are built rather than after.

3. Correct scaling of the significant areas of concern is much more important than the total number of potential concerns. By looking at different classes of conditions (steady state, transients, and accidents) and time phasing their occurrence during the scenarios of most interest, one "can avoid doing with more what can be done with less" [2.1.2].

4. Global scaling of the system is best done from a simplified, uniform top-down approach, but it needs to be complemented with bottom-up scaling assessments to capture the significant details and phenomena. If at all possible, an integral approach at the detailed level can improve scaling by reducing the number of nondimensional groups to be considered.

5. It is preferable to employ the same fluids, properties, and initial conditions when changes of phase or nonequilibrium conditions occur in the complex system.

6. Scaling uncertainties are reduced by employing as large a scale as practical and/or by performing counterpart tests.

7. Scaling distortions or biases are unavoidable and their impact must be evaluated as part of the scaling effort to ascertain that they are being kept at a low enough level not to jeopardize the validity of the test results. Some scaling biases become much less important if viewed at the overall system level and from the perspective of important behavioral parameters.

8. An analytical model validated by data from a well-scaled facility may be the only way to predict the complex system performance where test data could not be obtained.

9. The most essential and guiding principle in scaling complex systems is the need to understand the essential physics involved and to be able to preserve them from model to prototype.

10. Scaling efforts can be of little value without a strong parallel test instrumentation program to accurately obtain the necessary information and to improve the understanding of phenomena. Calibration of instrument, mass, and heat balances during the tests and steady-state measurements of test facility characteristics are essential to obtaining good test data.

REFERENCES

2.1.1 *An Integrated Structure and Scaling Methodology for Severe Accident Technical Issue Resolution*, NUREG/CR-5809, November 1991. Appendix D-A, "Hierarchical, Two-Tiered Scaling Analysis," by N. Zuber.

2.1.2 Zuber, N., "Scaling and Analysis of Complex Systems," Invited lecture, Ljubljana, Slovenia, April 1994.

2.1.3 Yadiroglu, G., et al., *Scaling of SBWR Related Tests*, GE Nuclear Energy Report NEDO-32288, Rev. 0, December 1995.

2.2.1 Barton, J. E., et al., *BWR Refill–Reflood Program; Task 4.4-30° SSTF; Description Document,* GE Report NUREG/CR-2133, May 1982.

2.2.2 *Compendium of ECCS Research for Realistic LOCA Analysis*, NUREG-1230, Upper Plenum Test Facility, Appendix A, pp. A-155 to A-159, December 1988.

2.2.3 Schwartzbeck, R. K., and Kocamustafaogullari, G., "Two-Phase Flow: Pattern Transition Scaling Studies," *ANS Proceedings of the 1988 National Heat Transfer Conference*, Houston, Texas, pp. 387-398, July 24–27, 1988.

2.2.4 Boucher, T. J., and Wolf, J. R., *Experiment Operating Specifications for Semiscale Mod-2C Feedwater and Steam Line Break Experiment Series*, EGG-SEMI-6625, May 1965.

2.4.1 Moody, F. J., *Introduction to Unsteady Thermofluid Mechanics*, Wiley, New York, 1990, pp. 87–89, 328–331.

2.4.2 Yadiroglu, G., et al., *Scaling of the SBWR Related Tests*, GE Nuclear Energy Report NEDO-32288, December 1995.

2.4.3 Wulff, W., personal communication, 1995.

2.4.4 Ishii, M., and Kataoka, I., *Similarity Analysis and Scaling Criteria for LWR's under Single-Phase and Two-Phase Natural Circulation*, NUREG/CR-3267 (ANL-83-32), 1983.

2.4.5 Kiang, R. L., "Scaling Criteria for Nuclear Reactor Thermal Hydraulics," *Nucl. Sci. Eng.*, **89**, 207–216, 1985.

2.4.6 Kocamustafaogullari, G., and Ishii, M., "Scaling Criteria for Two-Phase Flow Loops and Their Application to Conceptual 2 × 4 Simulation Loop Design," *Nucl. Technol.*, **65**, 146–160.

2.4.7 Reyes, J. N. Jr., "An Analytical Solution for Two-Phase Natural Circulation Flow Estimation," *Nucl. Eng. Des.*, **186**, 53–109, 1998. Also OSU-NE-9316, 1995.

2.4.8 Thom, R., *Structural Stability and Morphogenesis*, Advanced Book Program, Addison-Wesley, Reading, Mass., 1975.

2.4.9 Zeeman, E. C., *Catastrophe Theory, Selected Papers, 1972–1977*, Advanced Book Program, Addison-Wesley, Reading, Mass., 1977.

2.4.10 Reyes, J. N., Jr., "Scaling Single-State Variable Catastrophe Functions: An Application to Two-Phase Natural Circulation," *Nucl. Eng. Des.*, **151**, 41–48, 1994.

2.5.1 Tong, L. S., and Weisman, J., *Thermal Analysis of Pressurized Water Reactors*, 2nd ed., American Nuclear Society, La Grange Park, Ill., 1979.

2.5.2 Wulff, W., et al., "Quantifying Reactor Safety Margins, 3: Assessment and Ranging of Parameters," *Nucl. Eng. Des.*, **119**, 33–63, 1990.

2.5.3   Berenson, P. J., "Film Boiling Heat Transfer from a Horizontal Surface," *J. Heat Transfer*, **83**, 351–358, 1961.
2.5.4   Catton, I., et al., "Quantifying Reactor Safety Margins, 6: A Physically Based Method of Estimating PWR Large Break Loss of Coolant Accident PCT," *Nucl. Eng. Des.*, **119**, 109–117, 1990.
2.5.5   Peterson, P. F., et al., "Scaling for Integral Simulation of Mixing in Large Stratified Volumes," NURETH 6, *Proceedings of the 6th International Topical Meeting on Nuclear Reactor Thermal Hydraulics*, Grenoble, France, Vol. 1, pp. 202–211, 1993.
2.5.6   Peterson, P. F., "Scaling and Analysis of Mixing in Large Stratified Volumes," *Int. J. Heat Mass Transfer*, **37**(Suppl. 1), 97–106, 1994.
2.5.7   List, E. J., *Turbulent Jets and Plumes*, Pergamon Press, Elmsford, N.Y., 1982.
2.5.8   Eckert, E. R. G., and Jackson, T. W., *Analysis of Turbulent Free Convection Boundary Layer on Flat Plate*, NACA Report 1015, 1951.
2.5.9   Levy, S., "Integral Methods in Natural-Convection Flow," *J. Appl. Mech.*, **22**(4), 515–522, 1955.
2.5.10   Gebhart, B., et al., *Buoyancy-Induced Flows and Transport*, Hemisphere, New York, 1988.
2.5.11   Pilch, M., et al., *The Probability of Containment Failure by Direct Containment Heating in Zion*, NUREG/CR-6075, Suppl. 1, 1994.
2.6.1   Zuber, N., *Problems of Small Break LOCA*, NUREG-0724, 1980.
2.6.2   Taitel, Y., and Dukler, A. E., "A Model for Predicting Flow Regime Transition in Horizontal and Near Horizontal Gas–Liquid Flow," *A.I.Ch.E. J.*, **22**, 47, 1976.

CHAPTER 3

SYSTEM COMPUTER CODES

3.1 INTRODUCTION

A large number and wide variety of computer codes have been developed to analyze, design, and operate complex two-phase-flow systems. In this chapter the primary focus is on system computer codes, which as their name implies, have to deal with overall system behavior during numerous scenarios. System computer codes must be able not only to simulate several subsystems, many components and their couplings, but also the simultaneous occurrence of various phenomena and processes. The difficulties associated with such a large scope of interest are compounded by the complications produced by having to deal with two-phase rather than single-phase flow. In fact, the introduction of a second phase is felt in many different ways not encountered in single-phase flow.

1. In single-phase flow, all the fluid properties are known or can readily be calculated at every position along the flow channel. This is not true in two-phase flow. Let us, for instance, consider three of the properties that are essential to the solution of flow problems: density, thermal conductivity, and viscosity. As noted previously, in two-phase flow the liquid and gas phases usually do not have the same local velocity, and the density of the two-phase stream cannot be determined directly from the total gas and liquid flow rates specified. Similarly, no generally accepted expression exists for the viscosity or the thermal conductivity of a gas–liquid mixture. Thus, at least three new unknowns are introduced, and they greatly add to the complexity of two-phase-flow solutions.

2. A two-phase stream contains a very large number of gas–liquid interfaces. None exists in single-phase flow. At each liquid–gas interface, momentum, energy, and mass are transferred from one phase to the other. Such transfers can be expected

to play a major role in the predictions of two-phase flow; yet their understanding is far from complete because the transfer mechanisms and the interface areas over which they take place are difficult to specify or to measure experimentally. It is the author's opinion that the presence of interfaces and the difficult task of describing them has been, and will continue to be, the weakest element in developing reliable two-phase-flow system computer codes.

3. The two phases can arrange themselves in a large variety of flow patterns. The flow patterns can range from one of the phases being continuous and the other being present in the form of gas bubbles (*bubble flow*) or liquid drops (*dispersed flow*); or the two phases can be separated geometrically to produce *stratified flow* in a *horizontal pipe* or annular flow in a vertical channel; or two-phase flows could be in transition from one pattern to the other, producing what are referred to as *slug flow* and *churn flow* patterns. All two-phase-flow patterns are intermittent in their overall and local time behavior, and the variations with time are significantly larger than for a single phase. For example, in bubble flow the local gas or void fraction fluctuates intermittently from zero to unity, and in stratified and annular flow, waves of different amplitude and shape are present at the interface. In slug flow, the periodic changes in phase content tend to be quite large, while in churn flow, there can be even periodic changes in flow direction (i.e., a liquid film near the wall can be traveling upward part of the time and downward the remainder of the time). It should be apparent from this brief discussion that system computer codes cannot be expected to deal with the detailed or fine structure of two-phase-flow patterns and that they can treat them only on a time-averaged basis. In other words, two-phase flow has an added degree of freedom over single-phase flow: that of flow pattern. Because the values of fluid density, pressure drop, and heat transfer can be expected to vary with the flow pattern, it is necessary for a specified channel geometry to develop not a single solution but a family of approximate solutions to match all the possible types of flow pattern and their transition from one pattern to another.

4. Two-phase flow has several other degrees of freedom that do not occur in single-phase flow. For example,

- *Co-current and countercurrent flow can occur.* In co-current flow, the gas and liquid are flowing in the same direction; in countercurrent flow, they are flowing in opposite directions. The transfer mechanisms and behavior at the interfaces can be expected to be different whether the flow is co-current or countercurrent.
- *Evaporation, condensation, and mass transfer mechanisms can be present.* When one-component two-phase flow occurs, the stream contains the liquid and vapor forms of the same constituent and the possibility always exists that some of the liquid could be converted into vapor or that the vapor could be condensed into liquid. Evaporation and condensation phenomena are therefore an integral part of one-component two-phase-flow processes. In a two-component system, evaporation or condensation need not be taken into account, but mass transfer (i.e., transfer of material from and to the liquid by the gaseous phase) becomes important. This is true of most of the two-component processes, especially those found in rectifying and spray columns.

- *Thermal nonequilibrium can exist* between the two phases. When heat is being added or removed at a boundary surface, it is possible with surface temperatures above saturation temperature to find vapor bubbles being formed at the surface and flowing along it, even though the main stream is subcooled. If the temperature of the surface is high enough, there may be no rewetting by the liquid, and an inverted annular flow pattern results with superheated vapor next to the heated surface and a saturated or subcooled liquid core in contact with it. One can visualize other nonthermal equilibrium configurations, consisting of superheated dispersed vapors containing saturated liquid drops or, again, a subcooled or saturated condensing liquid film in contact with superheated steam. One anticipates that the transfer mechanisms, flow patterns, and interface behavior would be different for nonthermal equilibrium conditions than those for thermal equilibrium. Again, added complexities due to thermal nonequilibrium conditions have to be included in system computer codes.

5. Ideally, a system computer code should be able to cover not only two-phase-flow steady-state conditions but also the anticipated transients and the more severe prescribed accidents that need to be analyzed. The steady-state predictive capability is necessary to establish the initial and most likely conditions. The number of transient scenarios can be quite large and their time durations can range from being very slow (quasi-steady state) to extremely fast and reaching sonic velocity through the break in the case of a loss-of-coolant accident. Unsteady-state circumstances further complicate two-phase-flow behavior because the liquid and gas do not respond in the same way to time variations.

If all the preceding requirements have to be met, the system computer code is being asked to perform an "impossible" mission. History shows that, in fact, the earliest system computer codes were simple and approximate because they were the only analytical tools that could be put together and made to work at that time. They used uniform flow and thermal conditions and were referred to as *homogeneous models*. The next evolution of separated models recognized that the gas–liquid flow conditions could be different without dealing specifically with the interfaces or the flow patterns or any potential nonthermal equilibrium conditions. Most recent system computer codes employ two-fluid models and recognize the presence of flow patterns and exchanges at the gas–liquid interface. Averaging, in both time and in space (typically over an entire flow cross section or over a specified smaller volume element), is used in all models to arrive at tractable computer system codes.

During the evolution process of computer codes, different system computer codes were generated to deal with different events or different characteristics of the complex system. Some of this scope customizing still prevails in the industry today, but the number of variations has been reduced in recent years by employing a modular construction for system computer codes. This allows subsystems and components to be excluded when they have a minor role and/or to focus on the most essential phenomena.

All in all, the development of system computer codes is a remarkable achievement, but it was attained only after a large number of assumptions, simplifications, and validations against test data. The best way to illustrate this fact is to concentrate on one-dimensional flow and to follow the historical development starting from the simplest model and building up to the most complicated one. This approach has the advantage of highlighting both the merits and shortcomings of various system computer codes, which is the primary purpose of this chapter.

3.2 CONSERVATION LAWS IN ONE-DIMENSIONAL TWO-PHASE FLOW WITH NO INTERFACE EXCHANGES

The flow is one-dimensional when the fluid properties vary only in the direction of flow. In other words, fluid properties are uniform or averaged at every cross section along the axis of the channel. For the simplest possible case of separated or mixture models considered in this section, no interface interactions are considered and the solutions of the corresponding one-dimensional conservation equations are by definition approximate. They will yield information only on the way in which the average fluid properties vary along the channel and no information on the variation of properties normal to the flow direction or their gradients. There is, however, a wide range of steady-state and transient two-phase-flow problems which can be solved by excluding the gas–liquid interfaces and by considering the flow to be one-dimensional. Moreover, the one-dimensional equations, because of their simplified forms, give valuable insight about the controlling parameters of two-phase flow.

Consider the channel geometry shown in Figure 3.2.1. The flow is predominantly along the positive z axis. At any position z, the channel is inclined at an angle θ from the horizontal and has a cross-sectional-flow area A and a wetted perimeter P (defined as the contact perimeter between fluid and channel walls). Because the

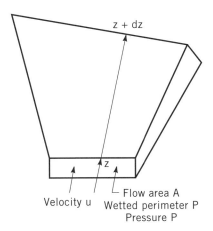

Figure 3.2.1 Flow channel.

flow is one-dimensional, the velocity vector has only one component u in the z direction, and the average velocity, $\bar{u}$, is

$$\bar{u} = \frac{1}{A} \int_A u \, dA \tag{3.2.1}$$

where the bar superscript is utilized to represent averaging over the channel cross section.

3.2.1 Conservation of Mass

If we follow the development of the one-dimensional laws by Meyer [3.2.1], the equation for conservation of mass in one dimension becomes

$$\frac{\partial}{\partial t} \left(\int_A \rho \, dA \right) + \frac{\partial}{\partial z} \left(\int_A \rho u \, dA \right) = 0 \tag{3.2.2}$$

where ρ is the fluid density and t represents the time. Equation (3.2.2) is general and applies to two- or single-phase flow. For two-phase flow, the relation can be simplified further by introducing the following definitions:

$$\bar{\rho} = \frac{1}{A} \int_A \rho \, dA = \text{average density at position } z \tag{3.2.3}$$

$$G = \frac{1}{A} \int_A \rho u \, dA = \text{average mass flow per unit area at } z \tag{3.2.4}$$

and the one-dimensional integral continuity relation becomes

$$\frac{\partial}{\partial t} (\bar{\rho} A) + \frac{\partial}{\partial z} (GA) = 0 \tag{3.2.5}$$

In two-phase flow, the average density $\bar{\rho}$ can be expressed in terms of the liquid and gas densities, ρ_L and ρ_G, and of the average gas weight-rate and volume fractions. Because the flow is one-dimensional, the local liquid and gas velocities are represented by their average values $\bar{u}_L$ and $\bar{u}_G$ over the liquid cross-sectional area A_L and the gas area A_G, or

$$\bar{u}_L = \frac{1}{A_L} \int_{A_L} u_L \, dA_L \qquad \bar{u}_G = \frac{1}{A_G} \int_{A_G} u_G \, dA_G \tag{3.2.6}$$

The average gas volume and weight-rate fractions are obtained from

$$\text{average gas volume fraction} = \bar{\alpha} = \frac{A_G}{A_L + A_G} = \frac{A_G}{A} \qquad (3.2.7)$$

$$\text{average gas weight-rate fraction} = \bar{x} = \frac{\text{weight flow of gas}}{\text{total weight flow}} = \frac{\rho_G \bar{u}_G \, \bar{\alpha}}{G} \qquad (3.2.8)$$

The average fluid density can now be evaluated:

$$\bar{\rho} = \frac{1}{A} \int_A \rho \, dA = \frac{1}{A} \int_{A_G} \rho_G \, dA_G + \frac{1}{A} \int_{A_L} \rho_L \, dA_L \qquad (3.2.9)$$

or

$$\bar{\rho} = \rho_G \frac{A_G}{A} + \rho_L \frac{A_L}{A} = \rho_G \bar{\alpha} + \rho_L (1 - \bar{\alpha}) \qquad (3.2.10)$$

3.2.2 Conservation of Momentum

The corresponding one-dimensional Meyer (3.2.1) integral momentum equation is

$$\frac{\partial}{\partial t}\left(\int_A \rho u \, dA\right) + \frac{\partial}{\partial z}\left(\int_A \rho u^2 \, dA\right)$$

$$= -A\frac{\partial p}{\partial z} - \int_P \tau_w \, dl - g\left(\int_A \rho \, dA\right) \sin\theta \qquad (3.2.11)$$

where p is the pressure, P the wetted perimeter, τ_w the wall shear stress, and g the gravitational constant. Equation (3.2.11) can be simplified further by introducing Eqs. (3.2.3) and (3.2.4) and by defining an average momentum density $\bar{\rho}_M$ and an average wall shear stress $\bar{\tau}_w$:

$$\frac{1}{\bar{\rho}_M} = \frac{1}{G^2}\left(\frac{1}{A}\int_A \rho u^2 \, dA\right) \qquad (3.2.12)$$

$$\bar{\tau}_w = \frac{1}{P}\int_P \tau_w \, dl \qquad (3.2.13)$$

The law of conservation of momentum becomes

$$\frac{\partial}{\partial t}(GA) + \frac{\partial}{\partial z}\left(\frac{G^2 A}{\bar{\rho}_M}\right) = -A\frac{\partial p}{\partial z} - \bar{\tau}_w P - \bar{\rho}g A \sin\theta \qquad (3.2.14)$$

The average momentum density $\bar{\rho}_M$ can again be expressed in terms of the average gas weight-rate and volume fractions:

$$\frac{1}{\bar{\rho}_M} = \frac{1}{G^2 A}\left(\int_{A_G} \rho_G u_G^2 \, dA_G + \int_{A_L} \rho_L u_L^2 \, dA_L\right) = \frac{1}{G^2}[\rho_G \bar{u}_G^2 \bar{\alpha} + \rho_L \bar{u}_L^2(1-\bar{\alpha})] \quad (3.2.15)$$

or

$$\frac{1}{\bar{\rho}_M} = \frac{1}{\rho_G}\frac{\bar{x}^2}{\bar{\alpha}} + \frac{1}{\rho_L}\frac{(1-\bar{x})^2}{1-\bar{\alpha}} \quad (3.2.16)$$

3.2.3 Conservation of Energy

The corresponding Meyer energy equation written in terms of fluid enthalpy H is

$$\frac{\partial}{\partial t}\left(\int_A \rho H \, dA\right) + \frac{\partial}{\partial z}\left(\int_A \rho u H \, dA\right) + \frac{\partial}{\partial t}\left(\int_A \frac{1}{2}\rho u^2 \, dA\right) + \frac{\partial}{\partial z}\left(\int_A \frac{1}{2}\rho u^3 \, dA\right)$$

$$= \int_{P_h} q'' \, dz + \int_A q''' \, dA + \frac{\partial}{\partial t}(pA) - g\sin\theta \int_A \rho u \, dA \quad (3.2.17)$$

where q'' is the heat flux over the heated perimeter P_h and q''' is the heat generation per unit volume. Equation (3.2.17) can be simplified by substituting Eqs. (3.2.4) and (3.2.12) and the following definitions of average flow-weighted enthalpy $\bar{H}$, density-weighted enthalpy $\bar{H}_\rho$, kinetic energy density $\bar{\rho}_{KE}$, average surface heat flux $\bar{q}''$, and average heat generation per unit volume $\bar{q}'''$:

$$\bar{H} = \frac{1}{G}\left(\frac{1}{A}\int_A \rho u H \, dA\right) \quad (3.2.18)$$

$$\bar{H}_\rho = \frac{1}{\bar{\rho}}\left(\frac{1}{A}\int_A \rho H \, dA\right) \quad (3.2.19)$$

$$\frac{1}{\bar{\rho}_{KE}} = \left[\frac{1}{G^3}\left(\frac{1}{A}\int_A \rho u^3 \, dA\right)\right]^{1/2} \quad (3.2.20)$$

$$\bar{q}'' = \frac{1}{P_h}\int_{P_h} q'' \, dz \quad (3.2.21)$$

$$\bar{q}''' = \frac{1}{A}\int_A q''' \, dA \quad (3.2.22)$$

The equation for conservation of energy becomes

$$\frac{\partial}{\partial t}(\bar{\rho}\bar{H}_p A) + \frac{\partial}{\partial z}(G\bar{H}A) + \frac{1}{2}\frac{\partial}{\partial t}\left(\frac{G^2 A}{\bar{\rho}_M}\right) + \frac{1}{2}\frac{\partial}{\partial z}\left(\frac{G^3 A}{\bar{\rho}_{KE}^2}\right)$$
$$= P_h\,\overline{q''} + A\,\overline{q'''} + \frac{\partial}{\partial t}(pA) - gGA\sin\theta \quad (3.2.23)$$

The properties $\bar{H}_p$, $\bar{H}$, and $\bar{\rho}_{KE}$ can again be expressed in terms of the average gas weight-rate and volume fractions:

$$\bar{H}_p = \frac{1}{\bar{\rho}}[\rho_G H_G \bar{\alpha} + \rho_L H_L (1 - \bar{\alpha})] \quad (3.2.24)$$

$$\bar{H} = \frac{1}{G}[\rho_G \bar{u}_G H_G \bar{\alpha} + \rho_L \bar{u}_L H_L (1 - \bar{\alpha})] = H_G \bar{x} + H_L (1 - \bar{x}) \quad (3.2.25)$$

$$\frac{1}{\bar{\rho}_{KE}} = \left[\frac{1}{\rho_G^2}\frac{\bar{x}^3}{\bar{\alpha}^2} + \frac{1}{\rho_L^2}\frac{(1-\bar{x})^3}{(1-\bar{\alpha})^2}\right]^{1/2} \quad (3.2.26)$$

3.3 DISCUSSION OF ONE-DIMENSIONAL TWO-PHASE CONSERVATION LAWS

3.3.1 Steady-State Flow

For steady-state conditions, the three basic conservation laws become

$$\frac{d}{dz}(GA) = 0 \quad (3.3.1)$$

$$\frac{d}{dz}\left(\frac{G^2 A}{\bar{\rho}_M}\right) = -A\frac{dp}{dz} - \bar{\tau}_w P - \bar{\rho}gA\sin\theta \quad (3.3.2)$$

$$\frac{d}{dz}(G\bar{H}A) + \frac{1}{2}\frac{d}{dz}\left(\frac{G^3 A}{\bar{\rho}_{KE}^2}\right) = P_h\overline{q''} + A\,\overline{q'''} - gGA\sin\theta \quad (3.3.3)$$

The continuity equation gives

$$GA = \text{constant} \quad (3.3.4)$$

and the momentum equation can be rewritten as

$$-\frac{dp}{dz} = G\frac{d}{dz}\left(\frac{G}{\bar{\rho}_M}\right) + \frac{\bar{\tau}_w P}{A} + \bar{\rho}g\sin\theta \quad (3.3.5)$$

According to Eq. (3.3.5), under steady-state conditions, the two-phase pressure drop is made up of three components.

$$\text{acceleration pressure drop} = G \frac{d}{dz}\left(\frac{G}{\overline{\rho}_M}\right)$$

$$\text{frictional pressure drop} = \frac{\overline{\tau}_w P}{A} \qquad (3.3.6)$$

$$\text{hydrostatic or head pressure drop} = \overline{\rho} g \sin\theta$$

Of the preceding three terms, the acceleration and head losses can be calculated if the average gas weight-rate and volume fractions, $\overline{x}$ and $\overline{\alpha}$, are known [see Eqs. (3.2.10) and (3.2.16)]. For steady-state conditions, the total two-phase pressure drop therefore can be obtained if G, $\overline{\tau}_w$, $\overline{x}$, and $\overline{\alpha}$ are known. Similarly, the energy equation can be written

$$\frac{d\overline{H}}{dz} + \frac{1}{2}\frac{d}{dz}\left(\frac{G^2}{\overline{\rho}_{KE}^2}\right) = \frac{P_h \overline{q''}}{GA} + \frac{\overline{q'''}}{G} - g\sin\theta \qquad (3.3.7)$$

and the change in enthalpy $\overline{H}$ can be computed if G, $\overline{x}$, $\overline{\alpha}$, $\overline{q''}$, and $\overline{q'''}$ are known.

3.3.2 The Many Densities of Two-Phase Flow

A striking novelty of the one-dimensional two-phase-flow equations is that they do not admit to a single density value. There are average density values for the head term $(\overline{\rho})$, for the acceleration term $(\overline{\rho}_M)$, and for the kinetic energy term $(\overline{\rho}_{KE})$. The existence of so many densities has not been recognized sufficiently by workers in the two-phase field. Many investigators have failed to use the correct form of density in defining some of the nondimensional groupings. One could, for instance, formulate a two-phase Reynolds number in terms of the channel hydraulic diameter (D_H), a mean two-phase density $(\overline{\rho}_{TP})$, velocity $\overline{u}_{TP}$, and viscosity $\overline{\mu}_{TP}$:

$$N_{RE_{TP}} = \frac{\overline{u}_{TP}\,\overline{\rho}_{TP}\, D_H}{\overline{\mu}_{TP}} \qquad (3.3.8)$$

As will be shown later, and as pointed out by Dukler et al. [3.3.1], the correct density to be used in Eq. (3.3.8) is not $\overline{\rho}_{TP} = \overline{\rho}$, as many workers have done in the past.

Another interesting result about the two-phase-flow densities can be obtained for the simplified case where the slip ratio of gas to liquid velocity $(\overline{u}_G/\overline{u}_L)$ is assumed to be a constant and independent of the gas weight fraction $\bar{x}$. The three densities $\bar{\rho}$, $\bar{\rho}_M$, and $\bar{\rho}_{KE}$ can be rewritten in terms of $\bar{u}_G/\bar{u}_L$:

$$\bar{\rho} = \frac{(1-\bar{x})(\bar{u}_G/\bar{u}_L) + \bar{x}}{(1/\rho_L)(\bar{u}_G/\bar{u}_L)(1-\bar{x}) + (1/\rho_G)\bar{x}}$$

$$\frac{1}{\bar{\rho}_M} = \left(\frac{\bar{x}}{\rho_G}\frac{\bar{u}_L}{\bar{u}_G} + \frac{1-\bar{x}}{\rho_L}\right)\left[1 + \bar{x}\left(\frac{\bar{u}_G}{\bar{u}_L} - 1\right)\right] \quad (3.3.9)$$

$$\frac{1}{\bar{\rho}_{KE}} = \left(\frac{\bar{x}}{\rho_G}\frac{\bar{u}_L}{\bar{u}_G} + \frac{1-\bar{x}}{\rho_L}\right)\left\{1 + \bar{x}\left[\left(\frac{\bar{u}_G}{\bar{u}_L}\right)^2 - 1\right]\right\}^{1/2}$$

Taking the derivatives of the expressions above with respect to $\bar{u}_G/\bar{u}_L$ shows that the density $\bar{\rho}$ decreases automatically with slip ratio. However, the other two densities, $\bar{\rho}_M$ and $\bar{\rho}_{KE}$, both exhibit a maximum. The momentum average density $\bar{\rho}_M$ reaches a maximum with respect to slip ratio when

$$\frac{\bar{u}_G}{\bar{u}_L} = \left(\frac{\rho_L}{\rho_G}\right)^{1/2} \quad (3.3.10)$$

and the average kinetic energy density $\bar{\rho}_{KE}$ maximizes at

$$\frac{\bar{u}_G}{\bar{u}_L} = \left(\frac{\rho_L}{\rho_G}\right)^{1/3} \quad (3.3.11)$$

As will be shown in Chapter 6, Eqs. (3.3.10) and (3.3.11) have been used to predict two-phase fluid densities, particularly in the case of two-phase critical flow.

3.3.3 Nondimensional Groupings

Many of the single-phase-flow correlations are expressed in terms of groupings obtained by nondimensionalizing the momentum and energy equations. Similar groupings can be formulated for two-phase flow. Let us consider the case of steady-state one-dimensional flow in a channel of constant cross-sectional area and inclined at an angle θ from the horizontal. The continuity equation reduces to the statement of $G = $ constant while the momentum equation becomes

$$G^2 \frac{d}{dz}\left(\frac{1}{\bar{\rho}_M}\right) = -\frac{dp}{dz} - \frac{\bar{\tau}_w P}{A} - \bar{\rho}g\sin\theta \quad (3.3.12)$$

Let us nondimensionalize Eq. (3.3.12) by introducing a reference length L, a reference average momentum density $(\bar{\rho}_M)_m$, and a reference average head density $(\bar{\rho})_m$. There results

$$\frac{d}{d(z/L)}\left[\frac{(\bar{\rho}_M)_m}{\bar{\rho}_M}\right] = \frac{-dp/dz}{G^2/L(\bar{\rho}_M)_m} - \frac{\bar{\tau}_w P}{AG^2/L(\bar{\rho}_M)_m} - g\left[\frac{\bar{\rho}}{(\bar{\rho})_m}\right]\frac{(\bar{\rho})_m}{G^2/L(\bar{\rho}_M)_m}\sin\theta \quad (3.3.13)$$

Equation (3.3.13) yields three dimensionless groupings, all on the right-hand side of the equation. The first grouping is called the *Euler number*. It gives the ratio of the axial pressure gradient forces to the acceleration forces. In single-phase flow, the Euler number is equal to one-half the friction factor f, when the pressure gradient is taken as the frictional pressure drop per unit length. Similarly, in two-phase flow, we shall set it equal to one-half the two-phase friction factor f_{TP} when we utilize the frictional two-phase pressure gradient $(dp/dz)_{FTP}$.

$$(N_{Eu})_{TP} = \frac{-(dp/dz)_{FTP}}{G^2/L(\bar{\rho}_M)_m} = \frac{1}{2}f_{TP} \quad (3.3.14)$$

The second grouping on the right-hand side of Eq. (3.3.13) can be rewritten by introducing the accepted definition of wall shear stress τ_w.

$$\tau_w = \mu\left.\frac{\partial u}{\partial y}\right|_{y=0} \quad (3.3.15)$$

where μ is the fluid viscosity and y is the distance perpendicular to the channel wall. Because the derivative in the relation (3.3.15) is taken at the wall, one can set the viscosity approximately equal to a wall viscosity μ_w. The shear stress expression can also be nondimensionalized by introducing the length L and the reference mean velocity $(u)_m$, and Eq. (3.3.15) becomes

$$\frac{\tau_w P}{AG^2/L(\bar{\rho}_M)_m} = 4\left[\frac{(u)_m \mu_w P}{4AG^2/(\bar{\rho}_M)_m}\right]\left[\frac{\partial (u/u_m)}{\partial y/L}\right]_{y=0} \quad (3.3.16)$$

The first term in brackets on the right-hand side of Eq. (3.3.16) is nondimensional. It is the inverse of the Reynolds number, which is equal to the ratio of acceleration to shear forces. If we define an equivalent hydraulic diameter D_H for the channel such that

$$D_H = \frac{4(\text{flow area})}{\text{wetted perimeter}} = \frac{4A}{P} \quad (3.3.17)$$

the *Reynolds number* in two-phase flow becomes

$$(N_{Re})_{TP} = \frac{G^2 D_H}{(\bar{\rho}_M)_m (u)_m \mu_w} \quad (3.3.18)$$

3.3 DISCUSSION OF ONE-DIMENSIONAL TWO-PHASE CONSERVATION LAWS

The last grouping on the right-hand side of Eq. (3.3.13) is called the *Froude number*. It gives the ratio of acceleration to gravity forces:

$$(N_{Fr})_{TP} = \frac{G^2}{gL(\bar{\rho}_M)_m(\bar{\rho})_m} \tag{3.3.19}$$

For the case of fully developed adiabatic flow, we can take $(\bar{\rho}_M)_m = \bar{\rho}_M$ and $(\bar{\rho})_m = \bar{\rho}$. We can also set $L = D_H$, and we can define the velocity $(u)_m$ as the true mean two-phase velocity, $\bar{u}_{TP}$ (i.e., total volumetric flow rate divided by cross-sectional area):

$$(u)_m = \bar{u}_{TP} = \bar{u}_L(1-\bar{\alpha}) + \bar{u}_G\bar{\alpha} = G\left(\frac{1-\bar{x}}{\rho_L} + \frac{\bar{x}}{\rho_G}\right) = \frac{G}{\bar{\rho}_H} \tag{3.3.20}$$

where $\bar{\rho}_H$ is the homogeneous density. Thus the preceding three dimensionless groupings become

$$(N_{Eu})_{TP} = \frac{1}{2}f_{TP} = \frac{-(dp/dz)_{FTP}}{G^2}D_H\bar{\rho}_M$$

$$(N_{Re})_{TP} = \frac{GD_H}{\mu_w}\frac{\bar{\rho}_H}{\bar{\rho}_M} \tag{3.3.21}$$

$$(N_{Fr})_{TP} = \frac{G^2}{gD_H\bar{\rho}_M\bar{\rho}}$$

Similar heat transfer groupings can be generated by employing Eq. (3.3.7) and introducing a reference length $D_t = 4A/P_h$ and a reference enthalpy increase ΔH_m. If we neglect the potential and kinetic energy by comparison to wall heat addition, and if we utilize accepted definitions for the heat flux $\bar{q}''$ and the heat transfer coefficient h_{TP}, there results

$$\frac{\bar{q}''}{G\,\Delta H_m} = \frac{(k_{TP}\,\partial T/\partial y)_{y=0}}{G\,\Delta H_m} = \frac{(k_w\,\partial T/\partial y)_{y=0}}{G\,\Delta H_m} = \frac{h_{TP}(T_w - T_b)}{G\,\Delta H_m} \tag{3.3.22}$$

where k_{TP} is the two-phase thermal conductivity, k_w its value at the wall, T_w the wall surface temperature, and T_b the bulk temperature. Equation (3.3.22) yields three nondimensional groupings:

$$\text{Nusselt number} = (N_{Nu})_{TP} = \frac{h_{TP}D_t}{k_w}$$

$$\text{Stanton number} = (N_{St})_{TP} = \frac{(N_{Nu})_{TP}}{(N_{Re})_{TP}(N_{Pr})_{TP}}\frac{D_H}{D_t}\frac{\bar{\rho}_H}{\bar{\rho}_M} \tag{3.3.23}$$

$$\text{Prandtl Number} = (N_{Pr})_{TP} = \frac{\mu_w(C_p)_{TP}}{k_w}$$

The most remarkable finding about the preceding nondimensional groupings is their very limited use in correlating two-phase flow or heat transfer data. Many other equations have been utilized instead when a much more consistent approach would have resulted by using similarity rules between single- and two-phase nondimensional groupings. For example, in fully developed single-phase flow, it has been found that the friction factor f is a function of the Reynolds number:

$$f = f(N_{Re}) \tag{3.3.24}$$

The similar relation in two-phase flow would be

$$f_{TP} = f_{TP}\,[(N_{Re})_{TP}] \tag{3.3.25}$$

with $(N_{Re})_{TP}$ obtained from Eq. (3.3.21).

One final comment should be made about Eq. (3.3.25). In contrast to single-phase flow, knowing the form of Eq. (3.3.25) does not specify the two-phase friction factor f_{TP} and the corresponding frictional pressure drop. One must also know the average gas volume fraction $\overline{\alpha}$ and the wall viscosity μ_w in order to calculate $(N_{Re})_{TP}$ and the pressure gradient.

3.4 HOMOGENEOUS/UNIFORM PROPERTY FLOW

3.4.1 Homogeneous Flow Conservation Equations

We shall define homogeneous flow as one where the two-phase fluid properties are constant and the liquid and gas velocities and temperatures are equal over the flow cross-sectional area. For one-dimensional flow, this means that

$$u_L = u_G = u_H \qquad \overline{u}_L = \overline{u}_G = \overline{u}_H \tag{3.4.1}$$

where u is the local velocity and the subscripts L, G, and H are used to represent the liquid, gas, and homogeneous conditions, respectively. In addition, it is assumed that all the fluid properties are constant across the cross section, whether they are the gas volume or weight-rate fraction or the thermal conductivity and viscosity.

Equating the average liquid and gas velocities gives

$$\overline{\alpha}_H = \frac{\rho_L \overline{x}}{\rho_L \overline{x} + \rho_G (1 - \overline{x})} \tag{3.4.2}$$

where $\overline{\alpha}_H$ is the average homogeneous volume fraction occupied by the gas, $\overline{x}$ the average gas weight flowing fraction, and ρ_L and ρ_G the liquid and gas densities. The corresponding effective fluid properties become

$$\text{specific volume} = \frac{1}{\bar{\rho}} = \frac{1-\bar{x}}{\rho_L} + \frac{\bar{x}}{\rho_G} \quad (3.4.3)$$

$$\text{density} = \bar{\rho} = \bar{\rho}_H = \bar{\rho}_M = \bar{\rho}_{KE} \quad (3.4.4)$$

$$\text{enthalpy} = \bar{H} = H_L(1-\bar{x}) + H_G \bar{x} = \bar{H}_H \quad (3.4.5)$$

$$\text{mass flow rate per unit area} = G = \bar{\rho}_H \bar{u}_H = [\rho_L(1-\bar{\alpha}_H) + \rho_G \bar{\alpha}_H] \bar{u}_H \quad (3.4.6)$$

Note that for homogeneous flow, the three densities of two-phase flow have collapsed into a single value, $\bar{\rho}_H$.

The conservation laws for homogeneous flow become

$$\frac{\partial}{\partial t}(\bar{\rho}_H A) + \frac{\partial}{\partial z}(\bar{\rho}_H \bar{u}_H A) = 0$$

$$\frac{\partial}{\partial t}(\bar{\rho}_H \bar{u}_H A) + \frac{\partial}{\partial z}(\bar{\rho}_H \bar{u}_H^2 A) = -A\frac{\partial p}{\partial z} - \bar{\tau}_w P - \bar{\rho}_H g A \sin\theta$$

$$\frac{\partial}{\partial t}(\bar{\rho}_H \bar{H}_H A) + \frac{\partial}{\partial z}(\bar{\rho}_H \bar{u}_H \bar{H}_H A) + \frac{1}{2}\frac{\partial}{\partial t}(\bar{\rho}_H \bar{u}_H^2 A) + \frac{1}{2}\frac{\partial}{\partial z}(\bar{\rho}_H \bar{u}_H^3 A)$$

$$= P_h \overline{q''} + A\overline{q'''} + \frac{\partial}{\partial t}(pA) - g\bar{\rho}_H \bar{u}_H A \sin\theta$$

(3.4.7)

where t represents the time, A the flow area, z the distance along the flow path, p the pressure, τ_w the wall shear stress, g the gravitational constant, and θ the flow channel inclination from the horizontal. P is the wetted perimeter, $\bar{\tau}_w$ the wall shear stress, P_h the heated perimeter, $\overline{q''}$ the average heat flux, and $\overline{q'''}$ the average heat generation per unit volume.

Equations (3.4.7) specify the variation of the three independent variables $\bar{u}_H$, $\bar{H}_H$, and p as a function of time and position z. These presume that the two phases are at the same temperature and see the same pressure p. It should be pointed out that Eqs. (3.4.7) are identical to the one-dimensional single-phase equations. This is not surprising since a homogeneous two-phase system with constant fluid properties can be expected to behave like a single-phase flow. Equations (3.4.7) can be solved if the constitutive equations or closure laws for $\bar{\tau}_w$ and $\overline{q''}$ can be expressed in terms of $\bar{u}_H$, $\bar{H}_H$, and the homogeneous fluid properties.

3.4.2 Closure Laws for Homogeneous Flow

The closure laws for homogeneous/uniform property flow can be derived readily from the methodology for single-phase flow. For example, for laminar flow in a circular pipe, the fully developed steady-state homogeneous momentum equation reduces to

$$\left(-\frac{dp}{dz}\right)_{FH} = -\frac{1}{r}\frac{d}{dr}\left(\mu_H r \frac{du_H}{dr}\right) \qquad (3.4.8)$$

where $(-dp/dz)_{FH}$ is the homogeneous frictional pressure drop in the direction z, u_H the homogeneous local velocity, and r the distance measured from the pipe centerline. If the homogeneous viscosity μ_H does not vary with radial position and if the velocity at the wall and the velocity gradient at the centerline are equal to zero, Eq. (3.4.8) can be integrated to yield

$$u_H = \left(-\frac{dp}{dz}\right)_{FH}\left(\frac{R^2 - r^2}{4\mu_H}\right) = 2\bar{u}_H\left(1 - \frac{r^2}{R^2}\right) \qquad (3.4.9)$$

In Eq. (3.4.9), R is the pipe radius and $\bar{u}_H$ the average homogeneous velocity over the cross-sectional flow area. At the centerline, $r = 0$, and Eq. (3.4.9) gives

$$\left(-\frac{dp}{dz}\right)_{FH} = \left(\frac{\rho_H \bar{u}_H^2}{2R\ 2}\right)\frac{64}{2R\ \bar{u}_H\rho_H/\mu_H} = f_H \frac{\rho_H\ \bar{u}_H^2}{2R\ 2} \qquad (3.4.10)$$

so that, by definition, the laminar homogeneous friction factor f_H is equal to

$$f_H = \frac{64}{2\ R\bar{u}_H\rho_H/\mu_H} = \frac{64}{N_{Re_H}} \qquad (3.4.11)$$

When the flow is turbulent, the homogeneous turbulent shear stress is defined as in single-phase flow; that is, it consists of a viscous component and a turbulent component expressed in terms of a homogeneous mixing length l_H:

$$\tau_H = \mu_H \frac{du_H}{dy} + \rho_H l_H^2 \left(\frac{du_H}{dy}\right)^2 \qquad (3.4.12)$$

An equivalent turbulent viscosity is sometimes introduced and

$$\tau_H = (\mu_H + \rho\epsilon_{MH})\frac{du_H}{dy}$$

where ϵ_{MH} represents the homogeneous eddy diffusivity for momentum. Except for positions very near the wall, the turbulent shear contribution overshadows the viscous term, and the turbulent shear stress becomes

$$\tau_H = \rho_H l_H^2 \left(\frac{du_H}{dy}\right)^2 = \rho_H \epsilon_{MH} \frac{du_H}{dy} \qquad (3.4.13)$$

The velocity distribution can be calculated from Eq. (3.4.13) if the mixing length l_H and the shear stress τ_H are specified. Various relations have been proposed for τ_H and l_H. The simplest combination has been advanced by Prandtl, who assumed that

$$\tau_H = \tau_{wH} \qquad (3.4.14)$$
$$l_H = Ky$$

K is the mixing-length constant and is taken usually equal to 0.4 for all fluids. Substitution of Eq. (3.4.14) into Eq. (3.4.13) and integration give for *smooth* pipe or parallel-plate flow,

$$u_H^+ = \frac{1}{K} \ln y^+ + 5.5 = 2.5 \ln y^+ + 5.5 \qquad (3.4.15)$$

In Eq. (3.4.15), the constant 5.5 is determined empirically, and the nondimensional velocity u_H^+ and distance y^+ are defined as follows:

$$u_H^+ = \frac{u_H}{\sqrt{\tau_{wH}/\rho_H}} \qquad (3.4.16)$$

$$y^+ = \frac{y\rho_H \sqrt{\tau_{wH}/\rho_H}}{\mu_H}$$

Very close to the channel wall, viscous effects must be included; and a buffer zone and a laminar sublayer are introduced. Their velocities are given, respectively, by

$$u_H^+ = \begin{cases} -3.05 + 5.0 \ln y^+ & 5 < y^+ < 30 \\ y^+ & 0 < y^+ < 5 \end{cases} \qquad (3.4.17)$$

Figure 3.4.1 shows the agreement of Eqs. (3.4.15) and (3.4.17) with the single-phase experimental data of Nikuradse [3.4.1] and Reichardt [3.4.2] in circular pipes. *These turbulent equations are equally applicable to homogeneous two-phase flow if we employ the homogeneous fluid properties.*

Once the velocity distribution is known, it can be integrated over the flow area, and a relation is obtained between the wall shear stress and the average fluid velocity.

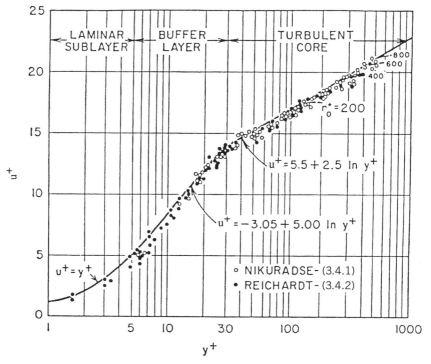

Figure 3.4.1 Universal turbulent velocity distribution in smooth pipes. (From Ref. 3.4.1.)

The wall shear stress then is expressed in terms of the frictional pressure drop, and an equation for the homogeneous friction factor f_H results:

$$\frac{1}{\sqrt{f_H}} = 2.0 \log \left(\frac{GD_H}{\mu_H} \sqrt{f_H} \right) - 0.8 \qquad (3.4.18)$$

Equation (3.4.18) is valid for homogeneous turbulent flow ($GD_H/\mu_H > 2000$) in a smooth channel. For simplification purposes it is often approximated by

$$f_H = \frac{0.184}{N_{Re_H}^{0.2}} \qquad (3.4.19)$$

When the flow channel is rough, the analytical solutions must be modified to account for the channel roughness. This is done by adjusting the constant 5.5 in Eq. (3.4.15) so that the friction factor depends not only on the Reynolds number but also upon a new parameter, ϵ/D_H, which describes the relative channel roughness. (The symbol ϵ represents the average height of the roughness projections from the wall, and D_H is the channel equivalent hydraulic diameter.) The most complete set of curves of single-phase friction factor in rough channels has been presented by

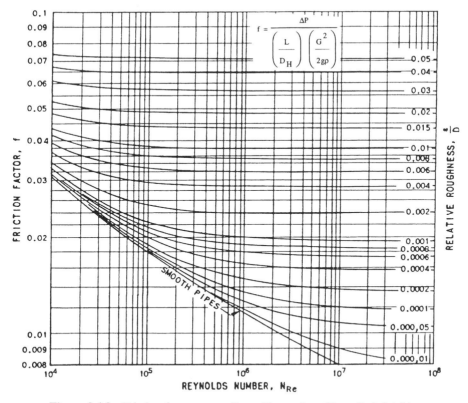

Figure 3.4.2 Friction factor versus Reynolds number. (From Ref. 3.4.3.)

Moody [3.4.3], and it is again applicable to two-phase homogeneous flow. It is reproduced in Figure 3.4.2. For approximation purposes, the homogeneous friction factor in rough channels can be calculated within ±5% from

$$f_H = 0.0055 \left[1 + \left(20,000 \frac{\epsilon}{D_H} + \frac{10^6}{GD_H/\mu_H} \right)^{1/3} \right] \qquad (3.4.20)$$

It should be noted that even though the foregoing turbulent equations were derived for pipe flow, they have been found to be applicable to noncircular geometries as long as the pipe diameter is replaced by an equivalent hydraulic diameter. This explains why Eqs. (3.4.18) and (3.4.20) and Figures 3.4.1 and 3.4.2 have already been formulated in terms of the hydraulic diameter D_H.

The preceding derivations for homogeneous laminar and turbulent flow friction factors were reproduced in detail in this section for two reasons. The first reason is to provide a reminder that single-phase *closure equations are obtained from steady, fully developed flow conditions and that generally they are presumed to apply to transient single-phase flow. This same assumption is utilized through all two-phase-flow complex system computer codes.* The second reason is to demonstrate that all required *closure equations for homogeneous two-phase flow are not new and can be deduced from single-phase flow.* Moreover, they are obtained from solutions of the velocity and temperature distributions across the flow direction. This conclusion again is valid for other channel geometries, for pressure losses across valves, entrance and exit configurations, and so on. This means that *the entire available set of analytical solutions or experimental correlations for single-phase flow apply to two-phase homogeneous/uniform property flows.* For example, the turbulent Dittus–Boelter correlations for turbulent heat transfer in a pipe [3.4.4] is valid for homogeneous two-phase flow and

$$(N_{\text{Nu}})_H = \frac{h_H D_H}{k_H} = 0.023 \left(\frac{G D_H}{\mu_H}\right)^{0.8} (N_{\text{Pr}})_H^{0.4} \qquad (3.4.21)$$

where $(N_{\text{Nu}})_H$ is the homogeneous Nusselt number and $(N_{\text{Pr}})_H$ is the corresponding Prandtl number, or

$$(N_{\text{Pr}})_H = \frac{\mu_H (C_p)_H}{k_H} \qquad (3.4.22)$$

Similarly, the analytical solutions for single-phase turbulent heat transfer by Martinelli [3.4.5] are applicable to homogeneous flow. In other words, the only new expressions needed for homogeneous/uniform property two-phase flow are those associated with the fluid properties and, in particular, the homogeneous viscosity and thermal conductivity.

3.4.3 Homogeneous Viscosity and Thermal Conductivity

3.4.3.1 Homogeneous Two-Phase Viscosity The expressions available for the viscosity of a liquid–gas mixture are mostly of an empirical nature. Because for homogeneous flow, the gas and liquid are presumed to be uniformly mixed. McAdams et al. [3.4.6] proposed that by analogy to the expression for the homogeneous density $\bar{\rho}_H$, the average two-phase homogeneous viscosity $\bar{\mu}_H$ be taken as

$$\frac{1}{\bar{\mu}_H} = \frac{1}{\mu_L}(1 - \bar{x}) + \frac{1}{\mu_G}\bar{x} \qquad (3.4.23)$$

It is interesting to note that Eq. (3.4.23) leads to the conclusion that the homogeneous Reynolds number is equal to the sum of the liquid and gas Reynolds numbers.

Bankoff [3.4.7] and others have chosen to define $\bar{\mu}_{TP}$ in terms of the average gas volume fraction $\bar{\alpha}$:

$$\bar{\mu}_{TP} = \mu_L(1 - \bar{\alpha}) + \mu_G \bar{\alpha} \qquad (3.4.24)$$

Others have preferred to weigh the liquid and gas viscosity according to the liquid and gas weight-rate fraction [3.4.8]:

$$\bar{\mu}_{TP} = \mu_L(1 - \bar{x}) + \mu_L(\bar{x}) \qquad (3.4.25)$$

The liquid and gas volumetric rate fractions [3.3.1] have also been used.

$$\bar{\mu}_{TP} = \mu_L(1 - \bar{x})\frac{\bar{\rho}_H}{\rho_L} + \mu_G \frac{\bar{x}\bar{\rho}_H}{\rho_G} \qquad (3.4.26)$$

An alternative approach sometimes used to determine the viscosity of a liquid–gas mixture is to employ relations developed for solid–fluid flow. The viscosity of solid–fluid mixtures $\bar{\mu}_{TP}$ has been correlated by expressions of the type

$$\bar{\mu}_{TP} = \mu_f(1 + C_1\bar{\alpha}_s + C_2\bar{\alpha}_s^2 + \cdots) \qquad (3.4.27)$$

where μ_f is the fluid viscosity and $\bar{\alpha}_s$ is the average volume fraction occupied by the solid particles. The constants $C_1, C_2, \ldots$ are established empirically and vary with the size and shape of the solid particles.

An often used and simplified version of Eq. (3.4.27) is the relation proposed by Einstein [3.4.9] for a dilute suspension of spherical solid particles in a fluid:

$$\bar{\mu}_{TP} = \mu_f(1 + 2.5\bar{\alpha}_s) \qquad (3.4.28)$$

Other forms applicable to solid particles [3.4.10] have been developed, including [3.4.11]:

$$\bar{\mu}_{TP} = \mu_f \frac{1}{(1 - \bar{\alpha}_s)^{2.5}} \qquad (3.4.29)$$

Equations (3.4.27), (3.4.28), and (3.4.29) have been applied to liquid–gas flow. Their limitations, however, should be recognized. The most striking aspect of the foregoing discussion is the number of equations that have been proposed for the two-phase viscosity.

3.4.3.2 Homogeneous Thermal Conductivity

Expressions have been developed for the thermal conductivity of solid materials containing small gas pores, including the early relation of Eucken described in Reference [3.4.12]:

$$\frac{k_{TP}}{k_s} = \frac{1 - \{1 - [3\,k_s/(2\,k_s + k_G)]\,(k_G/k_s)\}\,\overline{\alpha}_G}{1 + \{[3\,k_s/(2\,k_s + k_G)] - 1\}\,\overline{\alpha}_G} \qquad (3.4.30)$$

where the subscript s is used to represent the solid and G the gas. Bruggeman [3.4.13] proposed the following relation for liquid–solid mixtures:

$$\frac{k_L}{k_{TP}} = \left[\frac{1 - (k_L/k_{TP})\,(k_s/k_L)}{(1 - \overline{\alpha}_s)\,(1 - k_s/k_L)}\right]^{3/2} \qquad (3.4.31)$$

Merilo et al. [3.4.14] measured the resistivity of a liquid–gas pool and the corresponding homogeneous thermal conductivity was found to be for $\overline{\alpha}_G \leq 0.7$

$$\overline{k}_H = \overline{k}_L (1 - \overline{\alpha}_G)^{1.5} \qquad (3.4.32)$$

Equations (3.4.31) and (3.4.32) are equivalent if we replace the solid particles in Eq. (3.4.31) by gas bubbles and assume that the gas conductivity is small compared to that of the liquid. Even though Eq. (3.4.30) was derived only for a porous solid, its use might be advisable because it can be rewritten

$$\frac{\overline{k}_H}{\overline{k}_L} = \frac{1 - \{1 - [3\,\overline{k}_L/(2\,\overline{k}_L + \overline{k}_G)]\,(\overline{k}_G/\overline{k}_L)\}\,\overline{\alpha}_G}{1 + \{[3\,\overline{k}_L/(2\,\overline{k}_L + \overline{k}_G)]\,(\overline{k}_G/\overline{k}_L) - 1\}\,\overline{\alpha}_G} \qquad (3.4.33)$$

Equation (3.4.33) has the advantage of yielding $\overline{k}_H = \overline{k}_L$ when $\overline{\alpha}_G = 0$ and $\overline{k}_H = k_G$ when $\overline{\alpha}_G = 1$.

3.4.4 Summary: Homogeneous/Uniform Property Flow

1. By assuming equal gas and liquid velocity and temperature, homogeneous flow does not have to deal with gas–liquid interfaces. By employing uniform properties across the flow cross section, homogeneous/uniform property flow has another significant advantage: the ability to apply all available single-phase-flow analyses and empirical correlations. In that sense, no new closure equations are needed for a homogeneous/uniform property system computer code.

2. Homogeneous flow conditions are approached when the liquid and gas properties are similar, for example, close to the critical pressure of a single-component fluid. It is also applicable for very high velocities and pressure drops: that is, critical flow or flow across valves and other strong mixing devices. Finally, it has been used successfully to describe the dispersed flow of liquid drops in a gas stream.

3. As shown in Chapter 2, homogeneous flow can be employed at the top-down scaling level to obtain overall system and natural circulation scaling groups as long as they are not significantly affected by local velocity and temperature phenomena.

4. Homogeneous flow requires expressions for key physical properties and, in particular, its viscosity and thermal conductivity. In specifying such relations, preference should be given to those equations that satisfy the property values at 100% gas or liquid flow. Consideration should be given to the following set of consistent homogeneous properties:

$$\rho_H = \rho_L(1 - \overline{\alpha}_H) + \rho_G\overline{\alpha}_H$$
$$\mu_H = \mu_L(1 - \overline{\alpha}_H) + \mu_G\overline{\alpha}_H \quad (3.4.34)$$
$$k_H = k_L(1 - \overline{\alpha}_H) + k_G\overline{\alpha}_H$$

5. Consistency is necessary in employing the homogeneous/uniform property model. For example, it does not make sense to use a single-phase closure law based on uniform property conditions and to combine it with liquid properties at the bounding surfaces. Another frequent inconsistency is to get the average gas volume fraction from an empirical correlation that presumes different gas and liquid velocity and to employ it with homogeneous flow correlations for pressure drop and heat transfer.

6. A homogeneous/uniform property computer system code is an important default tool. It is the only tool available when closure laws are still to be developed for some aspects of the complex system. Homogeneous system computer codes are much easier to assemble and take considerably less time to run. Their principal shortcoming is presuming equal gas and liquid velocity and temperature. However, if they are applied judiciously, they can be very useful in estimating or scaling complex system behavior.

3.5 SEPARATED OR MIXTURE TWO-PHASE-FLOW MODELS WITH NO INTERFACE EXCHANGE

A multitude of separated or mixture models with no interface exchanges were developed initially to avoid the constraints of equal gas and liquid velocity. They were formulated in terms of area- and time-averaged parameters of the flow, and they assumed equal gas and liquid temperatures as in the case of the conservation laws derived in Section 3.2. Because no interfacial interactions were considered, the models had to rely upon semiempirically or empirically derived closure laws for the gas volume fraction and friction pressure drop and heat transfer at the boundary surfaces. In this section, single-phase models, which as their name implies are derived by analogy to single-phase flow, are covered first. All such single-phase models have the advantage of being inherently simple by assuming constant property values across the flow cross section and of being able to apply single-phase correlations directly. This is followed by a discussion of separated or mixture models, starting with the significant works of Lockhart and Martinelli [3.5.1] and Martinelli and Nelson [3.5.2] and finishing with other empirical relations that have been favored in recent years. Volumetric flux models and, in particular, the drift flux model are covered last.

3.5.1 Single-Phase Models

The earliest proposed single-phase model assumed that all the properties (including the density or gas volume fraction) were constant and uniform across the flow area. This permitted direct application of single-phase correlations to two-phase flow as illustrated below for frictional pressure drop. In the case of turbulent flow in a smooth pipe, the single-phase friction factor f can be written as

$$f = 0.184 N_{\text{Re}}^{-0.2} \tag{3.5.1}$$

The corresponding two-phase friction factor f_{TP} can be rewritten

$$f_{TP} = 0.184 (N_{\text{Re}})_{TP}^{-0.2} \tag{3.5.2}$$

If we use the definitions of f_{TP} and $(N_{\text{Re}})_{TP}$ given in Eqs. (3.3.21), one can form the ratio of the two-phase frictional pressure drop to that of the liquid flowing at the total mass flow rate per unit area, G, to get

$$\phi_{TPL_0}^2 = \frac{(dp/dz)_{FTP}}{(dp/dz)_{FL_0}} = \left(\frac{\rho_L}{\overline{\rho}_M}\right)^{0.8} \left(\frac{\rho_L}{\overline{\rho}_H}\right)^{0.2} \left(\frac{\mu_W}{\mu_L}\right)^{0.2} \tag{3.5.3}$$

which can be rewritten in terms of the gas average volume fraction $\overline{\alpha}$ and weight-rate fraction $\overline{x}$:

$$\phi_{TPL_0}^2 = \left[\frac{(1-\overline{x})^2}{1-\overline{\alpha}} + \frac{\rho_L}{\rho_G}\frac{\overline{x}^2}{\overline{\alpha}}\right]^{0.8} \left(1 - \overline{x} + \overline{x}\frac{\rho_L}{\rho_G}\right)^{0.2} \left(\frac{\overline{\mu}_{TP}}{\mu_L}\right)^{0.2} \tag{3.5.4}$$

In Eq. (3.5.4), μ_W was replaced by μ_{TP} to be consistent with the assumption of constant property across the flow cross section. By forming the ratio of two-phase friction pressure drop to the corresponding value for homogeneous flow, we get

$$\phi_{TPH}^2 = \left(\frac{\overline{\mu}_{TP}}{\mu_H}\right)^{0.2} \left(\frac{\overline{\rho}_H}{\overline{\rho}_M}\right)^{0.8} \tag{3.5.5}$$

Because of the 0.2 exponent on the viscosity ratio in Eq. (3.5.5), we can approximate Eq. (3.5.5) by

$$\phi_{TPH}^2 \approx \left(\frac{\overline{\rho}_H}{\overline{\rho}_M}\right)^{0.8} \quad (3.5.6)$$

and in so doing, we have developed an important scaling parameter $(\overline{\rho}_H/\overline{\rho}_M)$ to go from homogeneous to unequal gas and liquid velocity.

Similarly, one can form the ratio of two-phase heat transfer coefficient, h_{TPL}, to that of the coefficient for liquid flow at the total mass flow rate G, h_{L0}, to get

$$\frac{h_{TP}}{h_{L0}} = \left(\frac{k_{TP}}{k_L}\right)^{0.6} \left(\frac{\mu_L}{\mu_{TP}}\right)^{0.4} \left(\frac{C_{PTP}}{C_{PL}}\right)^{0.4} \left(\frac{\overline{\rho}_H}{\overline{\rho}_M}\right)^{0.8} \quad (3.5.7)$$

This single-phase approach offers a valuable substitute to the homogeneous model, where its applicability is questionable (e.g., $\rho_L/\rho_G \gg 1$). It has the advantage that it can be extended to other closure laws at the surface boundaries (i.e., heat transfer when unequal gas and liquid velocities are important). As in the case of homogeneous flow, this single-phase model requires expressions for the properties μ_{TP}, k_{TP}, and in addition, an empirical correlation for the average gas volume fraction $\overline{\alpha}$.

The single-phase models may not have received the credit they were due because of many inconsistencies in past applications. For example, many papers have employed the wrong density $\overline{\rho}$ instead of $\overline{\rho}_M$ in the Reynolds number or the wrong properties (e.g., $\overline{\mu}_L$ instead of $\overline{\mu}_{TP}$). It is the author's view that single-phase models provide a step improvement over the homogeneous model as a default and scaling model. They are inherently simple if they are coupled with an uncomplicated gas volume fraction relation [e.g., Eqs. (3.3.11)] because they have the ability to directly apply single-phase correlations. However, they suffer from the same shortcomings: utilizing a macroscopic representation of the two-phase stream and failing to consider its microstructure at the interfaces.

There have been many past attempts to include variations of properties and in particular of the density across the flow area. The models of Bankoff [3.4.7], who used a power law variation, and the model of Levy [3.5.3], who employed a mixing-length model, deserve being mentioned, even though they have not found meaningful application in system computer codes for complex systems.

3.5.2 Martinelli Separated Model

3.5.2.1 Lockhart–Martinelli Model The Lockhart–Martinelli model describes the flow of a gas–liquid mixture in a channel of cross-sectional area A and wetted perimeter P where the average liquid and gas velocities are equal to

$$\overline{u}_L = \frac{G_L}{(1-\overline{\alpha})\rho_L} \qquad \overline{u}_G = \frac{G_G}{\overline{\alpha}\rho_G} \quad (3.5.8)$$

and G_L and G_G are the mass flow rates of liquid and gas per total unit area. The basic premise of the Lockhart–Martinelli model is that the frictional losses in the liquid and gas phases can be calculated from relations of the single-phase type, so that

$$\left(\frac{dp}{dz}\right)_{LTP} = -f_{LTP}\frac{\rho_L \bar{u}_L^2}{2D_L} \qquad \left(\frac{dp}{dz}\right)_{GTP} = -f_{GTP}\frac{\rho_G \bar{u}_G^2}{2D_G} \qquad (3.5.9)$$

where D_L and D_G are equivalent hydraulic diameters for the liquid and gas phases. In a channel of area A and wetted perimeter P, the hydraulic diameter D_H is defined as

$$D_H = \frac{4A}{P} \qquad (3.5.10)$$

and in a similar manner, one can write[†]

$$D_L = \frac{4A_L}{P_L} = \frac{4A}{P}\frac{1-\alpha}{P_L/P} \qquad D_G = \frac{4A_G}{P_G} = \frac{4A}{P}\frac{\alpha}{P_G/P} \qquad (3.5.11)$$

where P_L and P_G are equivalent liquid and gas wetted perimeters. If the channel under consideration is smooth, one can express the friction factor f by means of the following single-phase relations:

$$f = \frac{c}{N_{Re}^n} \qquad (3.5.12)$$

If Eq. (3.5.12) is assumed to apply to f_{LTP} and f_{GTP}, there results

$$f_{LTP} = \frac{c_L}{(\bar{u}_L D_L \rho_L/\mu_L)^{n_L}} \qquad f_{GTP} = \frac{c_G}{(\bar{u}_G D_G \rho_G/\mu_G)^{n_G}} \qquad (3.5.13)$$

The constants c_L, c_G, n_L, and n_G are taken to be the same as in single-phase flow, and the values proposed by Lockhart and Martinelli[††] were

$$\frac{G_L D_H}{\mu_L} > 2000 \qquad c_L = 0.184 \qquad n_L = 0.2$$

$$\frac{G_L D_H}{\mu_L} < 2000 \qquad c_L = 64 \qquad n_L = 1.0$$

[†] Note that Lockhart and Martinelli chose to write $A_L = [(\pi/4) D_L^2]c'$ and $A_G = [(\pi/4) D_G^2]c''$ and introduced two new parameters, c' and c'', which add unnecessary complexity to the solution. The present approach not only simplifies the solution but also adds more credence to the final results.

[††] A more realistic approach in defining the transition from laminar to turbulent flow would be to employ the liquid and gas Reynolds numbers introduced into the friction factor equation.

3.5 SEPARATED OR MIXTURE TWO-PHASE-FLOW MODELS

$$\frac{G_G D_G}{\mu_G} > 2000 \qquad c_G = 0.184 \qquad n_G = 0.2$$

$$\frac{G_G D_G}{\mu_G} < 2000 \qquad c_G = 64 \qquad n_G = 1.0$$

Substitution of Eqs. (3.5.8), (3.5.11), and (3.5.13) into Eq. (3.5.9) gives

$$\left(\frac{dp}{dz}\right)_{LTP} = -\frac{c_L}{(G_L D_H/\mu_L)^{n_L}} \frac{G_L^2}{2\rho_L D_H} \frac{(P_L/P)^{1+n_L}}{(1-\overline{\alpha})^3}$$

$$\left(\frac{dp}{dz}\right)_{GTP} = -\frac{c_G}{(G_G D_H/\mu_G)^{n_G}} \frac{G_G^2}{2\rho_G D_H} \frac{(P_G/P)^{1+n_G}}{\overline{\alpha}^3} \qquad (3.5.14)$$

Lockhart and Martinelli next assumed that the static pressure drop in the liquid and gas phases were equal and that the hydrostatic losses were small with respect to friction (horizontal channel or large frictional pressure drop):

$$\left(\frac{dp}{dz}\right)_{LTP} = \left(\frac{dp}{dz}\right)_{GTP} = \left(\frac{dp}{dz}\right)_{FTP}$$

If we now consider the case of liquid or gas flowing alone in the same channel at the flow rate G_L and G_G, their frictional pressure drops would be

$$\left(\frac{dp}{dz}\right)_{FL} = -\frac{c_L}{(G_L D_H/\mu_L)^{n_L}} \frac{G_L^2}{2\rho_L D_H}$$

$$\left(\frac{dp}{dz}\right)_{FG} = -\frac{c_G}{(G_G D_H/\mu_G)^{n_G}} \frac{G_G^2}{2\rho_G D_H} \qquad (3.5.15)$$

and Eqs. (3.5.15) can be combined with Eqs. (3.5.13) and (3.5.14) to give

$$\phi_{TPL}^2 = \frac{(dp/dz)_{FTP}}{(dp/dz)_{FL}} = \frac{(P_L/P)^{1+n_L}}{(1-\overline{\alpha})^3}$$

$$\phi_{TPG}^2 = \frac{(dp/dz)_{FTP}}{(dp/dz)_{FG}} = \frac{(P_G/P)^{1+n_G}}{\overline{\alpha}^3} \qquad (3.5.16)$$

Division of ϕ_{TPL}^2 by ϕ_{TPG}^2 gives

$$\frac{\phi_{TPG}^2}{\phi_{TPL}^2} = \phi_{LG}^2 = \frac{(dp/dz)_{FL}}{(dp/dz)_{FG}} = \frac{(1-\overline{\alpha})^3}{\overline{\alpha}^3} \frac{(P/P_L)^{1+n_L}}{(P/P_G)^{1+n_G}} \qquad (3.5.17)$$

The grouping on the right side of Eq. (3.5.17) is only a function of the parameter

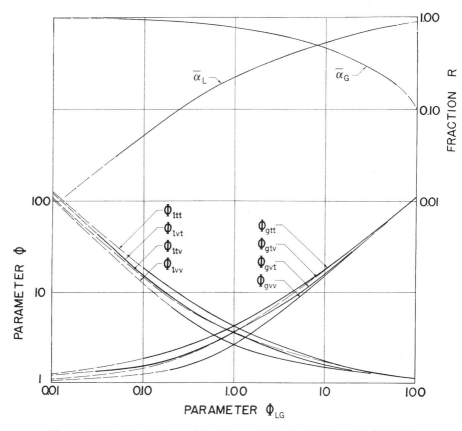

Figure 3.5.1 Correlations of Lockhart and Martinelli. (From Ref. 3.5.1.)

ϕ_{LG}^2. From this result, Lockhart and Martinelli inferred that each of the parameters $\overline{\alpha}$, P_L/P, and P_G/P, (i.e., ϕ_{TPL}^2 and ϕ_{TPG}^2) must depend only on the parameter ϕ_{LG}^2. They attempted to correlate the experimental measurements of ϕ_{TPL}^2, ϕ_{TPG}^2, and $\overline{\alpha}$ in terms of ϕ_{LG}^2, and the curves they derived are shown in Figure 3.5.1. Lockhart and Martinelli obtained one single curve for the average gas volumetric fraction $\overline{\alpha}$ and four different curves for the frictional groupings ϕ_{TPL}^2 and ϕ_{TPG}^2. The four curves in Figure 3.5.1 correspond to the four possible combinations of laminar or turbulent liquid and gas flow. The corresponding four parameters ϕ_{LG}^2 are defined as follows:

liquid laminar, gas laminar: $\quad \phi_{LG}^2 = \dfrac{G_L}{G_G} \dfrac{\rho_G}{\rho_L} \dfrac{\mu_L}{\mu_G}$

liquid laminar, gas turbulent: $\quad \phi_{LG}^2 = \dfrac{64}{0.184} \dfrac{G_L}{G_G} \dfrac{\rho_G}{\rho_L} \dfrac{\mu_L}{\mu_G} \left(\dfrac{G_G D_H}{\mu_G} \right)^{-0.8}$

3.5 SEPARATED OR MIXTURE TWO-PHASE-FLOW MODELS

liquid turbulent, gas laminar: $\quad \phi_{LG}^2 = \dfrac{0.184}{64} \dfrac{G_L}{G_G} \dfrac{\rho_G}{\rho_L} \dfrac{\mu_L}{\mu_G} \left(\dfrac{G_L D_H}{\mu_L}\right)^{-0.8}$

liquid turbulent, gas turbulent: $\quad \phi_{LG}^2 = \left(\dfrac{G_L}{G_G}\right)^{1.8} \dfrac{\rho_G}{\rho_L} \left(\dfrac{\mu_L}{\mu_G}\right)^{0.2}$

The Lockhart–Martinelli model clearly involves a multitude of assumptions; furthermore, it is an approximate empirical model because it requires experimental data to establish the curves presented in Figure 3.5.1; yet the model has been found repeatedly to yield acceptable and often good estimates of steady-state two-phase pressure drop and gas volumetric fraction as long as they fall within the range of the tests performed by Lockhart and Martinelli. *An interesting and advantageous feature of the Lockhart–Martinelli model is its use of liquid and gas fluid properties and its avoidance of unvalidated expressions for two-phase fluid properties.*

3.5.2.2 Martinelli–Nelson Model Martinelli and Nelson extended the correlation of Lockhart and Martinelli to the turbulent flow of steam–water mixtures. They found that the pressure drop curve plotted in Figure 3.5.1 compared satisfactorily with available steam–water results near atmospheric pressure, but that at high pressures, the predicted values fell above the experimental data. This is not surprising, since as the pressure increases and approaches the critical point, the vapor and liquid properties become equal and

$$\left(\frac{dp}{dz}\right)_{FTP} = -\frac{c_L}{(G_L D_H/\mu_L)^{n_L}} \frac{G^2}{2\rho_L D_H} \tag{3.5.18}$$

where for turbulent flow, $n_L = 0.2$ and $c_L = 0.184$. At the critical point, the parameters ϕ_{LG}^2 and ϕ_{TPL}^2 reduce to

$$\phi_{LG}^2 = \left(\frac{1-\bar{x}}{\bar{x}}\right)^{2-n_L} \qquad \phi_{TPL}^2 = \frac{1}{(1-\bar{x})^{2-n_L}} \tag{3.5.19}$$

and

$$\phi_{TPL}^2 = \left[\frac{1 + \phi_{LG}^{2/(2-n_L)}}{\phi_{LG}^{2/(2-n_L)}}\right]^{2-n_L} \tag{3.5.20}$$

If Eq. (3.5.20) were plotted on Figure 3.5.1, it would fall below the Lockhart–Martinelli curve. This fact led Martinelli–Nelson to draw a family of curves of ϕ_{TPL}^2 versus ϕ_{LG}^2 with pressure as a parameter. Since the Lockhart and Martinelli

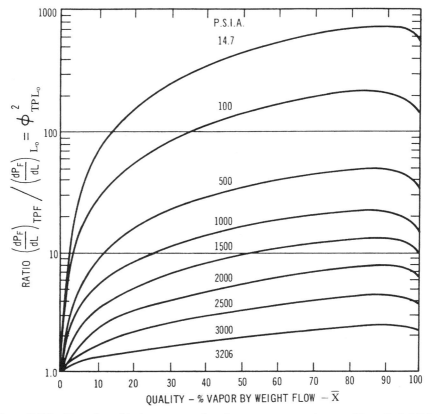

Figure 3.5.2 Two-phase friction pressure drop for steam–water mixtures. (From Ref. 3.5.2.)

curve was based mainly on air–water tests near atmospheric pressure, it was assumed to apply to steam–water mixtures which have similar properties (i.e., at 14.7 psia). Curves were drawn for intermediate pressures between the Lockhart–Martinelli curve and the curve corresponding to Eq. (3.5.20).

Figure 3.5.2 shows the correlations recommended by Martinelli–Nelson. To facilitate calculations, the curves are presented in terms of the parameters $\bar{x}$ and $\phi^2_{TPL_0}$ instead of ϕ^2_{LG} and ϕ^2_{TPL}. A parallel set of curves for the steam volumetric fraction $\bar{\alpha}$ is given in Figure 3.5.3. Except for the critical pressure curve (where $\bar{x} = \bar{\alpha}$) and the 14.7-psia curve that was reproduced from the Martinelli and Lockhart correlation, the remaining curves of Figure 3.5.3 were arbitrarily extrapolated by Martinelli–Nelson. A remarkable feature of the Martinelli–Nelson curves is that they have been in acceptable agreement with subsequent tests.

Another interesting result can be inferred from the Martinelli–Nelson work. It is that the Lockhart and Martinelli curves are not valid over all ranges of fluid properties. The Lockhart and Martinelli curves were found to give too-high frictional pressure drops when the liquid and gas properties become approximately

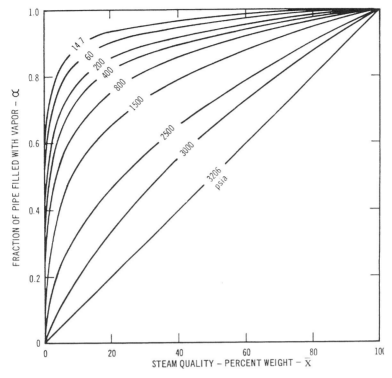

Figure 3.5.3 Steam volume fraction. (From Ref. 3.5.2.)

equal; they can also be expected to be too low when the liquid and gas properties become exceedingly different (e.g., under vacuum conditions). One final comment should be made about the Martinelli–Nelson results. If their proposed correlations were replotted as ϕ_{TPL}^2 versus $(1 - \overline{\alpha})$, one would find that the following expression is valid:

$$\phi_{TPL}^2 = \frac{1}{(1 - \overline{\alpha})^2} \qquad (3.5.21)$$

This is illustrated in Figure 3.5.4, where the Martinelli–Nelson curves at various pressures and the turbulent correlation of Lockhart and Martinelli are replotted together with some experimental data obtained by Isbin et al. (3.5.4) for steam–water mixtures at high flow rates.

The principal advantage of system computer codes based on separated or mixture models with no interface exchange is their simplification of the conservation equations. For example, for the Martinelli model, the gas volume fraction, $\overline{\alpha}$, is only a function of the fluid properties and the gas weight-rate fraction, $\overline{x}$, or $\overline{\alpha} = \overline{\alpha}(\overline{x})$. If we postulate that the fluid properties are constant, we can also write $\overline{\alpha} = \overline{\alpha}(H)$, and

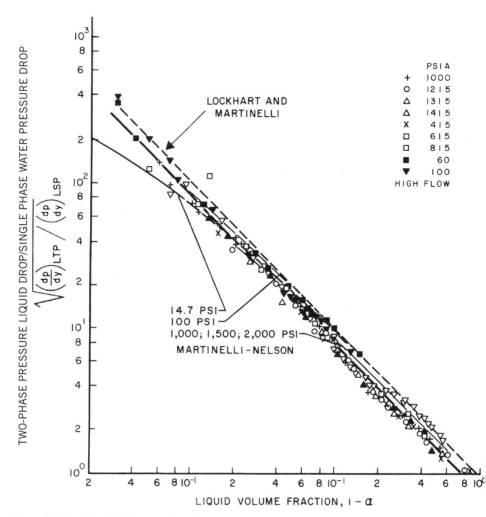

Figure 3.5.4 Simplified expression for pressure drop as a function of void fraction. (From Refs. 3.5.4 and 3.5.5.)

so on, for all the other average effective fluid properties ($\overline{\rho}_M$, $\overline{\rho}_{KE}$, $\overline{H}_\rho$, $\overline{\rho}$). Under such circumstances, the conservation laws become

$$\frac{d\overline{\rho}}{d\overline{H}} \frac{\partial}{\partial t} (\overline{H}A) + \frac{\partial}{\partial z} (GA) = 0 \qquad (3.5.22)$$

$$\frac{\partial}{\partial t} (GA) + \frac{d}{d\overline{H}} \left(\frac{1}{\overline{\rho}_M}\right) \frac{\partial}{\partial z} (G^2 A\overline{H}) = -A \frac{\partial p}{\partial z} - \overline{\tau}_w P - \overline{\rho}gA \sin\theta \qquad (3.5.23)$$

$$\frac{d\overline{\rho}}{d\overline{H}}\overline{H}_\rho + \overline{\rho}\frac{d\overline{H}_\rho}{d\overline{H}}\frac{\partial}{\partial t}(\overline{H}A) + \frac{\partial}{\partial z}(G\overline{H}A) + \frac{1}{2}\frac{d}{d\overline{H}}\left(\frac{1}{\overline{\rho}_M}\right)\frac{\partial}{\partial t}(G^2 A\overline{H})$$

$$+ \frac{1}{2}\frac{d}{d\overline{H}}\left(\frac{1}{\overline{\rho}_{KE}^2}\right)\frac{\partial}{\partial z}(G^3 A\overline{H}) = P_h \overline{q''} + A\overline{q'''} + \frac{\partial}{\partial t}(pA) - gGA\sin\theta \quad (3.5.24)$$

Equation (3.5.24) can be simplified further by introducing a new density $\overline{\rho}''$ such that

$$\overline{\rho}'' = \overline{\rho}\frac{d\overline{H}_\rho}{d\overline{H}} + (\overline{H}_\rho - \overline{H})\frac{d\overline{\rho}}{d\overline{H}} \quad (3.5.25)$$

and the energy equation becomes

$$\overline{\rho}''\frac{\partial}{\partial t}(\overline{H}A) + G\frac{\partial}{\partial z}(\overline{H}A) + \frac{1}{2}\frac{d}{d\overline{H}}\left(\frac{1}{\overline{\rho}_H}\right)\frac{\partial}{\partial t}(G^2 A\overline{H}) + \frac{1}{2}\frac{d}{d\overline{H}}\left(\frac{1}{\overline{\rho}_{KE}^2}\right)\frac{\partial}{\partial z}(G^2 A\overline{H})$$

$$= P_h \overline{q''} + A\overline{q'''} + \frac{\partial}{\partial t}(pA) - gGA\sin\theta \quad (3.5.26)$$

The new density $\overline{\rho}''$ can be shown to be equal to

$$\overline{\rho}'' = [\rho_G + (\rho_L - \rho_G)\overline{x}]\frac{d\overline{\alpha}}{d\overline{x}} \quad (3.5.27)$$

We can see that for the special case where $\overline{\alpha} = \overline{\alpha}(\overline{x})$, the three equations (3.5.22), (3.5.23), and (3.5.24) make it possible to calculate the three independent variables, G, $\overline{H}$, and p as a function of time and position. In other words, the two-phase conservation equations can be reduced to "analogous homogeneous" equations when the volume fraction $\overline{\alpha}$ is a known function of $\overline{x}$. For example, for the momentum equation, both the functions $\overline{\alpha}$ and the wall shear stress $\overline{\tau}_w$ can be specified by the Martinelli model.

System computer codes based on separated models with no interface exchange such as the Martinelli models can be very useful in predicting most complex system transients as well as several of the severe accidents. They are easy to use because of their reduced number of closure laws and they can yield acceptable answers as long as they are applied to conditions for which they are suitable (i.e., co-current flow with thermal equilibrium). Also, they must stay within the range of the tests used to develop the applied closure laws.

3.5.3 Other Separated Flow Models with No Interface Exchange

A multitude of analytical and empirical attempts have been made to modify or to extend the range of coverage of the Martinelli closure laws. They include:

- The analytical momentum exchange model by Levy [3.5.5] and the energy model by Gopalakrishman and Schrock [3.5.6], which provide expressions for

the gas volume fraction $\bar{\alpha}$ in terms of the gas weight rate fraction $\bar{x}$ and the density ratio ρ_L/ρ_G. These analytical expressions are simple in their form but not overly accurate.
- Empirical correlations, which attempt to cover all fluids and different flow directions. For example, Friedel [3.5.7] developed an empirical correlation for the frictional pressure drop parameter $\phi^2_{TPL_0}$, which is applicable to any fluid as long as $\mu_L/\mu_G < 1000$. It is still considered to be the best generally available correlation. Similarly, the CISE correlation [3.5.8] was considered to be the most accurate correlation for the gas volume fraction $\bar{\alpha}$. In recent years an EPRI correlation developed by Chexal and Lellouche [3.5.9] is being preferred and it will be covered under Section 3.5.4. Both the Friedel and CISE or EPRI correlations are good default expressions to be employed when data are lacking for the complex system being investigated.
- Customized expressions for $\bar{\alpha}$ and $\phi^2_{TPL_0}$, developed from tests for the conditions of interest to the complex system being evaluated. Such relations are the most accurate and if they are available, they are and should be preferred to other analytical or generalized correlations.

In summary, separated models with no interface exchanges are useful because of their simplicity. Separated models that do not involve special two-phase fluid property relations are preferable. They are not very accurate and they should not be applied beyond their range of applicability. Their biggest failure is their empirical nature and their avoidance of dealing with the detailed structure of two-phase flow.

3.5.4 Drift Flux Model

The drift flux model is another type of separated or mixture flow model that relies on the volumetric flux (or superficial velocity) of each phase instead of their actual velocity. It was originally developed by Wallis [3.5.10] and embellished upon by Zuber and Findlay [3.5.11]. The-time averaged flux parameters j, j_L, and j_G are defined from

$$j_L = u_L(1 - \alpha) \qquad j_G = u_G \alpha \qquad j = u_{TP} = j_L + j_G \qquad (3.5.28)$$

If we next write the identity

$$\alpha u_G = u_{TP} \alpha + \alpha(u_G - u_{TP}) \qquad (3.5.29)$$

and integrate it over the channel flow area, there results

$$\overline{\alpha u_G} = \overline{u_{TP} \alpha} + \overline{\alpha(u_G - u_{TP})} \qquad (3.5.30)$$

3.5 SEPARATED OR MIXTURE TWO-PHASE-FLOW MODELS

where the bars are used to represent average values over the flow area. By definition,

$$\overline{j_G} = \overline{\alpha u_G} = \frac{G_G}{\rho_G} \qquad \overline{j} = \overline{u_{TP}} = \frac{G_G}{\rho_G} + \frac{G_L}{\rho_L} \qquad (3.5.31)$$

and Eq. (3.5.30) becomes

$$\frac{\overline{j_G}}{\overline{j}} = \overline{\lambda} = \frac{G_G/\rho_G}{(G_L/\rho_L) + (G_G/\rho_G)} = \left[\frac{\overline{u_{TP}\alpha}}{\overline{u_{TP}}\overline{\alpha}} + \frac{\overline{\alpha(u_G - u_{TP})}}{\overline{u_{TP}}\overline{\alpha}}\right]\overline{\alpha} \qquad (3.5.32)$$

where $\overline{\lambda}$ is the average volumetric flow fraction equivalent of the weight fraction $\overline{x}$. In Eq. (3.5.32), the grouping on the right side which multiplies the average volumetric fraction $\overline{\alpha}$ consists of two terms. The first one accounts for the nonuniform distribution of the volumetric fraction α and of the two-phase velocity u_{TP}. The second term considers the drift velocity of the gaseous phase with respect to the two-phase velocity $(u_G - u_{TP})$ and the nonuniform distributions of $u_G - u_{TP}$, and α.

Zuber and Findlay have attempted to specify the two groupings that multiply $\overline{\alpha}$. The first grouping was determined, as suggested by Bankoff [3.4.7], by assuming power law variations for u_{TP} and α so that at a radial position r in a pipe of diameter D,

$$u_{TP} = (u_{TP})_c \left[1 - \left(\frac{2r}{D}\right)^{1/m'}\right] \qquad \alpha = \alpha_c \left[1 - \left(\frac{2r}{D}\right)^{1/n'}\right] \qquad (3.5.33)$$

where the subscript c is used to represent conditions at the pipe centerline and

$$\frac{\overline{\alpha u_{TP}}}{\overline{\alpha}\, \overline{u_{TP}}} = \frac{1}{K_z} = 1 + \frac{2m'n'}{m' + n' + 2m'n'} \qquad (3.5.34)$$

For nearly all reasonable profiles, the constant $1/K_z$ was found to vary from 1.0 to 1.5, which agrees with the range prescribed to the comparable Bankoff's constant.

The second grouping on the right side of Eq. (3.5.32) is more difficult to specify because the local drift velocity $(u_G - u_{TP})$ depends on the stress fields in the two phases and the momentum and energy transfer at the liquid–gas interfaces. Expressions for the drift velocity have been obtained for specific flow patterns, and some typical relations are presented in Chapter 4. Let us consider here the simplified case where $u_G - u_{TP}$ is uniform over the entire cross-sectional area. Equation (3.5.32) becomes,

$$\frac{\overline{j_G}}{\overline{j}} = \frac{G_G/\rho_G}{G_G/\rho_G + G_L/\rho_L}\left[\frac{1}{K_z} + (u_G - u_{TP})\frac{1}{G_G/\rho_G + G_L/\rho_L}\right]\overline{\alpha} \qquad (3.5.35)$$

Equation (3.5.35) often has been written in the form

$$\overline{\alpha} = \frac{\overline{j_G}}{C_0\overline{j} + \overline{U_{Gj}}} \qquad (3.5.36)$$

where C_o is the flux concentration parameter and is equivalent to $1/K_z$ and $\overline{U}_{Gj}$ is the drift velocity or $\overline{u_G - u_{TP}}$.

As a first approximation, we can take for the drift velocity, $\overline{U}_{Gj}$ or $\overline{u_G - u_{TP}}$, the following expressions suggested by Zuber and Findlay [3.5.11] for slug and bubble flow and by Ishii [3.5.12] for annular flow:

$$\overline{u_G - u_{TP}} = 0.35 \left[\frac{g(\rho_L - \rho_G)(D)}{\rho_L} \right]^{1/2} \quad \text{and} \quad C_0 = 1.2 \quad \text{for slug flow}$$

$$\overline{u_G - u_{TP}} = 1.53 \left[\frac{\sigma g(\rho_L - \rho_G)}{\rho_L^2} \right]^{1/4} \quad \text{and} \quad C_0 = 1.2 \quad \text{for bubble flow} \quad (3.5.37)$$

$$\overline{u_G - u_{TP}} = \frac{(1 - \overline{\alpha})^{3/2}}{\overline{\alpha} + a} \sqrt{\frac{(\rho_L - \rho_G)g D_H}{0.015 \rho_L}} \quad \text{for annular flow with}$$

$$a = \sqrt{\frac{1 + 75(1 - \overline{\alpha})}{\sqrt{\overline{\alpha}}}} \sqrt{\frac{\rho_G}{\rho_L}} \quad \text{and} \quad C_0 = 1 + \frac{1 - \overline{\alpha}}{\overline{\alpha} + a}$$

Equations (3.5.36) and (3.5.37) make it possible to calculate the average volume fraction $\overline{\alpha}$ in terms of the two-phase stream properties. This, without doubt, is the major usefulness of the drift flux models. However, to calculate $\overline{\alpha}$ from the models, one must select an empirical or arbitrary value for C_0 or $1/K_z$. A value of $C_0 = 1.2$ is used frequently. Furthermore, in the case of Eq. (3.5.35), it is necessary to use relations for $\overline{U}_{Gj}$ of the type (3.5.37), which imply knowledge of the prevailing flow pattern and of its transition mode from one pattern to another.

In general, both C_0 and $\overline{U}_{Gj}$ are functions of flow pattern, degree of thermal equilibrium, channel characteristics, and total mass flow. A large number of formulations have been proposed for C_0 and $\overline{U}_{Gj}$ to cover many flow patterns and flow directions. A very simple formulation was developed by Spore and Shiralkar [3.5.13], who used the bubbly flow relation of Eq. (3.5.37) but varied its constant 1.53 and the concentration parameter C_0 as a function of the gas volume fraction $\overline{\alpha}$ as shown in Figure 3.5.5. At the gas volume fraction $\overline{\alpha}_t = 0.65$, which corresponds to a flow regime transition to annular flow, Spore and Shiralkar employed

$$C_0 = 1.2 \quad \text{and} \quad \overline{U}_{Gj} = 2.05 \left[\frac{\sigma g(\rho_L - \rho_G)}{\rho_L^2} \right]^{1/4} \quad (3.5.38)$$

Also, they allowed the two constants 1.2 and 2.05 to decrease linearly to 1.0 and zero from $\overline{\alpha}_t = 0.65$ to $\overline{\alpha} = 1.0$, respectively. This approach ensures that the correct gas superficial velocity is obtained when $\overline{\alpha} \rightarrow 1.0$. The formulation of Figure 3.5.5 was found to give reasonable predictions when combined with a modified Martinelli correlation to recognize the impact of the total flow rate G.

3.5 SEPARATED OR MIXTURE TWO-PHASE-FLOW MODELS

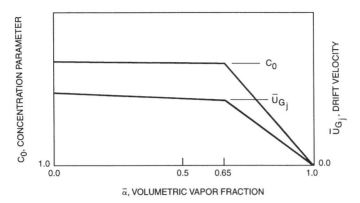

Figure 3.5.5 $C_0 - \overline{U}_{Gj}$ dependence on volumetric vapor fraction. (From Ref. 3.5.13.)

The most extensive correlation for predicting drift flux gas volume fraction was developed by EPRI [3.5.9]. It is supposed to cover all flow patterns and flow directions and it has been incorporated in one of the latest computer system codes, RELAP5/MOD3 [3.5.14]. While it deals with other fluids and gases, only its form for steam–water mixtures will be given here. Its distribution parameter, C_0, is calculated from

$$C_0 = \frac{L(\overline{\alpha},p)}{K_0 + (1 - K_0)\overline{\alpha}^r} \tag{3.5.39}$$

$$L(\overline{\alpha},p) = \frac{1 - \exp(-C_1\overline{\alpha})}{1 - \exp(-C_1)} \tag{3.5.40}$$

$$C_1 = \frac{4p_{\text{crit}}^2}{p(p_{\text{crit}} - p)} \tag{3.5.41}$$

p_{crit} = critical pressure

$$K_0 = B_1 + (1 - B_1)\left(\frac{\rho_G}{\rho_L}\right)^{1/4} \tag{3.5.42}$$

$$B_1 = \min(0.8, A_1) \tag{3.5.43}$$

$$A_1 = \frac{1}{1 + \exp(-N_{\text{Re}}/6000)} \tag{3.5.44}$$

$$N_{\text{Re}} = \begin{cases} N_{\text{Re}_G} & \text{if } N_{\text{Re}_G} > N_{\text{Re}_L} \text{ or } N_{\text{Re}_G} < 0 \\ N_{\text{Re}_L} & \text{if } N_{\text{Re}_G} \leq N_{\text{Re}_L} \end{cases} \tag{3.5.45}$$

N_{Re_L} = local liquid superficial Reynolds number

$$= \frac{\rho_L \overline{j}_L D_H}{\mu_L} \tag{3.5.46}$$

N_{Re_G} = local vapor superficial Reynolds number

$$= \frac{\rho_G \bar{j}_G D_H}{\mu_G} \tag{3.5.47}$$

$\bar{j}_L$ = liquid superficial velocity

$$= \bar{\alpha}_L \bar{u}_L \tag{3.5.48}$$

$\bar{j}_G$ = vapor superficial velocity

$$= \bar{\alpha}_G \bar{u}_G \tag{3.5.49}$$

$$r = \frac{1 + 1.57(\rho_G/\rho_L)}{1 - B_1} \tag{3.5.50}$$

The sign of $\bar{j}_k$ is positive if phase k flows upward and negative if it flows downward. This convention determines the sign of N_{Re_G}, N_{Re_L}, and N_{Re}.

The vapor drift velocity, $\bar{U}_{Gj}$, is calculated from

$$\bar{U}_{Gj} = 1.41 \left[\frac{(\rho_L - \rho_G)\sigma g}{\rho_L^2} \right]^{1/4} C_2 C_3 C_4 C_9 \tag{3.5.51}$$

where

$$C_2 = \begin{cases} 1 & \text{if } C_5 \geq 1 \\ \dfrac{1}{1 - \exp(-C_6)} & \text{if } C_5 < 1 \end{cases} \tag{3.5.52}$$

$$C_5 = \left[150 \left(\frac{\rho_G}{\rho_L} \right) \right]^{1/2} \tag{3.5.53}$$

$$C_6 = \frac{C_5}{1 - C_5} \tag{3.5.54}$$

$$C_4 = \begin{cases} 1 & \text{if } C_7 \geq 1 \\ \dfrac{1}{1 - \exp(-C_8)} & \text{if } C_7 < 1 \end{cases} \tag{3.5.55}$$

$$C_7 = \left(\frac{D_2}{D_H} \right)^{0.6} \tag{3.5.56}$$

$D_2 = 0.09144$ (normalizing diameter)

3.5 SEPARATED OR MIXTURE TWO-PHASE-FLOW MODELS

$$C_8 = \frac{C_7}{1 - C_7} \qquad (3.5.57)$$

$$C_9 = \begin{cases} (1 - \bar{\alpha})^{B_1} & \text{if } N_{Re_G} \geq 0 \\ \min[0.7, (1 - \bar{\alpha})^{0.65}] & \text{if } N_{Re_G} < 0 \end{cases} \qquad (3.5.58)$$

The parameter C_3 depends on the directions of the vapor and liquid flows:

Upflow (both $\bar{j}_G$ and $\bar{j}_L$ are positive):

$$C_3 = \max\left[0.50, 2 \exp\left(-\frac{|N_{Re_G}|}{60,000}\right)\right] \qquad (3.5.59)$$

Downflow (both g_G and j_L are negative):

$$C_3 = 2\left(\frac{C_{10}}{2}\right)^{B_2} \qquad (3.5.60)$$

$$B_2 = \frac{1}{1 + 0.05 \left|N_{Re_L}/35,000\right|^{0.4}} \qquad (3.5.61)$$

$$C_{10} = 2 \exp\left[\left(\frac{|N_{Re_L}|}{350,000}\right)^{0.4}\right] - 1.75|N_{Re_L}|^{0.03} \exp\left[-\frac{|N_{Re_L}|}{50,000}\left(\frac{D_1}{D}\right)^2\right]$$
$$+ \left(\frac{D_1}{D_H}\right)^{0.25} |Re_L|^{0.001} \qquad (3.5.62)$$

$$D_1 = 0.0381 \text{ (normalizing diameter)} \qquad (3.5.63)$$

Countercurrent flow ($\bar{j}_G$ is positive, $\bar{j}_L$ is negative):

$$C_3 = \max\left[0.50, 2 \exp\left(\frac{-|N_{Re_L}|}{60,000}\right)\right] \qquad (3.5.64)$$

For co-current horizontal, the concentration parameter is

$$C_0 = \frac{1 + \bar{\alpha}^{0.05}(1 - \bar{\alpha}^2)L}{K_0 + (1 - K_0)\bar{\alpha}^r} \qquad (3.5.65)$$

The EPRI correlation is useful because it is the result of an intensive effort to fit available data over a number of gas–liquid flow conditions. It has the important advantage of not having to consider various flow patterns. However, it is primarily an empirical correlation.

The drift flux model has several advantages over other separated models:

- The drift flux model relies on the *local* phase velocity difference $u_G - u_{TP}$ averaged over the cross section or $\overline{u_G - u_{TP}}$ rather than on the difference of average phase velocities (e.g., $\bar{u}_G - \bar{u}_j$). Physically, the drift flux formulation is more appropriate and correct.
- The drift flux model recognizes to a limited extent the flow distribution within the channel through the parameter C_0. It also incorporates some knowledge about flow patterns. The more flow pattern information it includes, the more complicated the formulation becomes.
- Drift flux models can inherently handle co-current (up or down) flow and countercurrent flow.
- The drift flux model has been particularly successful in dealing with countercurrent flow limitations (see Chapter 5) and it has been extended to deal with interface exchanges in the system computer code TRAC-BWR [3.5.15], which was developed to analyze transients and accidents in boiling water nuclear reactors.

However, the drift flux model requires closure laws for the frictional pressure drop and heat transfer at the flow boundaries over and above the drift flux relations. These closure laws are often obtained from previously described separated models without assuring their consistency with the drift flux–derived gas volume fraction.

The choice between the drift flux model and other separated two-phase flow models appears to be more a matter of user and code developer preference. In this author's opinion, the drift flux model makes it easier to specify the interface exchanges, as discussed in Chapter 4. However, the most important rule for making a selection among the various available separated system computer codes should be the degree to which the code has been validated against a whole range of test data.

3.6 TWO-FLUID MODELS INCLUDING INTERFACE EXCHANGE

Two-fluid models incorporating interface exchanges are being used widely at the present time in complex system computer codes because they can permit different gas and liquid velocities and flow directions as well as different liquid and gas temperatures. These models still rely on one-dimensional formulations and upon time and spatial averaging. In other words, they do not consider fluid variation of properties (e.g., velocity and temperature) and their gradients at the interfaces or at the wall boundaries. Such two-fluid models therefore require closure laws to deal with the interfaces as well as the wall boundary conditions. Because these closure laws are known to depend on flow patterns, flow regime maps must also be specified. In this section we first describe the conservation equations for two-fluid models, including interfacial exchanges, and discuss their limitations. This is followed by a review of typical flow pattern maps incorporated in corresponding system computer codes. Closure laws are covered next and are followed by other pertinent remarks and summary comments about two-fluid-system computer codes.

3.6.1 Two-Fluid-Model Conservation Laws with Interface Exchange

The following two-fluid conservation laws are based on the set proposed by Yadigaroglu and Lahey (3.6.1). The continuity equations for each phase are

$$\frac{\partial}{\partial t}[\rho_L(1-\overline{\alpha})] + \frac{1}{A}\frac{\partial}{\partial z}[\rho_L(1-\overline{\alpha})\overline{u}_L A] = -\Gamma \qquad (3.6.1)$$

$$\frac{\partial}{\partial t}[\rho_G(1-\overline{\alpha})] + \frac{1}{A}\frac{\partial}{\partial z}[\rho_G(1-\overline{\alpha})\overline{u}_G A] = \Gamma \qquad (3.6.2)$$

Equations (3.6.1) and (3.6.2) allow for some of the liquid to be transferred (or converted) into the gas (or vapor) phase at the rate Γ per unit volume. Some of that transfer can occur at the wall and the rest at the interface. In the case of one-component flow and thermodynamic equilibrium, there is no interface heat transfer and all the transfer takes place at the wall or $\Gamma = \Gamma_w$. It should be noted that addition of Eqs. (3.6.1) and (3.6.2) reproduces Eq. (3.2.5) for a constant area A.

The corresponding phasic momentum equations are

$$\frac{\partial}{\partial t}[\rho_L(1-\overline{\alpha})\overline{u}_L] + \frac{1}{A}\frac{\partial}{\partial z}[\rho_L(1-\overline{\alpha})\overline{u}_L^2 A] = -(1-\overline{\alpha})\frac{\partial p}{\partial z} - g\rho_L(1-\overline{\alpha})\sin\theta$$
$$-\frac{P_{wL}\tau_{wL}}{A} + \frac{P_i\tau_i}{A} - \Gamma\overline{u}_{iL} + C\overline{\alpha}\,\overline{\rho}\,(1-\overline{\alpha})\frac{\partial(\overline{u}_G - \overline{u}_L)}{\partial t} \qquad (3.6.3)$$

$$\frac{\partial}{\partial t}(\rho_G\overline{\alpha}\,\overline{u}_G) + \frac{1}{A}\frac{\partial}{\partial z}(\rho_G\overline{\alpha}\,\overline{u}_G^2 A) = -\overline{\alpha}\frac{\partial p}{\partial z} - g\rho_G\overline{\alpha}\sin\theta - \frac{P_{WG}\tau_{WG}}{A}$$
$$-\frac{P_i\tau_i}{A} + \Gamma\overline{u}_{iG} - C\overline{\alpha}\,\overline{\rho}(1-\overline{\alpha})\frac{\partial(\overline{u}_G - \overline{u}_L)}{\partial t} \qquad (3.6.4)$$

The terms on the right side of Eqs. (3.6.3) and (3.6.4) are the forces on phase k (where k can represent the liquid phase L or the gas phase G). The first term on the right side of Eqs. (3.6.3) and (3.6.4) is the net pressure force acting on phase k; the second term is the gravitational force. The third and fourth terms are the shear stresses acting on the phase at the wall and at the interface, and they are denoted, respectively, by τ_{wk} and τ_i. The parameter P_{wk} is the part of the wall perimeter wetted by phase k. The next terms on the right side represent the momentum addition into phase k by mass exchange at the interface. The mass entering phase k has an interface velocity $\overline{u}_{ik}$. The last term represents the virtual mass term and it was not included in the conservation laws without interface exchanges. It represents the force required to accelerate the apparent mass of the surrounding phase when the relative velocity between phases changes. There is continued debate about the form of the virtual mass term. Its simplest form, as employed in the RELAP system computer codes, is utilized here. The constant C varies with flow pattern. It has a value of 0.5 for bubble or dispersed flow and it is set to zero for separated or stratified flows. The various proposed formulations for virtual mass are discussed by Ishii and Mishima [3.6.2]. However, it should be mentioned here that the virtual mass term is important only for very fast changes in the flow (e.g., critical flow) and that its primary impact is to stabilize the numerical solution of the conservation equations.

Here again, it should be noted that addition of Eqs. (3.6.3) and (3.6.4) eliminates the interfacial terms and yields Eq. (3.2.14). Moreover, Eqs. (3.6.3) and (3.6.4) presume that the liquid and gas phases are at the same pressure p. The corresponding phasic energy equations were written in terms of total enthalpy H_k^0, or

$$H_k^0 = H_k + \frac{\overline{u}_k^2}{2} - gz \sin \theta \tag{3.6.5}$$

After manipulation of some of the terms, Yadigaroglu and Lahey obtained

$$\frac{\partial}{\partial t}[\rho_L(1-\overline{\alpha})\overline{H}_L^0] + \frac{1}{A}\frac{\partial}{\partial z}[\rho_L(1-\overline{\alpha})\overline{H}_L^0 \overline{u}_L A]$$

$$= q_L'''(1-\overline{\alpha}) + \frac{q_{iL}'' P_i}{A} + \frac{q_{wL}'' P_{hL}}{A} - \Gamma \overline{H}_{iL}^0 + (1-\overline{\alpha})\frac{\partial p}{\partial t} - \xi \frac{P_i}{A} \tau_i \overline{u}_{iL} \tag{3.6.6}$$

$$\frac{\partial}{\partial t}(\rho_G \overline{\alpha} \overline{H}_G^0) + \frac{1}{A}\frac{\partial}{\partial z}(\rho_G \overline{\alpha} \overline{H}_G^0 \overline{u}_G A)$$

$$= q_G''' \overline{\alpha} + \frac{q_{iG}'' P_i}{A} + \frac{q_{wG}'' P_{hG}}{A} + \Gamma \overline{H}_{iG}^0 + \overline{\alpha}\frac{\partial p}{\partial t} + \xi \frac{P_i}{A} \tau_i \overline{u}_{iG} \tag{3.6.7}$$

The first term on the right-hand side of Eqs. (3.6.6) and (3.6.7) is the internal heat generation due to a volumetric source q_k'''. The second and third terms are the sensible heat inputs from the interfacial perimeter P_i and from the heated portion of the perimeter wetted by phase k, P_{hk}. The fourth term accounts for energy addition to phase k due to interfacial mass transfer with $\overline{H}_{ik}^0$ being the total enthalpy characteristic of this exchange. The fifth term accounts for work due to expansion or contraction of the phases, and the last term is related to the interfacial energy dissipation. The parameter ξ represents the fraction of the energy dissipated at the interface that gets transferred to the gas phase. Here again it is worth noting that the same interfacial perimeter P_i is used for the momentum and energy equations and that the addition of Eqs. (3.6.6) and (3.6.7) reproduces Eq. (3.2.23).

The conservation laws tacitly assumed that the two phases are at the same pressure p. In most cases this is a plausible assumption because the pressure differences between the two phases are expected to be small. However, in the case of stratified flow, a pressure difference exists between the two phases and it must be taken into account to determine the interfacial instability and the transition to slug flow. Similarly, the equal pressure rule can be invalid when the flows are oscillatory. If one considers the special case of oscillatory flow with a zero time-averaged flow, the time-averaged wall shear stress does not vanish (because it is proportional to the fluctuating velocity squared). Under those circumstances, wall shear closure laws of the type discussed in Sections 3.2 to 3.5 are not valid. In other words, the proposed conservation laws may not be adequate when the flow pattern is intermittent, particularly when the periodic changes with time are large as in the case of churn flow.

The two-fluid model requires the knowledge of several exchange terms, including the volumetric mass exchange, Γ; the wall shear force applied to each phase and the interfacial shear force; the heat supplied from the wall to each phase; and the two interfacial energy transfer rates. In fact, it is necessary to provide a total of 13 expressions for Γ, P_{wL}, P_{wG}, τ_{wL}, τ_{wG}, P_{hL}, P_{hG}, q''_{wL}, q''_{wG}, P_i, τ_i, q''_{iL}, and q''_{iG}. Some of the variables are interrelated, such as

$$P_{wL} + P_{wG} = P_w \quad \text{and} \quad P_{hL} + P_{hG} = P_h \tag{3.6.8}$$

In addition, one can write the following heat balance at the interface:

$$\Gamma(\overline{H}_{iG} - \overline{H}_{iL}) + \frac{P_i}{A}(q''_{iG} + q''_{iL}) = 0 \tag{3.6.9}$$

This means that a total of 10 closure laws must be provided. Because the closure laws depend on flow patterns, the total number of closure laws must be multiplied by the number of flow patterns. This total number has continued to increase over the years and it can be quite large, as attempts are made to reproduce more and more of the physics of two-phase flow. This proliferation of closure laws is illustrated clearly in Sections 3.6.2 and 3.6.3.

The six conservation laws can be simplified and reduced in number with simplifying assumptions. As noted already, if interface exchanges are neglected, the number of conservation laws decrease from 6 to 3. If one specifies a relation between the liquid and gas velocity (e.g., drift flux model), there is a need for only a mixture momentum equation. Similarly, if thermal equilibrium exists in one-component flow with the two phases at saturation temperature, there is no need to use separate energy equations. Under many nonequilibrium conditions, only one of the phases does not stay at saturation temperature and one of the phase energy equations can be eliminated, and so on.

The basic equations vary from one system computer code to another. As noted, the virtual mass term is different because it is sometimes written in terms of the total derivative rather than the partial time derivative used in Eqs. (3.6.3) and (3.6.4). Also, the equations have employed alternative flow variables (e.g., RELAP employs internal energy instead of the total enthalpy). However, these differences at the conservation laws level are not significant, but they become larger when dealing with flow pattern maps and closure laws and their impact is not easy to assess.

3.6.2 Flow Pattern Maps for Computer System Codes

To ease the numerical computations, system computer codes tend to employ simplified flow pattern maps, expressed primarily in terms of the gas volume fraction $\bar{\alpha}$. For example, in the case of the drift flux system computer code TRAC-BF1/MOD1 [3.5.15], the same flow regime map is assumed to be valid for both horizontal and vertical flow. Furthermore, the flow regime is rather simple and consists of two distinct patterns, a liquid continuous pattern at low gas volume fractions, $\bar{\alpha}$, and a vapor-continuous regime at high values of $\bar{\alpha}$, with a transition zone in between. The liquid continuous regime applies to single-phase liquid flow, bubbly, churn, and inverted annular flows. The vapor continuous regime applies to the dispersed droplet flow and single-phase vapor flow. The transition regime involves annular-droplet and film flow situations. The flow determination logic in TRAC-BF1/MOD1 is:

$$\begin{aligned} &\text{bubbly/churn flow} &&\text{for} &&\bar{\alpha} < \bar{\alpha}_{tran} \\ &\text{annular flow} &&\text{for} &&\bar{\alpha}_{tran} < \bar{\alpha} < \bar{\alpha}_{tran} + 0.25 \\ &\text{dispersed droplet flow} &&\text{for} &&\bar{\alpha} > \bar{\alpha}_{tran} + 0.25 \end{aligned} \quad (3.6.10)$$

where according to Ishii [3.5.12],

$$\bar{\alpha}_{tran} = \left(1 + 4\sqrt{\frac{\rho_G}{\rho_L}}\right)\frac{1}{C_0} - 4\sqrt{\frac{\rho_G}{\rho_L}} - 0.15 \quad (3.6.11)$$

with

$$C_0 = C_\infty - \sqrt{\frac{\rho_G}{\rho_L}}(C_\infty - 1) \quad (3.6.12)$$

In the shear model, C_∞ is taken as

$$C_\infty = 1 + 0.2 \left(\frac{\rho_L}{G}\sqrt{gD_H}\right)^{0.5} \quad (3.6.13)$$

while in the heat transfer model it is

$$C_\infty = 1.393 - 0.0155 \log \frac{GD_H}{\mu_L} \quad (3.6.14)$$

The simplicity of the flow pattern map in TRAC-BF1/MOD1 is rather striking; so is the use of different correlations for C_∞ in defining the transition from bubbly/churn to annular flow at the interfacial shear and heat transfer surfaces.

One of the most extensive flow regime maps is provided in the system computer code RELAP5/MOD3. The employed vertical flow regime map is shown in Figure

3.6 TWO-FLUID MODELS INCLUDING INTERFACE EXCHANGE

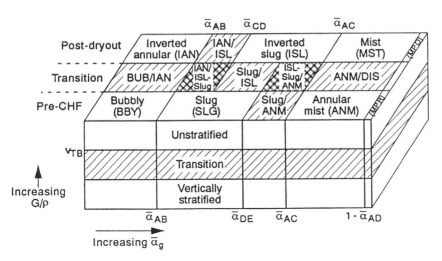

Figure 3.6.1 Schematic of vertical flow regime map with hatchings indicating transitions (From Ref. 3.5.14.)

3.6.1 and it includes a total of nine regimes, four for precritical heat flux (CHF) heat transfer (which corresponds to a liquid or a boiling liquid at the heated surface), four for post-CHF heat transfer (which corresponds to vapor blanketing at the vapor surface), and one for vertical stratification. For pre-CHF conditions, the regimes modeled are bubbly (BBY), slug (SLG), annular mist (ANM), and mist (MPR). For post-CHF, the bubbly, slug, and annular mist regimes are transformed into the inverted annular (IAN), inverted slug (ISL), and mist (MST) regimes. An additional post-CHF mist (MPO) regime is added to match the pre-CHF MPR regime. The ninth regime included in Figure 3.6.1 is vertically stratified flow for sufficiently low mixture velocity. Transition regions between flow pattern are provided in the code and are shown in Figure 3.6.1. The boundaries for transition from one regime to another in RELAP5/MOD3 are as follows:

- Bubbly flow does not exist in tubes of small diameter:

$$D_H < 19 \left[\frac{g(\rho_L - \rho_G)}{\sigma} \right]^{-1/2} \quad (3.6.15)$$

- If the diameter exceeds the value of Eq. (3.6.15), the transition from bubbly to slug takes place when

$$\overline{\alpha}_{AB} = \begin{cases} 0.25 & \text{for } G < 2000 \text{ kg/m}^2 \cdot \text{s} \\ 0.25 + 0.001\,(G - 2000)(0.25) & \text{for } 2000 < G < 3000 \text{ kg/m}^2 \cdot \text{s} \\ 0.5 & \text{for } G > 3000 \text{ kg/m}^2 \cdot \text{s} \end{cases}$$

$$(3.6.16)$$

The slug-to-annular mist transition occurs when

$$\bar{\alpha}_{AC} = \min[\bar{\alpha}^f_{crit}, \alpha^e_{crit}]$$

$$\bar{\alpha}^f_{crit} = \begin{cases} \dfrac{1}{\bar{u}_G}\left[\dfrac{gD_H(\rho_L - \rho_G)}{\rho_G}\right]^{1/2} & \text{for upflow} \\ 0.75 & \text{for downflow and countercurrent flow} \end{cases} \quad (3.6.17)$$

$$\bar{\alpha}^e_{crit} = \dfrac{3.2}{\bar{u}_G}\left[\dfrac{g\sigma(\rho_L - \rho_G)}{\rho_G^2}\right]^{1/4}$$

where $\bar{\alpha}^f_{crit}$ is controlling the transition in small tubes due to flow reversal and $\bar{\alpha}^e_{crit}$ is controlling the transition in large tubes due to droplet entrainment. The transition gas volume fraction $\bar{\alpha}_{AC}$ is constrained between 0.56 and 0.9. For the transition to full mist flow, $1 - \bar{\alpha}_{AC} = 10^{-4}$.

At low mass fluxes, the possibility of vertical stratification exists. A volume is considered stratified if the difference in gas volume fractions of the volumes above and below it exceeds 0.5 and if the average mass flux in the volume in question is less than that corresponding to the Taylor bubble rise velocity, or $\bar{u}_{TB}$, or

$$G \leq \bar{\rho}\bar{u}_{TB} \leq 0.35\,\bar{\rho}\left[\dfrac{gD(\rho_L - \rho_G)}{\rho_L}\right]^{0.5} \quad (3.6.18)$$

In RELAP5-MOD3, a different flow regime map is employed in horizontal flow and it is shown in Figure 3.6.2. It is similar to the vertical flow pattern map except that the post-CHF regimes are not included and a horizontal stratification regime replaces the vertical one. For horizontal flow, the bubbly-to-slug transition is a constant, or

$$\bar{\alpha}_{AB} = 0.25 \quad (3.6.19)$$

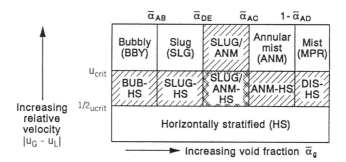

Figure 3.6.2 Schematic of horizontal flow regime map with hatchings indicating transition regions (From Ref. 3.5.14.)

and the slug-to-annular mist transition is

$$\overline{\alpha}_{AC} = 0.85 \tag{3.6.20}$$

Horizontal stratified flow will occur when the gas velocity falls below a limiting value developed by Taitel and Dukler [3.6.3]:

$$\overline{u}_G = \frac{1}{2}\left[\frac{(\rho_L - \rho_G)g\overline{\alpha}A}{\rho_G D \sin v}\right]^{1/2}(1 - \cos v) \tag{3.6.21}$$

with the angle v being related to the gas volume fraction $\overline{\alpha}$ and the water level H in the circular pipe of diameter D by

$$H = \frac{D}{2}(1 + \cos v) \tag{3.6.22}$$

$$\pi\overline{\alpha} = v - \sin v \cos v$$

where v is half the angle subtended from the top of the pipe to the horizontal level of the liquid.

RELAP5/MOD3 finally recognizes a high mixing flow map which applies to such components as pumps. It consists of a bubble regime for $\overline{\alpha} < 0.5$, a mist regime for $\overline{\alpha} \geq 0.95$, and a transition regime for $0.5 < \overline{\alpha} < 0.95$. This transition region is modeled as a mixture of bubbles dispersed in liquid and droplets dispersed in vapor.

It is apparent that computer system codes such as RETRAN are capable of incorporating flow regime maps to match various anticipated pattern configurations and their transitions. They tend to be based on flow pattern studies; however, the flow regime maps utilized are simplified considerably to ease the computer code numerical calculations burden even if it is at the expense of physical reality. More complex and accurate flow patterns and transition criteria are available (see Chapter 4), but their utilization cannot be justified due to other approximations inherent in system computer codes. For example, to avoid sudden changes in flow regime that may be deleterious to numerical stability, transition regions covering a span $\Delta\overline{\alpha}$ of at least 0.1 are included in computer system codes and that required computational flexibility overshadows any accuracy that might be derived from more exact flow regime maps.

3.6.3 Closure Laws for Two-Fluid Models with Interfacial Exchange

The primary purpose of this section is to describe the process employed to specify closure laws in system computer codes with interfacial exchanges. It is not meant to be an all-inclusive discussion of the various closure laws employed or their preferred versions. Instead, the goal is to look at the complexity of the process and its potential uncertainties. Interface shear is covered first, followed by wall shear. Finally, similar summary discussions for interface and wall heat transfer are provided.

3.6.3.1 Interface Shear
To predict the interfacial shear, it is necessary to specify the interfacial area and the interfacial friction factor.[†]

3.6.3.1.1 Interfacial Shear Area
In most system computer codes, the interfacial area density P_i/A or interfacial surface area per unit volume is defined from a simplified geometric representation of the prevailing flow pattern. The expressions depend on the average gas volume fraction $\bar{\alpha}$, and on characteristic lengths of the flow pattern such as the diameter of bubbles and droplets in bubble, slug, and dispersed flows and the film thickness in annular flows. For example, in RELAP5/MOD3 the bubbles or droplets in bubble or dispersed flow are assumed to be spherical particles with a size distribution of the Nukiyama–Tanasawa form. If $\bar{d}'_b$ is the most probable bubble diameter, the interfacial area per unit volume is

$$\frac{P_i}{A} = \frac{2.4\bar{\alpha}}{\bar{d}'_b} = \frac{3.6\bar{\alpha}}{\bar{d}_b} \qquad (3.6.23)$$

where $\bar{d}_b$ is the average bubble diameter and is equal to $1.5\,\bar{d}'_b$.

The maximum bubble diameter $\bar{d}_{b,\max}$ is related to the critical Weber number, $N_{We,c}$:

$$N_{We,c} = \frac{\bar{d}_{b,\max}\rho_L(u_G - u_L)^2}{\sigma} = \frac{\bar{d}_{b,\max}\rho_L \overline{U}_{Gj}^2}{\sigma(1 - \bar{\alpha})^2} \qquad (3.6.24)$$

By assuming that $\bar{d}_{b,\max} = 2\bar{d}_b$ and employing a critical Weber number of 10, RELAP5/MOD3 gets for bubble flow

$$\frac{P_i}{A} = 0.72 \frac{\bar{\alpha}}{(1 - \bar{\alpha})^2} \frac{\rho_L \overline{U}_{Gj}^2}{\sigma} \qquad (3.6.25)$$

A similar expression was derived for droplet flow by employing a critical Weber number of 3.

In the case of slug flow, the flow is modeled as a series of Taylor bubbles separated by liquid slugs containing small gas bubbles. The diameter of the Taylor bubble is approximated by the channel diameter D_H, and its surface/volume ratio is taken as $4.5\,D_H$. The channel average void fraction $\bar{\alpha}$ is given by

$$\bar{\alpha} = \bar{\alpha}_{TB} + \bar{\alpha}_{GS}(1 - \bar{\alpha}_{TB}) \quad \text{or} \quad \bar{\alpha}_{TB} = \frac{\bar{\alpha} - \bar{\alpha}_{GS}}{1 - \bar{\alpha}_{GS}} \qquad (3.6.26)$$

where $\bar{\alpha}_{TB}$ is the fraction of the cell occupied by the Taylor bubble and $\bar{\alpha}_{GS}$ is the

---

[†]An alternative approach for obtaining the interfacial shear is described in Section 3.6.3.2. It was developed for drift flux models.

remaining gas volume fraction in the liquid slug and film. The value of $\overline{\alpha}_{GS}$ is obtained in RELAP5/MOD3 by interpolation between the transition from bubble to slug flow, $\overline{\alpha}_{AB}$, and from slug to annular flow, $\overline{\alpha}_{AC}$, as shown in Figure 3.6.1.

An exponential relation was adopted such that $\overline{\alpha}_{GS} = \overline{\alpha}_{AB}$ at the bubble-to-slug flow transition and approaches zero at the slug-to-annular transition, or

$$\overline{\alpha}_{GS} = \overline{\alpha}_{AB} \exp\left(-8\frac{\overline{\alpha} - \overline{\alpha}_{AB}}{\overline{\alpha}_{AC} - \overline{\alpha}_{AB}}\right) \qquad (3.6.27)$$

The interfacial area can now be calculated. It is made up of Taylor bubbles with an interface of $(4.5/D_H)$ and of smaller spherical gas bubbles with diameter $\overline{d}_b$ in the liquid so that using Eq. (3.6.23) there results for slug flow

$$\frac{P_i}{A} = \frac{4.5}{D_H}\overline{\alpha}_{TB} + 3.6\frac{\overline{\alpha}_{GS}(1 - \overline{\alpha}_{TB})}{\overline{d}_b} \qquad (3.6.28)$$

In Eq. (3.6.28), $\overline{\alpha}_{GS}$ is derived from Eq. (3.6.27), $\overline{\alpha}_{TB}$ is obtained from Eq. (3.6.26), and $\overline{d}_b$ is calculated from Eq. (3.6.24) and the specified critical Weber number of 10. It should be recognized that under this formulation the Taylor bubble and the smaller gas bubbles are presumed to behave in the same manner at their interfaces.

In annular flow, the interfacial area per unit volume can easily be obtained for the case of no liquid entrainment in the gas core and a liquid-film thickness δ:

$$\overline{\alpha} = \left(1 - \frac{2\delta}{D_H}\right)^2 \quad \text{or} \quad \frac{P_i}{A} = \frac{4}{D_H}\sqrt{\overline{\alpha}} \qquad (3.6.29)$$

In the case of annular flow with entrainment or annular mist flow, one can define the liquid volume fraction of the liquid film along the wall as $\overline{\alpha}_{LF}$. The liquid droplet volume fraction in the gas core is $1 - \overline{\alpha} - \overline{\alpha}_{LF}$. The interfacial area in RELAP5/MOD3 was then obtained by adding the interfacial area at the liquid–gas core perimeter to that of the liquid droplets. Using Eqs. (3.6.29) and (3.6.23), there results

$$\frac{P_i}{A} = \frac{4}{D_H}\sqrt{1 - \overline{\alpha}_{LF}} + 3.6\frac{1 - \overline{\alpha} - \overline{\alpha}_{Lf}}{\overline{d}_p} \qquad (3.6.30)$$

where $\overline{d}_p$ is the average diameter of the core liquid droplets and its value is derived from a critical Weber number of 3.

In RELAP5/MOD3, an exponential function is proposed to calculate the film volume fraction $\overline{\alpha}_{Lf}$:

$$\overline{\alpha}_{Lf} = (1 - \overline{\alpha})\exp\left[-7.5(10^{-5})\frac{\overline{\alpha}\,\overline{u}_G}{\overline{\alpha}\,\overline{u}_{G,e}}\right] \qquad (3.6.31)$$

where $\overline{\alpha}\overline{u}_{G,e}$ is the critical gas velocity to suspend a liquid droplet. In vertical flow it is given by

$$\overline{\alpha}\overline{u}_{G,e} = 3.2\frac{[g\sigma(\rho_L - \rho_G)]^{1/4}}{\rho_G^{1/2}} \qquad (3.6.32)$$

The preceding discussion about the interfacial area density was detailed on purpose. It was done to highlight the many geometric and other assumptions involved. For example, they include the concepts of spherical bubbles and droplets with a specified size distribution, of different critical Weber numbers for bubbles and droplets and of a critical gas velocity to suspend liquid droplets. Also, when one of the phases assumes two forms (e.g., liquid film and liquid drops), their interfaces are added arithmetically, which implies that the two forms behave the same way at their interfaces. Moreover, slowly changing exponential functions were introduced to assure a smooth transition in the interface areas from one flow pattern to another. Finally, some of the prescriptions are influenced as we change flow regimes (see Figure 3.6.2) or flow direction, and the number of closure laws can be expected to multiply with any such variations.

3.6.3.1.2 Interfacial Shear Stress Once the interfacial area is specified, the corresponding interfacial shear τ_i is obtained from

$$\tau_i = -\frac{C_D}{8}\rho_c\left|\overline{u}_G - \overline{u}_L\right|(\overline{u}_G - \overline{u}_L) \qquad (3.6.33)$$

where C_D is the interface drag coefficient and ρ_c is the density of the continuous phase. As pointed out before, $\overline{u_G - u_L}$ would be a more appropriate velocity difference in Eq. (3.6.33) and the drift flux model is superior in that respect. The values of the drag coefficient C_D are again dependent on the flow regime, and the following expressions developed by Ishii and Chawla [3.6.4] are typical of those found in system computer codes: For bubbly or dispersed flows,

$$C_D = 24\frac{1 + 0.1N_{Re_p}^{0.75}}{N_{Re_p}} \qquad (3.6.34)$$

where N_{Re_p} is the Reynolds number of the gas bubble or liquid droplet:

$$N_{Re_p} = \frac{(\overline{u}_G - \overline{u}_L)d_p\rho_c}{\mu_m} \qquad (3.6.35)$$

where μ_m is the mixture viscosity and $\mu_m = \mu_L/(1 - \overline{\alpha})$ for bubbles and $\mu_m = \mu_G/\overline{\alpha}^{2.5}$ for droplets. For slug flow,

$$C_D = 9.8(1 - \overline{\alpha}_{TB})^3 \qquad (3.6.36)$$

where $\bar{\alpha}_{TB}$ is obtained from Eqs. (3.6.26) and (3.6.27). For annular flow, C_D is replaced by the friction factor f_i at the liquid film/gas core interface. Generally, the friction factor has been expressed in terms of the liquid-film thickness $\delta \approx \bar{\alpha}_{Lf} D_H / 4$ with $\bar{\alpha}_{Lf}$ obtained from Eq. (3.6.31).

According to Bharathan [3.6.5], f_i is obtained:

$$f_i = 4 [0.005 + A(\delta^*)^B] \quad (3.6.37)$$

with

$$D_H^* = D_H \left[\frac{(\rho_L - \rho_G)g}{\sigma} \right]^{1/2} \quad \text{and} \quad \delta^* = \delta \left[\frac{(\rho_L - \rho_G)g}{\sigma} \right]^{1/2} \quad (3.6.38)$$

$$\log A = -0.56 + \frac{9.07}{D_H^*} \quad \text{and} \quad B = 1.63 + \frac{4.74}{D_H^*} \quad (3.6.39)$$

The corresponding drag on the water droplets can be obtained from Eq. (3.6.34).

It should be noted that additional and different formulations have been proposed for the drag coefficient C_D because:

- Some of the additions are necessary to take into account different flow direction and flow pattern maps. For example, in the case of horizontal stratified flow, RELAP5/MOD3 employs

$$f_i = \frac{0.3164}{N_{\text{Re}_L}^{0.25}} \quad (3.6.40)$$

with

$$N_{\text{Re}_L} = \frac{D_i \rho_G (\bar{u}_G - \bar{u}_L)}{\mu_G} \quad \text{and} \quad D_i = \frac{\bar{\alpha} \pi D_H}{\nu + \sin \nu} \quad (3.6.41)$$

where the nomenclature of Eq. (3.6.22) is utilized.

- Some of the changes result from trying to be more representative of configurations of the complex system. For instance, TRAC-BF2/MOD1 uses a different correlation for $\overline{U}_{Gj}$ in a multirod fuel assembly. The 0.35 constant in Eq. (3.5.37) was reduced to 0.188 to be representative of multirod experimental data.

- Many other differences come about from efforts to improve the physical picture. Instead of employing an empirically defined exponential, Eq. (3.6.31), to calculate the amount of liquid entrained, TRAC-BWR versions have employed a modified version of Ishii and Mishima's correlation, [3.6.6] where the liquid entrainment, E, is given

$$E = tgh\left[75(10^{-7}) j_G^{*2.5} D_H^{*1.25} N_{\text{Re}_L}^{0.25}\right] \quad (3.6.42)$$

where

$$j_G^* = \left[\frac{j_G}{[\sigma(\rho_L - \rho_G)g/\rho_G^2](\rho_G/\rho_L)^{0.667}}\right]^{0.25}$$

(3.6.43)

$$D_H^* = D_H\sqrt{\frac{g(\rho_L - \rho_G)}{\sigma}} \qquad N_{Re_L} = \frac{\rho_L j_L D_H}{\mu_L}$$

- Finally, some of the shortcomings arise from the simplified approach employed to represent flow pattern interfacial areas and the interfacial drag coefficient. For instance, single rather than clusters of particles are considered in the proposed mist flow formulation. Yet it is known that interactions between particles can become important. In the dispersed flow pattern, the interactions between drops tend to become significant for $\overline{\alpha} > 0.92$. Also, a substantial increase in the drag coefficient has been observed for large liquid drop concentrations or where the fluid particles are distorted.

In summary, a comprehensive and accepted set of models for the shear forces and interfacial areas does not exist for all flow regimes, and most of those available are for "idealized" flow regimes. Also, the closure laws and other supporting correlations for the interfacial areas and shear forces are based on correlations derived from adiabatic steady-state test conditions. The inherent assumption is made that they are applicable to transient and heated conditions. Despite all the preceding shortcomings, system computer codes employing two-fluid models with interfacial exchanges have performed the best of all system computer codes because they make a genuine effort to represent the physical behavior of two-phase flows.

3.6.3.2 Wall Shear Different methodologies have been proposed to calculate the wall shear. They are all based on closure laws derived from fully developed steady-state two-phase-flow correlations. For example in TRAC-BF1/MOD1, the wall shear is obtained by multiplying the single-phase wall shear by a two-phase multiplier as originally recommended by Lockhart–Martinelli [3.5.1] and Martinelli–Nelson [3.5.2]. The multiplier proposed by Hancox and Nicoll [3.6.7] is used. The remaining issue is how to apportion the wall friction between the two phases. If the interfacial shear is known, this distribution can be done accurately from the steady-state fully developed momentum, Eqs. (3.6.3) and (3.6.4), which simplify to

$$-(1 - \overline{\alpha})\frac{\partial p}{\partial z} + g\rho_L(1 - \overline{\alpha})\sin\theta - \frac{P_{WL}\tau_{WL}}{A} + \frac{P_i\tau_i}{A} = 0 \qquad (3.6.44)$$

$$-\overline{\alpha}\frac{\partial p}{\partial z} + g\rho_L\overline{\alpha}\sin\theta - \frac{P_{WG}\tau_{WG}}{A} + \frac{P_i\tau_i}{A} = 0 \qquad (3.6.45)$$

3.6 TWO-FLUID MODELS INCLUDING INTERFACE EXCHANGE

If we multiply Eq. (3.6.44) by $\bar{\alpha}$ and Eq. (3.6.45) by $(1 - \bar{\alpha})$ and subtract them, one gets

$$\frac{P_{WL}\tau_{WL}}{A}\bar{\alpha} - \frac{P_{WG}\tau_{WG}}{A}(1 - \bar{\alpha}) = \frac{P_i\tau_i}{A} + g(\rho_L - \rho_G)\bar{\alpha}(1 - \bar{\alpha})\sin\theta \quad (3.6.46)$$

One can also write

$$P_{WG} + P_{WL} = P \quad (3.6.47)$$

$$\frac{P_{T_W}}{A} = \frac{P_{WL}\tau_{WL}}{A} + \frac{P_{WG}\tau_{WG}}{A}$$

so that combining Eqs. (3.6.46) and (3.6.47) yields

$$\frac{P_{WG}\tau_{WG}}{A} = \frac{P_{T_W}}{A}\bar{\alpha} - \frac{P_i\tau_i}{A} - g(\rho_L - \rho_G)\bar{\alpha}(1 - \bar{\alpha})\sin\theta$$

$$\frac{P_{WL}\tau_{WL}}{A} = \frac{P_{T_W}(1 - \bar{\alpha})}{A} + \frac{P_i\tau_i}{A} + g(\rho_L - \rho_G)\bar{\alpha}(1 - \bar{\alpha})\sin\theta \quad (3.6.48)$$

If the total wall shear, P_{T_W}/A, the gas volume fraction, $\bar{\alpha}$, and the interface shear, $P_i\tau_i/A$, are known, the separate phasic frictional pressure losses can be calculated from Eqs. (3.6.48). It is unfortunate that this approach [i.e., Eqs. (3.6.47) and (3.6.48)] has not found increased use in computer system codes instead of arbitrarily assuming P_{WL} and P_{WG}.

Conversely, if the distribution of the wall shear between the two phases is known, the interface shear can be calculated. For certain flow regimes, this can be readily done. For example, for bubbly flow $P_{WG} = 0$ and $P_{WL} = P$, so that

$$\frac{P_i\tau_i}{A} = \frac{P_{T_W}}{A}\bar{\alpha} - g(\rho_L - \rho_G)\bar{\alpha}(1 - \bar{\alpha})\sin\theta \quad (3.6.49)$$

In fact, Eq. (3.6.49) is generally applicable to slug and annular flow since $P_{WL} \approx P$ and $P_{WG} \approx 0$ for those flow regimes. This method has been particularly successful with drift flux models, where it helps reduce the complexity of the interfacial calculations. One final simplified approach is worth mentioning. One can rewrite Eq. (3.6.47) with τ_G and τ_L representing the single-phase gas and liquid wall shear

$$\frac{\tau_W}{\tau_L} = \frac{P_{WL}}{P}\frac{\tau_{WL}}{\tau_L} + \frac{P_{WG}}{P}\frac{\tau_{WG}}{\tau_L}\frac{\tau_G}{\tau_L} \quad (3.6.50)$$

If we use the Lockhart–Martinelli nomenclature and the same 0.2 power variation for the friction factor, Eq. (3.6.50) becomes

$$\phi_{TPL}^2 = \frac{P_{WL}}{P} \frac{1}{(1-\overline{\alpha})^{1.8}} + \frac{P_{WG}}{P} \phi_{GL}^2 \frac{1}{\overline{\alpha}^{1.8}} \qquad (3.6.51)$$

The wall shear can be calculated from Eq. (3.6.51) if the wetted perimeter ratios P_{WL}/P and $P_{WG}/P = (1 - P_{WL}/P)$ are known. Alternatively, if the wetted perimeter ratios can be established from flow patterns, the wall shear can be specified. For instance, with $P_{WL}/P = 1$ and $P_{WG} = 0$, there results

$$\phi_{TPL}^2 = \frac{1}{(1-\overline{\alpha})^{1.8}} \qquad (3.6.52)$$

which is similar to Lockhart–Martinelli Eq. (3.5.21).

In summary, the wall shear prediction in system computer codes with interfacial exchanges relies upon steady-state empirical correlations which are not always consistent with the interfacial perimeter and gas volume fraction calculated values. Also, the empirical correlations are not always consistent and tend to vary from one system computer code to another.

3.6.3.3 Interfacial and Wall Heat Transfer As in the case of shear, the system computer code has to deal with both interfacial and wall heat transfer. The difficulties are compounded by having to consider several heat transfer regimes:

- Single-phase liquid or gas forced or natural convection
- Gas–liquid mixture forced or natural convection
- Nucleate boiling up to the point of critical heat flux where the heated surface stops being fully wetted
- Transition boiling where the heated surface is alternately covered by liquid and vapor
- Film boiling where the surface is blanketed with vapor and where thermal radiation from the wall can become significant
- Condensation where vapor is converted back to liquid

As shown in Figure 3.6.1, the flow patterns can vary with the prevailing heat transfer regime and only an illustration of the heat transfer methodology will be provided here for a few cases:

- Wall heat transfer to gas–liquid mixture
- Wall heat transfer to gas–liquid mixtures without mass transfer
- Interfacial heat transfer during dispersed patterns with mass transfer
- Wall nucleate, transition, and film boiling

3.6.3.3.1 Wall Heat Transfer to Gas–Liquid Mixtures The process is similar to wall shear and consists of apportioning the heated surface areas to the liquid and gas phases and applying the corresponding phasic heat transfer coefficients over those portions of the heated perimeter, that is,

$$h_{TP} P_h = h_{LTP} P_{hL} + h_{GTP} P_{hG} \tag{3.6.53}$$

or

$$\frac{h_{TP}}{h_L} = \frac{h_{LTP}}{h_L} \frac{P_{hL}}{P_h} + \frac{h_{GTP}}{h_G} \frac{h_G}{h_L} \frac{P_{hG}}{P_h} \tag{3.6.54}$$

where h_L and h_G are the liquid and gas heat transfer coefficients for single-phase flow. For bubble flow without mass transfer, we can set $P_{hL}/P_H = 1.0$ and $P_{hG}/P_H = 0$, so that for turbulent flow

$$\frac{h_{TP}}{h_L} = \frac{h_{LTP}}{h_L} = (1 - \overline{\alpha})^{-0.8} \tag{3.6.55}$$

In RELAP5/MOD-3, the turbulent heat transfer coefficient is taken according to Collier [3.6.8] as

$$h_{TP} = h_{L0}(1 - \overline{\alpha})^{-0.8} \tag{3.6.56}$$

where h_{L0} is the heat transfer coefficient calculated for the total mass flux with liquid properties.

A similar expression is utilized for slug flow, while for annular flow a forced convection correlation from Collier [3.6.8] is employed, or

$$h_{TP} = 0.029 \frac{k_L}{D_H} N_{Re}^{0.87} N_{Pr_L}^{0.33} \tag{3.6.57}$$

where k_L is the liquid thermal conductivity and N_{Pr_L} the liquid Prandtl number. The Reynolds number in Eq. (3.6.57) is based on the total mass flux and the homogeneous viscosity of Eq. (3.4.23). A gaseous heat transfer coefficient is also recognized when $\overline{\alpha} > 0.99$ and it is set equal to the single-phase gas heat transfer coefficient multiplied by $100(\overline{\alpha} - 0.99)$.

This very brief example shows that some liberty is taken with the physics or the fluid properties in specifying the applicable wall heat transfer coefficients. Another assumption implied in the preceding equation is that the liquid and gas phases are well mixed and relatively at the same temperature.

3.6.3.3.2 Interfacial Heat Transfer to Gas Bubbles and Liquid Drops In all such interfacial cases, the heat transfer area is calculated as for interfacial shear by assuming spherical geometries and by determining the bubble or liquid drop diameter from a critical Weber number (see Section 3.6.3.1). The heat transfer process to gas bubbles and liquid drops is usually controlled by the interface, which has the most difficulty delivering heat to the opposite interface. For simplification purposes, the resistance to heat transfer at the opposite interface is often arbitrarily presumed to be negligible. This is illustrated below for three important cases involving mass transfer.

Let us first consider interfacial heat transfer from a superheated liquid to vapor bubbles of radius R. If the number of bubbles per unit volume is N, the gas volume fraction is $\bar{\alpha} = 4\pi R^3 N/3$ and the corresponding interfacial heat transfer area, P_{hi}/A, is

$$\frac{P_{hi}}{A} = 4\pi R^2 N = \frac{3\bar{\alpha}}{R} = 3\bar{\alpha}^{2/3}(4\pi N)^{1/3} \qquad (3.6.58)$$

The corresponding volumetric vapor generation rate, in terms of time t, is

$$\Gamma = N\rho_G \cdot 4\pi R^2 \frac{dR}{dt} \qquad (3.6.59)$$

The rate of bubble growth is controlled by conduction from the superheated liquid and it has been determined by Plesset and Zwick [3.6.9] to be analytically

$$\frac{dR}{dt} = \frac{\rho_L C_{PL} \Delta T_{sup}}{\rho_G H_{LG}} \sqrt{\frac{3k_L}{\pi t \rho_L C_{PL}}} \qquad (3.6.60)$$

where ΔT_{sup} is the superheat of the liquid, H_{LG} the heat of vaporization, and C_{PL} the specific heat of the liquid.

According to Berne [3.6.10], the vapor continuity equation can be approximated by

$$\rho_G \frac{D\bar{\alpha}}{Dt} = \Gamma$$

where D/Dt is the total derivative with respect to time, so that

$$\Gamma = \frac{18}{\sqrt{\pi}} \left(\frac{4\pi}{3}\right)^{2/3} \bar{\alpha}^{1/3} N^{2/3} \rho_G \frac{k_L}{\rho_L C_{PL}} \left(\frac{\rho_L C_{PL} \Delta T_{sup}}{\rho_G H_{LG}}\right)^2 = \frac{Nq''_{iL} \cdot 4\pi R^2}{H_{LG}} \qquad (3.6.61)$$

where q''_{iL} is the heat flux from the water to the interface. The interfacial heat transfer rate is then

$$\frac{P_{hi}}{A} q''_{iL} = \frac{3\bar{\alpha}}{R} \frac{6}{\pi R} \frac{k_L \rho_L C_{PL}}{\rho_G H_{LG}} \Delta T^2_{sup} \qquad (3.6.62)$$

Equation (3.6.62) is similar to an expression found in RELAP5/MOD3, except for replacing $\bar{\alpha}$ by a value of $\frac{1}{6}$. In this particular case, the vapor is at the saturation temperature and it offers no resistance to the water heat transfer.

Let us next consider the case of vapor bubbles in a subcooled liquid. The process involves condensation at the vapor–liquid interface. Empirical correlations for that mode of condensation have been developed and they include the Lahey and Moody [3.6.11] expression used in BWR-TRAC.

$$\frac{P_{hi}}{A} q''_{iL} = \frac{H_0 H_{LG}}{1/\rho_G - 1/\rho_L} \bar{\alpha} \, \Delta T_{sub} \tag{3.6.63}$$

with $H_0 = 150$ (hr $-°$F)$^{-1} = 0.075$ (s $\cdot$ °C)$^{-1}$, or the Ünal [3.6.12] correlation used in RELAP for bubbles of radius R:

$$h_{iL} = \frac{C \phi H_{LG} R}{1/\rho_G - 1/\rho_L} \tag{3.6.64}$$

with

$$C = \begin{cases} 0.25(10^{10}) p^{1.418} & \text{for } 1 < p \le 17.7 \text{ MPa} \\ 65 - 5.69(10^{-5})(p - 10^5) & \text{for } 0.1 \le p \le 1 \text{ MPa} \end{cases}$$

and

$$\phi = \begin{cases} \left(\dfrac{\bar{u}_L}{0.61}\right)^{0.47} & \text{for } \bar{u}_L > 0.61 \text{ m/s} \\ 1 & \text{for } \bar{u}_L \le 0.61 \text{ m/s} \end{cases}$$

For the vapor side, very large heat transfer coefficients are presumed to exist and they range from 1000 to 10,000 W/m^2 $\cdot$ K, depending on whether the vapor is superheated or saturated.

Let us, finally, consider the case of liquid droplets in superheated vapor. The process is controlled by gaseous convection to the liquid, which is specified from the correlation of Lee and Ryley [3.6.13]:

$$\frac{h_{GTP} d_d}{k_G} = 2 + 0.74 \left[\frac{d_d (\bar{u}_G - \bar{u}_L) \rho_G}{\mu_G}\right]^{0.5} N_{Pr_G}^{1/3} \tag{3.6.65}$$

where d_d is the droplet diameter. In this case the heat transfer on the liquid side is by conduction augmented by recirculation within the drop. In most cases, the heat transfer within the drops is made large enough to be negligible with respect to the gaseous heat transfer.

As in the case of interfacial shear, the three preceding examples of interfacial heat transfer are seen to employ idealized geometries, empirical correlations, and not to take into account the interactions between bubbles or liquid drops. However,

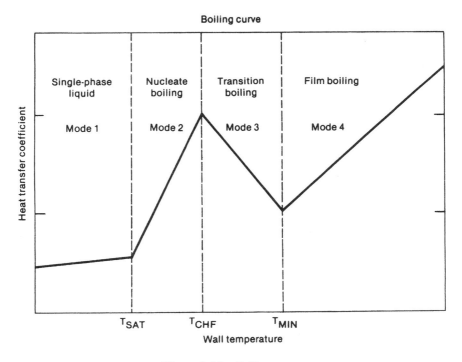

Figure 3.6.3 Boiling curve.

they have the important advantage of being able to recognize that the gas and liquid phases are not at the same temperature.

3.6.3.3.3 Wall Nucleate, Transition, and Film Boiling This set of heat transfer mechanisms was selected for a brief discussion here because there is no comparable pressure drop or shear behavior. Experimental studies have shown that heat transfer from a solid wall to a single-component fluid varies as the temperature difference between the wall and the bulk of the fluid passes through certain limits. The rate of heat transfer or heat flux varies dramatically with the temperature difference between the wall and fluid. These variations are illustrated in Figure 3.6.3, which is known as the *boiling curve.*

At the left most part of the curve there is single-phase natural or forced convection to the liquid and it prevails until the wall temperature T_w exceeds the saturation temperature T_{sat}. As the wall temperature rises enough above saturation, boiling occurs. If the liquid is subcooled, subcooled/nucleate boiling takes place, with bubbles being formed at the wall and being condensed at the bubble–subcooled liquid interface. If the bulk liquid is saturated, saturated nucleate boiling will exist and increased vaporization will take place until the wall temperature T_{CHF} corresponding to critical heat flux (CHF) is reached. At the critical heat flux condition, the wall is no longer fully wetted and its temperature rises sharply because of the

TABLE 3.6.1 Logic for Selection of Wall Heat Transfer

Fluid Condition[a]	Wall Conditions			
	No Boiling Transition		Boiling Transition	
Flow regime	$T_w < T_{sat}$	$T_{sat} < T_w < T_{CHF}$	$T_{CHF} < T_w < T_{min}$	$T_{min} < T_w$
$\bar{\alpha} = 0$	Liquid convection	Liquid convection	N/A	N/A
$0 < \bar{\alpha} < \bar{\alpha}_{tran}$	Liquid convection	Subcooled/nucleate boiling	Transition boiling	Film boiling
$\bar{\alpha}_{tran} < \bar{\alpha} < 1$	Condensation	Forced convection vaporization	Transition boiling	Film boiling
$\bar{\alpha} = 1$	N/A	Vapor convection	Vapor convection	Vapor convection

[a] $\bar{\alpha}_{tran}$ corresponds to transition to annular flow; CHF, critical heat flux.

ensuing heat transfer degradation. Beyond the point of CHF, transition boiling occurs. (In a constant or heat-increasing flux environment, this regime would not be encountered.) The degree of surface being wetted during transition boiling continues to decrease as well as the heat transfer rate. The transition boiling regime is complete when the minimum stable film temperature T_{min} is reached. The entire boiling surface is blanketed with a vapor film and the film boiling regime prevails beyond T_{min}. The film boiling wall temperature increases as the heat flux is raised and radiation effects become significant. All the preceding modes of heat transfer are listed in Table 3.6.1 except for the need to recognize the benefits of axial heat conduction during quenching or reflood of a surface in film or transition boiling.

Each of the heat transfer modes shown in Table 3.6.1 requires a correlation or closure law. Selection of the appropriate correlation depends on the values of the wall temperature T_w and of the void fraction $\bar{\alpha}$. In Table 3.6.1, $\bar{\alpha}_{tran}$ corresponds to transition to annular flow. The applicable heat transfer correlations usually have all been determined experimentally, and they generally vary with the geometry and whether the flow is in natural or forced circulation. The correlations become less precise as the geometry differs from that of the tests and as the flow goes from forced to natural convection. Prototypic tests have been used often (e.g., for quenching or critical heat flux to reduce the prevailing uncertainties). Still, the remaining number of involved phenomena and closure laws is rather large and so are the uncertainties in their representations and predictions in complex system computer codes.

3.7 OTHER ASPECTS OF SYSTEM COMPUTER CODES

3.7.1 Numerical Solution of System Computer Codes

Once the differential equations and closure laws have been formulated for a two-phase-flow complex system, they need to be solved numerically on a computer. The numerical technique employed needs to be accurate or the predictions of the computer program will depend on both the mathematical model and the numerical solution procedure. Under such circumstances, comparisons of the system computer code are of little value because the agreement (or disagreement) could be due to inadequate representation of the complex physical system or to inaccurate numerical techniques, or a combination of both. It is therefore important to assure or to be able to assess the accuracy of the numerical procedure. Considerable efforts have been made toward that objective. However, they fall beyond the scope of this book and only a few summary comments are offered here.

1. The accepted numerical solution scheme is based on replacing the differential equations with a system of finite difference equations. The derivatives in the differential equations are replaced by spatial or temporal differences, which turn the solution of the differential equations problem into an algebraic one. Depending on the spatial and temporal nodalization, the numerical solutions may not be stable and not converge to the solution of the partial differential equations. This possibility is easy to visualize in the case of a pressure wave propagating downstream. The time step employed in the numerical solution must be limited in size so that the pressure wave cannot propagate beyond one spatial mesh in one time step. This restriction can be illustrated for the simplified differential equation

$$\frac{\partial \rho}{\partial t} + \frac{\partial \rho u}{\partial z} = 0 \tag{3.7.1}$$

which for a constant velocity $u > 0$ can be approximated by the finite difference approximation

$$\frac{\rho_j^{n+1} - \rho_j^n}{\Delta t} + u \frac{\rho_j^n - \rho_{j-1}^n}{\Delta z} = 0 \tag{3.7.2}$$

The stability condition for this finite difference equation is

$$\frac{u \, \Delta t}{\Delta z} \leq 1.0 \tag{3.7.3}$$

where the quantity $u \, \Delta t / \Delta z$ is often called the *Courant number*. In the case where u represents the wave propagation velocity or the speed of sound, the time step must be very small and the numerical solutions can generate truncation errors and take a significant amount of computer time. Similar stability limitations can be derived for other derivative terms in the conservation laws of two-phase flow, but they generally tend to be less restrictive than the Courant number.

2. A truncation error in time and space is produced by the typical finite difference scheme of Eq. (3.7.2). By using Taylor series expansion, Roache [3.7.1] showed that the partial differential equation being solved by Eq. (3.7.2) is closer to

$$\frac{\partial \rho}{\partial t} + u \frac{\partial \rho}{\partial z} = \frac{u \, \Delta z}{2} \left(1 - \frac{u \, \Delta t}{\Delta z} \right) \frac{\partial^2 \rho}{\partial z^2} \tag{3.7.4}$$

than Eq. (3.7.1). The last term in Eq. (3.7.4) can be viewed as an artificial numerical

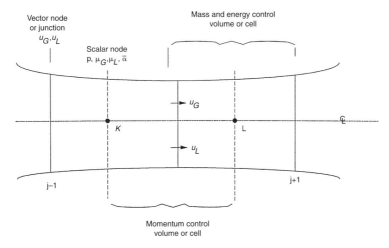

Figure 3.7.1. Difference equation nodalization schematic. (From Ref. 3.5.14.)

diffusion term. This numerical diffusion can have a substantial effect on the solutions obtained from finite difference schemes. For example, it plays an important role for thermal stratification or liquid-level evaluations where the finite difference scheme tends to smear the prevailing sharp interface.

3. The finite difference method of Eq. (3.7.2) is referred to as an *explicit* scheme because it permits the calculation of ρ_j at the new time step $n + 1$ from the values of the flow variables at the old time step n. Such explicit schemes as noted previously are subject to the Courant time step restriction for stability reasons. To avoid stability restrictions, the differential equations can be written in terms of the new time step. For example, Eq. (3.7.2) could be replaced by

$$\frac{\rho_j^{n+1} - \rho_j^n}{\Delta t} + u \frac{\rho_j^{n+1} - \rho_{j-1}^{n+1}}{\Delta z} = 0 \qquad (3.7.5)$$

in which ρ_j^{n+1} cannot be solved because of the involvement of ρ_{j-1}^{n+1}. There results an implicit finite difference scheme which produces a time-advancing matrix to be solved by inversion. Most of the two-phase system computer codes avoid fully explicit formulations. They range from being fully implicit in the case of the French code, CATHARE, to being semi-implicit in the case of the TRAC or RELAP codes. The semi-implicit scheme evaluates only some of the terms in the differential equations implicitly in time, and the terms evaluated implicitly are defined as the scheme is developed.

In addition to selective implicit evaluation of spatial gradients at the new time, the difference equations are based on the concept of a staggered spatial control volume (or mesh cell) in which the scalar properties (pressure, energies, and gas volume fraction) of the flow are defined at cell centers, and vector quantities (velocities) are defined at the cell boundaries. Figure 3.7.1 illustrates the resulting one-dimensional spatial noding. The term *cell* means an increment in the spatial variable z, corresponding to the mass and energy control volume. In Figure 3.7.1 the mass and energy equations are integrated with respect to the spatial variable z from the junction at z_j to z_j+1. The momentum equations are integrated with respect to the spatial variables from one cell center to the adjoining cell center or from z_K to z_L in Figure 3.7.1. This approach is employed to avoid the instability associated with a fully centered scheme.

Several other means (such as integrating algorithms) are employed to improve the stability of the numerical scheme for mesh sizes of practical interest. Furthermore, Mahaffy [3.7.2] suggested a two-step approach which includes a stabilizing step prior to the semi-implicit step. Under any circumstances, the well posedness of the final numerical scheme as well as its accuracy must be demonstrated by extensive numerical testing during development.

4. Numerical instabilities can also be produced by discontinuities in the formulation of the closure laws or flow pattern maps with the numerical solutions oscillating back and forth across the discontinuities. This has led to modifications of the closure laws or to transition zones in the flow patterns being provided at the discontinuities. Such changes are, in fact, departures from the physical situation, and they need to be assessed to ascertain that their physical impact is minimal.

5. It is essential to assess the accuracy of the utilized numerical scheme. This can be done by comparing the numerical results to available exact solutions or to very accurate benchmark results obtained by other solutions, such as the method of characteristics. While considerable progress has been made in recent years in the numerical solutions of system computer codes, it should not detract from validating their accuracy.

3.7.2 Nodalization

There are other reasons beyond good numerical performance for selecting the cell size or the nodalization for the prototype two-phase-flow complex system. First, the prototype computer model must be nodalized enough to capture the important design characteristics of the system and relevant subsystems and important components. Second, the nodalization must be able to represent the phenomena judged to be significant by the ranking and scaling processes discussed in Chapters 1 and 2. Third, the predicted system performance results cannot be overly sensitive to the nodalization. As a general guide, there is merit to repeat a few of the prototype system calculations with one half and double the chosen nodalization to assure that they yield comparable results. Fourth, and not least, it is important that the prototype computer calculations be performed in a time- and cost-effective manner. Put all together, the preceding requirements mean that experience is key to determining the final nodalization selection of the prototype complex system computer code. Furthermore, if all the requirements above are satisfied, there is no reason why the same nodalization cannot be employed for the separate effects, component, and integral tests utilized to validate the system computer code. This approach eliminates having to assess the impact of nodalization if it is changed from test to test or from test to prototype.

It is recognized that in a few cases, a smaller mesh size may be necessary to better understand the prevailing phenomena in some tests. However, nodalization cannot and should not be used to get agreement with the test data. Also, the predictions for all such necessary cases should be repeated with the nodalization selected for the prototype system to decide whether the finer test mesh is necessary in the prototype model or to assign an uncertainty factor to keeping the prototypic cell size in those few cases. It is recognized that some have suggested nodalization sensitivity studies to quantify nodalization as an independent contributor to computer systems code uncertainty. However, such a quantification can be expected to be highly user dependent and very costly when it is applied to the multitude of scenarios, closure laws, and tests to be considered. As discussed in Reference 1.1.4, the proposed path selects the prototype system nodalization for reasons of experience, accuracy, and efficiency and applies the same nodalization to validate the system computer code from the selected assessment matrix of tests. This approach eliminates the freedom of manipulating nodalization and it reduces the need to evaluate its contribution to system computer code uncertainty.

3.7.3 Subchannel Representation

In many sections of the prototype two-phase-flow system, the exclusive use of one-dimensional flow is not adequate. The flow cross section needs to be subdivided into subchannels and to take into account interchanges or mixing, and differences in pressure and heat inputs among the subchannels. Several system computer codes have been developed to deal with subchannels and they include COBRA-IV [3.7.3], ASSERT-4 [3.7.4], and VIPRE [3.7.5]. The COBRA computer code was developed first and it relies on the approach utilized for mixing in single-phase turbulent flow. It is based on a one-dimensional separated two-phase-flow model with different gas and liquid axial velocities between adjacent subchannels. VIPRE is a subsequent improved version of COBRA. ASSERT-4 is based on a drift flux model. It employs equal volume exchange between subchannels containing gas–liquid mixtures of different average densities.

The primary purpose of this section is not to provide a detailed discussion of subchannel codes but rather to highlight their differences and to assess their limitations and uncertainties. As noted before, the flow cross section is subdivided into discrete subchannels and one-dimensional time- and flow-averaged conditions are used in each subchannel. No interface exchanges are considered within the subchannels. In single-phase flow, the fluctuating equal mass flux exchange between subchannels i and j is represented by G'_{ij} and the momentum flux from one subchannel to the other is given by

$$G'_{ij}(\overline{u}_j - \overline{u}_i) = \overline{\rho}\epsilon \left(\frac{d\overline{u}}{dy}\right)_{ij} = \overline{\rho}\epsilon \left(\frac{\overline{u}_j - \overline{u}_i}{l}\right) \quad (3.7.6)$$

where $\overline{\rho}$ is the average density of adjacent channels, ϵ the eddy diffusivity, y the distance measured in the transverse direction, and l the effective mixing length between the subchannels. If the gap clearance between subchannels is s, the fluctuating mass cross flow per unit length of interconnection is W'_{ij}, which is equal to

$$W'_{ij} = sG'_{ij} = \frac{\overline{\rho s \epsilon}}{l} = \frac{\overline{\rho \epsilon}}{Gl} s\overline{G} = \beta s \overline{G} \qquad (3.7.7)$$

where β or $\overline{\rho \epsilon}/Gl$ is referred to as the mixing coefficient and $\overline{G}$ is the average mass flow rate per unit area or $\overline{G} = \frac{1}{2}(G_i + G_j)$.

The single-phase mixing coefficient β has been shown to have the form [3.7.6]

$$\beta = K \frac{D_H}{s} N_{\text{Re}}^{-0.1} \qquad (3.7.8)$$

where K is related to flow geometry and is obtained from experimental data. A comparable approach is employed by COBRA-IV in two-phase flow. It consists of adding to the mass conservation equation (3.2.6) the term $\sum_{j=1}^{n} W_{ij}$ to account for the mass cross-flow from all the potential adjoining channels. Similarly, the axial momentum conservation is identical to Eq. (3.2.14) except for recognizing the following additional momentum fluxes across the interface of the subchannels:

$\sum_{j=1}^{n} W'_{ij}(\overline{u}_i - \overline{u}_j)$ corresponding to the fluctuating momentum component

$\sum_{j=1}^{n} W_{ij}\overline{u}^*$ representing the donor cell (shown by *) momentum transfer

For the conservation of transverse momentum, one gets

$$\frac{\partial}{\partial t} W_{ij} + \frac{\partial}{\partial z} W_{ij}\overline{u}^* = \frac{s}{l}(p_i - p_j) - \frac{s}{l} F_{ij} \qquad (3.7.9)$$

where a pressure difference between cells is recognized and F_{ij} is the transverse friction and form pressure loss.

To resolve these equations, several new closure laws are required:

- The mixing coefficient β must be specified to calculate W'_{ij}.
- The mixing length l, which is taken equal to the distance between subchannel centroids or given by a suitable mixing-length correlation.
- The friction and form transverse pressure loss relation, which is written in terms of W_{ij}^2.

The biggest uncertainty is associated with the mixing coefficient β, which is generally determined empirically from tests. For example, for a channel with a 3×3 square array of rods of 10 mm in diameter, 13.3-mm rod pitch, and 1-m heated length, Wang and Cao [3.7.7] found that for single-phase flow at 14.7 MPa,

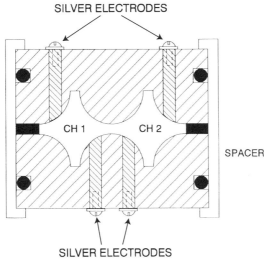

Figure 3.7.2. Test section of Tapucu et al. (Reprinted from Ref. 3.7.8 with permission from Elsevier Science.)

Geometric parameters of test section

Rod radius	8.8 ± 0.1 mm
Gap clearance	1.7 ± 0.05 mm
Cross-sectional area	
Subchannel 1	116.6 ± 2 mm²
Subchannel 2	116.6 ± 2 mm²
Hydraulic diameters	
Subchannel 1	7.6 ± 0.2 mm
Subchannel 2	7.6 ± 0.2 mm
Centroid-to-centroid distance	18.7 ± 0.1 mm
Interconnection length	1321 ± 5 mm

$$\beta = 0.0056 \frac{D_H}{s} N_{\text{Re}}^{-0.1} \quad (3.7.10)$$

For subcooled boiling conditions, COBRA-IV was used to fit the data and to obtain

$$\beta = 0.015 \frac{D_H}{s} N_{\text{Re}}^{-0.1} \quad (3.7.11)$$

As pointed out by Wang and Cao, subcooled boiling increases the mixing between subchannels, but only a few subcooled tests exist and their results are far from being consistent, due to differences in geometry and operating conditions. Similarly, Tapucu et al. [3.7.8] measured mixing between two subchannels of two-phase flow in a multirod configuration as shown in Figure 3.7.2 and again they determined

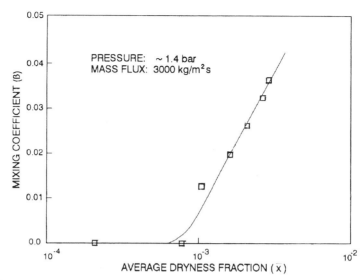

Figure 3.7.3. COBRA mixing coefficient as a function of the dryness fraction. (Reprinted from Ref. 3.7.8 with permission from Elsevier Science.)

the COBRA mixing coefficient by adjusting it to obtain the best prediction of the experimentally determined void fractions. The relationship they found between the mixing coefficient and the average gas weight fraction is reproduced in Figure 3.7.3, which again shows a strong dependence of the mixing coefficient upon the gas content.

In the drift flux model of ASSERT-4, an equal volume exchange occurs between subchannels that contain mixtures of different void fractions and the transverse fluctuating velocities are assumed to be equal for the liquid and gas phases. The net fluctuating transverse flow W''_{ij} is written as

$$W''_{ij} = (W''_{ij})_L + (W''_{ij})_G \tag{3.7.12}$$

with

$$(W''_{ij})_L = \frac{\rho_L s \epsilon}{l}(\overline{\alpha}_j - \overline{\alpha}_i) \quad \text{and} \quad (W''_{ij})_G = \frac{\rho_G s \epsilon}{l}(\overline{\alpha}_i - \overline{\alpha}_j) \tag{3.7.13}$$

According to Eqs. (3.7.13), the gas and liquid flow in opposite direction and their volume exchange $(W''_{ij})_L/\rho_L$ and $(W''_{ij})_G/\rho_G$ is the same. According to Tapucu et al., this trend was observed in their tests, with the liquid flow decreasing in the channel of low gas volume content while its gas volume fraction increased. This is shown in Figure 3.7.4.

Here again, the transverse friction and pressure losses, as well as a relation for the mixing coefficient β, were adjusted to obtain a satisfactory comparison with

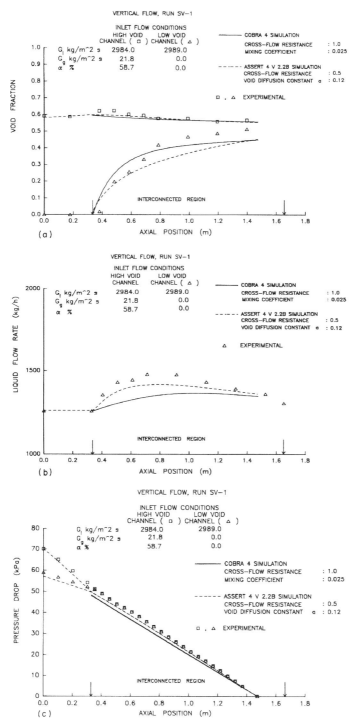

Figure 3.7.4. (*a*) Comparison of ASSERT-4 and COBRA-IV predictions: void fraction; (*b*) comparison of ASSERT-4 and COBRA-IV predictions: mass flow; (*c*) comparison of ASSERT-4 and COBRA-IV predictions: pressure. (Reprinted from Ref. 3.7.8 with permission of Elsevier Science.)

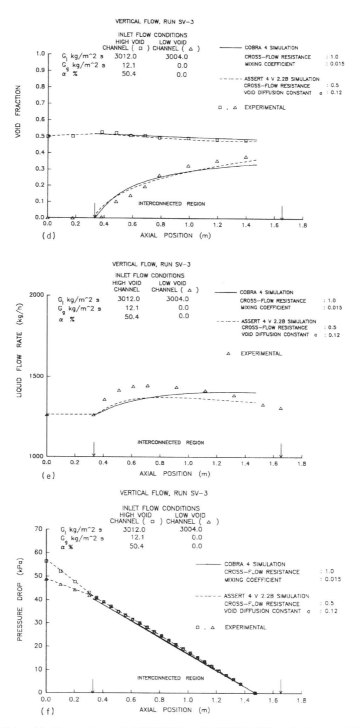

Figure 3.7.4. (*d*) Comparison of ASSERT-4 and COBRA-IV predictions: void fraction; (*e*) comparison of ASSERT-4 and COBRA-IV predictions: mass flow; (*f*) comparison of ASSERT-4 and COBRA-IV predictions: pressure. (Reprinted from Ref. 3.7.8 with permission of Elsevier Science.)

the test data. In addition, there was a need to recognize in the drift flux model a *void drift*, the tendency of the gas to shift to the higher-velocity channels. This void drift was included by replacing $\overline{\alpha}_i - \overline{\alpha}_j$ in Eq. (3.7.13) by

$$(\overline{\alpha}_i - \overline{\alpha}_j) - (\overline{\alpha}_i - \overline{\alpha}_j)_{EQ} \text{ with}$$

$$(\overline{\alpha}_i - \overline{\alpha}_j)_{EQ} = \frac{\overline{\alpha}_i A_i + \overline{\alpha}_j A_j}{A_i + A_j} \frac{A_i + A_j}{G_l A_i + G_j A_j} (G_i - G_j) \quad (3.7.14)$$

In summary, subchannel codes are available to predict diversion cross flow due to pressure differences between subchannels, turbulent mixing between them, and even void drift. However, they require a number of assumptions and are dependent on empirical adjustments to match the data. This is illustrated in Figure 3.7.4, comparing COBRA-IV and ASSERT-4 to Tapucu et al., experimental results. It is observed that the COBRA mixing coefficient changes from Figure 3.7.4*a–c* to Figure 3.7.4*d–f* and that the cross-flow resistance in the ASSERT-4 analysis has twice the value of that used in COBRA. One can imply from that comparison that our state of knowledge for subchannel conditions is not as good as for one-dimensional two-phase flow and needs considerably more verification in terms of channel geometry, pressure, and two-phase-flow conditions.

3.7.4 Validation of System Computer Codes

System computer codes require many assumptions, approximations, and a very large number of closure laws. System computer codes may even be missing physics in a few areas. For all those reasons, it is essential that they be tested against a wide and complete set of separate effects, components, and integral tests. In fact, *a system computer code is only as good as the validation and verification to which it has been subjected.* The degree of checking performed must be documented, scrutable, and kept up to date. Also, because system computer codes evolve and change over time, the specific version of the code used in the verification process must be identified and the need to repeat previous comparisons to tests dealt with.

The validation of a system computer code involves several steps. These are illustrated using TRAC-PF1/MOD as an example.

- A code manual and programmer's guide is needed. For TRAC-PF1/MOD1, it consists of NUREG/CR-3858, An Advanced Best Estimate Computer Program for Pressurized Water Reactors Thermal-Hydraulic Analysis.
- A user's guide is necessary, as provided by NUREG/CR-4442.
- The closure laws and their range of applicability must be described as done in NUREG/CR-4278, TRAC-PF1/MOD1 Models and Correlations.
- The code assessments must be documented as in NUREG/CR-4278, TRAC-PF1/MOD1 Development Assessment. An additional 38 assessment reports are listed in NUREG/CR-5249, Appendix F.

This process involves a very large amount of work and it takes a significant length of time. Completeness is essential to the success and validity of system computer codes The following are a few practices to enhance those chances:

- The number of scenarios, phenomena, and conditions to be covered by system computer codes has increased continuously over time. This has led to additions and changes to the original codes, and even though the costs may be high, the new code versions must be subjected to strong validation and verification efforts as well as rechecking the impact on original version elements of the codes.
- The usual structure of system computer codes is such that it does not lead to an easy evaluation of a small change or a new addition without running the entire system computer code. It would be desirable to have code structures permit sensitivity assessments of the merits and benefits of such changes.
- Sometimes, flexibility is provided in system computer codes by including alternative closure laws and leaving the choice to the users without evaluating the inconsistency they might generate when coupled to other closure laws. Comparative evaluation of such alternatives would be very helpful.
- A common misconception is that the use of more complicated closure laws, more detailed flow pattern maps, or multidimensional or finer nodalization can substitute for the lack of information at the interfaces or as a result of the averaging process. In fact, it is not clear that such strategies do not decrease the accuracy of the predictions. For that reason, the cost/benefit of their implementation deserves a closer look.
- System computer codes have been referred to as empirically based tools with hundreds of adjustable constants and correlations to fit the test data. This fitting process has raised doubts about the ability to extrapolate their predictions beyond available testing. In some cases, large-scale and expensive three-dimensional two-phase-flow studies had to be carried out to reduce those elements of doubt [3.7.9]. Alternative, less costly scaling strategies to check the system computer codes deserve consideration.
- In the author's view, system computer codes are the only tools available to predict the performance of complex systems. Their empirical nature comes about not by choice but rather from the lack of understanding of two-phase flow or from missing information about the geometry of the interfaces and the interfacial processes of mass, momentum, and energy transfer. System computer codes are a remarkable development and achievement, but their empirical nature should be a constant reminder of their shortcomings and of the need to have an intensive validation and verification program. Because no major scientific or fundamental improvement can be forecasted for several years, the criticism of system computer codes may be justifiable, but that energy is better directed to finding ways to improve them and build up confidence in their application to predict the performance of two-phase-flow complex systems.

3.8 SUMMARY

1. Considerable progress has been made in developing system computer codes. Their numerical methods are quite robust and their latest models are much more based on a physical basis. However, they continue to rely on a multitude of empirical information and correlations obtained under steady-state conditions. This is due to inadequate descriptions of the gas–liquid interfacial geometries and their motion and deformation with time and location. To overcome these shortcomings in knowledge, system computer codes employ (a) time and flow cross-section averaging; (b) closure laws for the mass, momentum, and energy transfer at the interfaces to replace the missing interfacial information; (c) empirically based wall process models; and (d) flow regime maps because they affect the closure laws. Until understanding of the topology of interfaces improves considerably (and it may take many years), the present formulation of system computer codes may be the only and best available means of predicting the performance of complex two-phase-flow systems.

2. A large variety of two-phase-flow system computer codes are available. They include:

- Uniform homogeneous models that presume equal gas and liquid velocity and require expressions for the homogeneous viscosity and thermal conductivity but can employ all single-phase closure laws. These models offer an important default solution to complex two-phase-flow problems and they have been especially useful in top-down scaling assessments.
- Separated two-fluid models with no interface exchanges. They can take into account unequal gas and liquid velocity, and they can provide adequate solutions to two-phase-flow complex systems under normal and less severe transient conditions. Separated models that employ the liquid and gas properties are preferable.
- Drift flux and separated two-fluid models incorporating interface exchanges that require flow regime maps and closure laws to define interfacial areas and transfers of mass, momentum, and energy. They provide the most accurate predictions of two-phase-flow complex systems because they best represent the prevailing physics.

3. The number of closure laws and empirical information grows nearly exponentially as one proceeds from homogeneous models, to separate models with no interface exchanges, to two-fluid models with interfacial transfers. The number of approximations and inconsistencies in the respective system computer codes grows correspondingly, and the associated validation and verification programs need to be much more detailed and intensive.

4. Subchannel system computer codes have been developed to deal with multidimensional maldistributions. However, they still rely on time and space averaging and require additional empirical information and closure laws. Here again, they cannot compensate for the basic missing interfacial information.

5. In the author's opinion:

- System computer codes may be at the point of diminishing returns. In fact, there may be gains to be made by simplifying them and eliminating the prevailing inconsistencies rather than making their flow pattern maps and closure laws more precise or more complex.
- If new versions of system computer codes are to be developed, a joint team of computer code developers and two-phase-flow technical experts should be put together to carry it out. This will lead to a better balance between two-phase-flow physics and their computer simulation. The new code versions should also employ a structure that makes it easy to perform sensitivity studies on potential changes and new significant phenomena.
- The primary concern with system computer codes is to ascertain that they have not missed a significant phenomenon. Once phenomena are identified, they tend to be adequately described in the computer code.
- The methodologies of Chapters 1 and 2 are best applied before tests and system computer codes are implemented rather than the other way. Considerable savings in costs and schedules would result.
- The next major improvement in system computer codes will come from advances in the topology of the gas–liquid interfaces and the interfacial transfers of mass, momentum, and energy. They will require basic experimental studies with new and sophisticated instruments and new physical basic model development. Their use in practical applications is many years away, but this should not stop such studies from being carried out.

REFERENCES

3.2.1 Meyer, J. E., "Conservation Laws in One-Dimensional Hydrodynamics," *Bettis Tech. Rev.*, **20**, 61–72, 1960.

3.2.2 Dukler, A. E., Wicks, M., III, and Cleveland, R. G., "Frictional Pressure Drop in Two-Phase Flow, B: An Approach through Similarity Analysis," *A.I.Ch.E. J.*, **10**(1), 38–43, 1964.

3.4.1 Nikuradse, J., *Gesetzmässigkeiten der turbulenten Strömung in glätten Rohren*, VDI Forschungsheft, Vol. 356, 1932; Vol. 361, 1933.

3.4.2 Reichardt, M., "Complete Representation of Turbulent Velocity Distribution in Smooth Pipe," *Z. Angew. Math. Mech.*, **31**, 1951.

3.4.3 Moody, L. F., "Friction Factors for Pipe Flow," *Trans. ASME*, **66**, 671–684, 1944.

3.4.4 McAdams, W. H., *Heat Transmission*, 3rd ed., McGraw-Hill, New York, 1954.

3.4.5 Martinelli, R. C., "Heat Transfer to Molten Metals," *Trans. ASME*, **69**, 947–959, 1947.

3.4.6 McAdams, W. M., Woods, W. K., and Heroman, L. C., "Vaporization inside Horizontal Tubes, II: Benzene–Oil Mixtures," *Trans. ASME*, **64**, 1942.

3.4.7 Bankoff, S. G., "A Variable Density Single Fluid Model for Two-Phase Flow with a Particular Reference to Steam–Water Flows," *Trans. ASME*, Ser. C, **82**, 265, 1960.

3.4.8 Jaspari, G. P., Lombardi, C., and Peterlongo, G., *Pressure Drop in Steam–Water Mixtures*, Cèntro Informazióni Stùdi Esperienze, Report R-83, 1964.

3.4.9   Einstein, A., "New Determination of Molecular Dimension," *Ann. Phys.* **19**, 289, 1906.

3.4.10 Beattie, D. R. H., and Whalley, P. B., "A Simple Two-Phase Frictional Pressure Drop Calculation Method," *Int. J. Multiphase Flow*, **8**, 83–87, 1981.

3.4.11 Roscoe, R., *Br. Appl. Phys.*, **3**, 267, 1952.

3.4.12 Jakob, M., *Heat Transfer*, Wiley, New York, 1949, p. 86.

3.4.13 Bruggeman, D. A. G., "Berechnung Verscheideiner physikalischer Konstanten von heterogenen Substanzan," *Ann. Phys. Leipzig*, **24**, 636, 1935.

3.4.14 Merilo, M., Dechene, R. L., and Cichowlas, W. M., "Void Fraction Measurement with a Rotating Electric Field Conductance Gauge," *Trans.* ASME, **99**, 330, 1977.

3.5.1   Lockhart, R. W., and Martinelli, R. C., "Proposed Correlation of Data for Isothermal Two-Phase Two-Component Flow in Pipes," *Chem. Eng. Prog.*, **45**, 34–48, 1949.

3.5.2   Martinelli, R. C. and Nelson, D. B., "Prediction of Pressure Drop during Forced Circulation Boiling of Water," *Trans. ASME*, **70**, 695–702, 1948.

3.5.3   Levy, S., "Prediction of Two-Phase Pressure Drop and Density Distribution from Mixing Length Theory," *Trans. ASME*, Ser. C, **82**, 137, 1963.

3.5.4   Isbin, H. S., Moen, R. H., Wickey, R. O., Mosher, D. R., and Larson, H. C., "Two-Phase Steam–Water Pressure Drops," Reprint 147, Nuclear Engineering and Science Conference, Chicago, March 1958.

3.5.5   Levy, S., "Steam Slip-Theoretical Prediction from Momentum Model," *J. Heat Transfer*, **82C**(1), 113–124, 1960.

3.5.6   Gopalakrishnan, A., and Schrock, V. E., *Void Fraction from the Energy Equation*, Heat Transfer and Fluid Mechanics Institute, Stanford University Press, Palo Alto, 1964.

3.5.7   Friedel, L., "Improved Friction Pressure Drop Correlations for Horizontal and Vertical Two-Phase Flow," paper presented at the European Two-Phase Flow Group Meeting, Ispra, Italy, 1979. Also, P. B. Whalley, *Boiling, Condensation and Gas–Liquid Flow*, Oxford Science Publications, Oxford, 1987, Appendix B.

3.5.8   Premoli, A., Francesco, D., and Prina, A., "An Empirical Correlation for Evaluating Two-Phase Mixture Density under Adiabatic Conditions," paper presented at the European Two-Phase Flow Group Meeting, Milan, Italy, 1970. Also, P. B. Whalley, *Boiling, Condensation and Gas–Liquid Flow*, Oxford Science Publications, Oxford, 1987, Appendix C.

3.5.9   Chexal, B., and Lellouche, G., *A Full Range Drift Flux Correlation for Vertical Flows*, EPRI Report NP-3989-SR, Rev. 1, 1986.

3.5.10 Wallis, G. B., *One-Dimensional Two-Phase Flow*, McGraw-Hill, New York, 1969.

3.5.11 Zuber, N. and Findlay, J. A., "Average Volumetric Concentration in Two-Phase Flow Systems," *J. Heat Transfer*, **87**, 453, 1965.

3.5.12 Ishii, M., *One-Dimensional Drift Flux Model and Constitutive Equations for Relative Motion between Phases on Various Two-Phase Flow Regimes*, ANL 77-47, October 1977.

3.5.13 Spore, J. W., and Shiralkar, B. S., *A Generalized Computational Model for Transient Two-Phase Thermal Hydraulics in a Single Channel*, NR-13, pp. 71–76, Sixth Int. Heat Transfer Conference, Toronto, 1978.

3.5.14 Carlson, K. E., et al., *RELAP5/Mod 3 Code Manual*, NUREG/CR-5535, June 1990.

3.5.15 Borkowski, J. A., et al., *TRAC-BF1/MOD1 Models and Correlations*, NUREG/CR-4391, EGG-2680, Rev. 4, 1992.

3.6.1   Yadigaroglu, G., and Lahey, R. T., Jr., "On the Various Forms of the Conservation Equations in Two-Phase Flow," *Int. J. Multiphase Flow*, **2**, 477–494, 1987.

3.6.2 Ishii, M., and Mishima, K., "Two-Fluid Model and Hydrodynamic Constitutive Relations," *Nucl. Eng. Des.*, **82**, 107–126, 1984.

3.6.3 Taitel, Y., and Dukler, A. E., "A Model for Predicting Flow Regime Transitions in Horizontal and Near Horizontal Gas–Liquid Flow," *A.I.Ch.E. J.*, **22**, 47–55, 1976.

3.6.4 Ishii, M., and Chawla, T. C., *Local Drag Laws in Dispersed Two-Phase Flow*, NUREG-CR-1230, ANL 79-105, 1979.

3.6.5 Bharathan, D., *Air–Water Countercurrent Annular Flow*, EPRI Report, NP-1165, 1979.

3.6.6 Ishii, M., and Mishima, K., *Correlation for Liquid Entrainment in Annular Two-Phase Flow of Low Viscous Fluid*, ANL/KAS/LWR 81-2, March 1981.

3.6.7 Hancox, W. T., and Nicoll, W. B., "Prediction of Time Dependent Diabatic Two-Phase Water Flows," *Prog. Heat Mass Transfer*, **6**, 119–125, 1972.

3.6.8 Collier, J. G., *Convective Boiling and Condensation*, McGraw-Hill, New York, 1972.

3.6.9 Plesset, M. S. and Zwick, S. A., "The Growth of Bubbles in Superheated Liquids," *J. Appl. Phys.*, **25**, 443–500, 1954.

3.6.10 Berne, P., *Analyse critique des modèles de taux d'autovaporisation utiliser dans le calcul des écoulements diphasiques en conduite*, CEA Report R-5205, 1983.

3.6.11 Lahey, R. T., Jr., and Moody, F. J., *The Thermal-Hydraulics of a Boiling Water Nuclear Reactor*, American Nuclear Society, La Grange, Ill., 1977.

3.6.12 Ünal, H. C., "Maximum Bubble Diameter, Maximum Bubble-Growth Time and Bubble-Growth Rate during the Subcooled Nucleate Flow Boiling of Water up to 17.7 MN/m^2," *Int. J. Heat Mass Transfer*, **19**, 643–649, 1976.

3.6.13 Lee, K., and Ryley, O. J., "The Evaporation of Water Droplets in Superheated Steam," *J. Heat Transfer*, **10**, 445–451, 1968.

3.7.1 Roache, P., *Computational Fluid Dynamics*, Hermosa Publishers, Albuquerque, N. Mex., 1976.

3.7.2 Mahaffy, J. H., "A Stability-Enhancing Two-Step Method for Fluid Flow Calculations", *J. Comp. Phys.*, **66**, 329–341, 1982.

3.7.3 Stewart, C., et al., *COBRA-IV: The Model and the Method*, BNWL-2214, NRC-4, 1977.

3.7.4 Tye, A., et al., "Fundamental Derivation of the Equations Used in the ASSERT-4 Subchannel Code," paper presented at the 17th Annual CNS Simulation Symposium on Reactor Dynamics and Plant Control, Kingston, Ontario, Canada, 1992.

3.7.5 Srikantial, G. S., "VIPRE: A Reactor Core Thermal-Hydraulics Analyses Code for Utility Applications," *Nucl. Technol.*, **100**, 216–227, 1992.

3.7.6 Rogers, J. T., and Todreas, N. E., "Coolant Interchange Mixing in Reactor Fuel Rod Bundles Single Phase Coolants," in *Heat Transfer in Rod Bundles*, ASME booklet, 1968.

3.7.7 Wang, J., and Cao, L., "Experimental Research for Subcooled Boiling Mixing between Subchannels in a Bundle," *NURETH 6 Proc.*, **1**, 663–670, 1993.

3.7.8 Tapucu, A. et al., "The Effect of Turbulent Mixing Models on the Predictions of Subchannel Codes," *Nucl. Eng. Des.*, **149**, 221–231, 1994.

3.7.9 Damerell, P. S. and Simons, J. W., *2D/3D Program*, NUREG/1A-0126 and 1A-0127, 1993.

CHAPTER 4

FLOW PATTERNS

4.1 INTRODUCTION

A significant improvement was made in system computer codes when flow patterns were introduced because they allowed increased physics of the two-phase-flow process to be incorporated in the simulation of complex systems. Also, closure laws based on flow patterns could be employed to describe the transfer of mass, momentum, and energy at the gas–liquid interfaces. Ideally, it would have been desirable to have:

- An accurate description of the flow regimes and an understanding of how they evolve from one flow pattern to another
- Physically based models for each flow regime which took into account the effects of geometry, scale, variable properties, and developing flow conditions
- Direct derivation of the closure laws from the analytical descriptions of the flow patterns
- Clear statements about the limitations of the selected recipes in the system computer codes and, in particular, when they may not represent the actual situation adequately

Unfortunately, such goals could not be satisfied and incorporated into present system computer codes, and it is unlikely that they will be met for many years. The simple reason is that two-phase-flow patterns and their behavior are highly complex and far from being well understood. It is the purpose of this chapter to describe our present state of knowledge about flow patterns to highlight the complications, shortcomings, and differences in our understanding and to cover recent physically

160 FLOW PATTERNS

based advances. The chapter begins with a brief description of the potential flow regimes followed by a more detailed discussion of each of the major type of flow patterns. Emphasis is put on mass and momentum exchanges based on the premise that energy exchanges can be treated similarly. Experimental data as well as proposed analytical models are described for each flow pattern. A special effort is made to cover variations in available empirical correlations and closure laws and to provide conclusions pertinent to two-phase flow in complex systems and their simulation by computer codes.

4.2 DESCRIPTION OF FLOW PATTERNS

When two phases flow co-currently in a channel, they can arrange themselves in a number of different configurations, called *flow patterns* or *flow regimes*. Each flow pattern is characterized by a relatively similar distribution of the two phases and of their interfaces. Transition from one flow pattern to another takes place whenever a major change occurs in geometry of the gas–liquid interface. Figure 4.2.1 shows the flow patterns observed in a vertical circular duct, and Figure 4.2.2 illustrates typical flow regimes reported for horizontal flow in a pipe. These include:

· *Bubble flow.*	The liquid is the continuous phase and the gas is dispersed in the liquid in the form of bubbles of variable shape and size.
· *Slug flow.*	Slug flow is characterized by large gas bubbles almost filling the channel and separated by slugs of liquid.
· *Annular flow.*	Annular flow consists of an annular liquid film and of a gas core with or without drops of liquid in it.
· *Spray, mist, dispersed, or fog flow.*	The gas occupies most of the cross-sectional area and the liquid is in the form of small droplets dispersed in the gas.

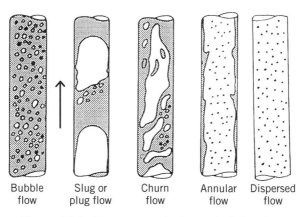

Figure 4.2.1 Flow patterns in the vertical direction.

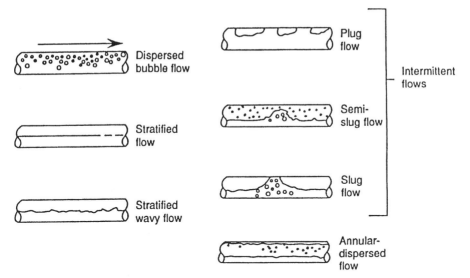

Figure 4.2.2 Flow patterns in the horizontal direction.

- *Stratified flow.*   Because of gravity forces the liquid flows along the bottom of a horizontal or inclined channel, and the gas flows above it.

A variety of other flow regimes have been reported in the literature, some of which are illustrated in Figures 4.2.1 and 4.2.2. In many cases, new names have been introduced to better define the distribution of the two phases. For instance, the terms *wavy annular* or *wavy stratified* have been used to identify the presence of waves at the gas–liquid interface. Similarly, *plug* and *semiannular* wording have been offered to describe transitional flow patterns, such as between bubble and slug or again between slug and annular flow. We shall here include all such variations within the discussion of a specific flow pattern, and we therefore need to recognize only the five major groups of flow regimes already identified: bubble, slug, stratified, annular, and dispersed (fog).

In discussing each of the major types of flow patterns, it is necessary not only to specify such properties as the pressure drop and the volumetric gas fraction, but also the conditions that determine the transition from one flow pattern to another. Furthermore, the interface areas and the transfer processes at the interfaces need to be covered. Also, because the geometry at the interface is quite complex, it is often necessary to simplify it considerably to obtain an analytical representation. Finally, it would be naive to expect that a single mathematical expression or empirical correlation could adequately describe the many existing variations of each flow regime. As a matter of fact, several expressions have been developed to describe the conditions encountered within a single flow pattern and it can be quite difficult to choose among them.

4.3 BUBBLE FLOW

Bubble flow is of importance to the chemical process industry, where the rise of bubbles through a liquid, both individually and in swarms (clusters), has received considerable attention. Properly speaking, bubbly flow is not a fully developed flow regime because given enough time or distance, the bubbles may collide with each other; and their agglomeration could lead to the formation of large bubbles or slug flow. In some instances, where proper care is taken in their generation, the bubbles present in the stream are small enough that they will touch rarely, and bubble flow will persist for a significant distance. The occurrence of this type of bubble regime and its transition to other forms of bubbly flow have been reported by several investigators.

4.3.1 Types of Bubble Flow

Single gas bubbles rising through liquids have been studied extensively and it has been found that when the bubbles are very small, surface tension forces make them spherical and they tend to preserve that configuration as long as their rising velocity or Reynolds number remains small. In most practical circumstances, the single gas bubbles are not spherical and the gas–liquid interface can vary over time and distance for the bubbles to assume different forms ranging from spherical to ellipsoidal to dimpled and cap spherical or ellipsoidal shapes. The bubble shape depends not only on the bubble size and rising velocity but also on the density of the gas and the viscosity, surface tension, and density of the liquid.

Similarly, many variations in bubble shape have been reported where the flow involves several bubbles except that now their behavior is complicated further by their ability to interact, collide, and combine. Zuber and Hench [4.3.1] have observed bubble flow in a batch process (no net liquid flow) in a 10 cm by 10 cm Plexiglas tube filled with water and equipped with various types of orifice plates through which air could be introduced at the bottom. They measured the average gas volume fraction $\bar{\alpha}$ as they increased the gas flow rate, and their results are reproduced in Figure 4.3.1. As shown in that figure, they distinguished three types of bubbly flow. In the *ideal bubbly flow regime*, the volumetric gas fraction and the bubble emission frequency increased with gas flow rate. In this regime the bubbles are nearly equal in diameter and are uniformly distributed across the test section. They rise at a uniform velocity, and their rise does not interfere with that of the surrounding bubbles. The bubbles do not have a wake, and the liquid is relatively undisturbed except in the immediate vicinity of the bubble. It was found that in order to operate in this ideal regime, the gas had to be introduced through uniformly distributed holes of small diameter.

Beyond the ideal bubble regime, the bubble emission frequency remains constant as the gas flow is increased and the bubble diameter becomes larger. The bubbles are no longer uniform in shape or diameter and they generate wakes that promote coalescence. Large bubbles start to appear in the center of the channel, and the liquid flows upward, particularly in the region of large bubbles.

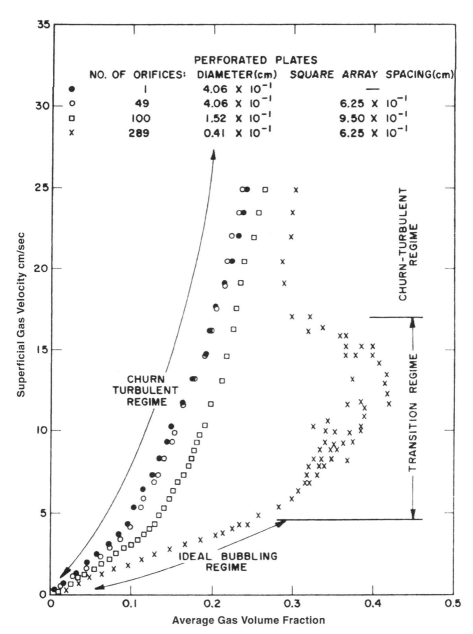

Figure 4.3.1 Void fraction in bubbly flow. (From Ref. 4.3.1.)

This *transition regime* is characterized by a large scatter of the data as illustrated in Figure 4.3.1.

In the *churn-turbulent regime*, the bubbles tend to concentrate in the central region. These bubbles transport liquid in their wakes, and in a batch process the upward motion of the liquid in the core is compensated by downward liquid flow along the channel walls. Large eddies are present, and large slugs of air were observed. However, due to the large size of the channel used in the test, the slugs were not stable and shattered under the action of Taylor hydrodynamic instability. For a smaller channel size, the churn-turbulent flow would take the form of slug flow, and this type of regime is discussed later. As shown in Figure 4.3.1, the churn-turbulent regime can occur at very low gas flow rates with large orifice holes. It is also characterized by a reduced rate of gas volumetric fraction change with increased gas flow rate.

In order to analytically describe the various bubbly regimes discussed above, it is necessary first to consider the problem of a single bubble rising through a liquid. Once the rise of a single bubble is known, the conditions corresponding to a swarm of bubbles can be inferred by considering a conglomeration of single bubbles, and eventually their interaction across the flow cross section.

4.3.2 Single-Bubble Rise in Infinite Medium

The rise of single bubbles in an infinite medium has been studied experimentally by Haberman and Morton [4.3.2], Peebles and Garber [4.3.3] and Harmathy [4.3.4]. Figure 4.3.2 shows the measured terminal bubble rise velocity in terms of bubble size as reported in Reference 4.3.2 for air bubbling through water. The curve has an odd shape due to bubble deformation by drag and circulation. In regimes AB and BC, the bubbles are spherical and have a straight-line or streamline motion. In region CD, the bubbles are spheroidal or ellipsoidal and flat, and they move in a zigzag manner. In region DE the bubbles are deformed. They have a spherical or ellipsoidal cap shape, and their travel is highly irregular. As the bubble diameter grows beyond about 1.4 mm, its rising velocity decreases according to Figure 4.3.2. The change in shape from spherical to wobbling to spherical cap continually increases the drag forces, and the rising velocity tends to level off. The values of the drag coefficient C_D and of the terminal velocity u_t proposed by Haberman and Morton for the various regions after improvement for region CD are as follows:

Region	C_D	u_t
AB	$24\left(\dfrac{\mu_L}{\rho_L u_t d}\right)$	$\dfrac{d^2 g(\rho_L - \rho_G)}{18\mu_L}$
BC	$18.3\left(\dfrac{\mu_L}{\rho_L u_t d}\right)^{0.68}$	$\left[\dfrac{2g(\rho_L - \rho_G)}{27\rho_L}\left(\dfrac{\rho_L d}{\mu_L}\right)^{2/3}\right]^{3/4}$
CD	$1.46 g\left(\dfrac{d}{2}\right)^2 \dfrac{\rho_L - \rho_G}{\sigma}$	$1.35\left(\dfrac{2\sigma}{\rho_L d}\right)^{1/2}$
DE	2.6	$1.05\left[\dfrac{g(\rho_L - \rho_G)}{\rho_L} r_e\right]^{1/2}$

(4.3.1)

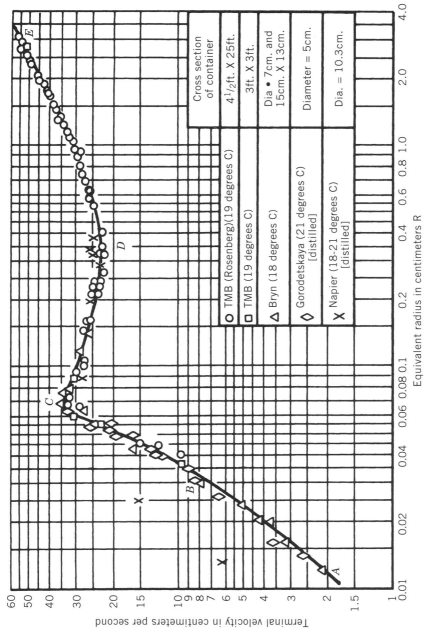

Figure 4.3.2 Terminal velocity of air bubbles in filtered or distilled water as a function of bubble size. (From Ref. 4.3.2.)

166 FLOW PATTERNS

where d is the bubble diameter, ρ_L and ρ_G the liquid and gas densities, σ the surface tension, g the gravitational constants, and μ_L the liquid viscosity. The symbol r_e is an equivalent bubble radius defined by

$$r_e = \left| \frac{3}{4\pi} V_b \right|^{1/3} \tag{4.3.2}$$

where V_b is the bubble volume. Subsequently, Peebles and Garber and Harmathy suggested that for regions *CD* and *DE*, where

$$\frac{\rho_L u_t d}{\mu_L} > 500 \quad \text{and} \quad \frac{g(\rho_L - \rho_G)d^2}{\sigma} < 13$$

the following relations be used for the drag coefficient and terminal velocity:

Peebles–Garber: $\quad C_D = 0.95 \left[\dfrac{g(\rho_L - \rho_G)d^2}{\sigma} \right]^{1/2} \quad u_t = 1.18 \left[\dfrac{\sigma g(\rho_L - \rho_G)}{\rho_L^2} \right]^{1/4}$

$$\tag{4.3.3}$$

Harmathy: $\quad C_D = 0.575 \left[\dfrac{g(\rho_L - \rho_G)d^2}{\sigma} \right]^{1/2} \quad u_t = 1.53 \left[\dfrac{\sigma g(\rho_L - \rho_G)}{\rho_L^2} \right]^{1/4}$

Satisfactory agreement of Eqs. (4.3.3) with the test data was reported.

4.3.3 Analysis of Single-Bubble Rise in Infinite Medium

When a bubble rises in an infinite medium, its velocity is determined by the bubble buoyancy and by the opposing forces due to liquid inertia, surface tension, and viscosity. The ratio of the bubble buoyancy to the three liquid forces can be expressed in terms of the following dimensionless groups:

$$\frac{\rho_L u_t^2}{gd(\rho_L - \rho_G)} \quad \frac{\sigma}{gd^2(\rho_L - \rho_G)} \quad \frac{u_t \mu_L}{gd^2(\rho_L - \rho_G)} \tag{4.3.4}$$

where d is the bubble diameter and where the viscous force on the bubble was derived from Stokes drag law. The first of the terms above controls when inertia effects are dominant, while the second and third terms are important when the role of surface tension and viscosity becomes significant. In the most general case, all three dimensionless parameters have to be considered. In the practical case of churn-turbulent bubbly flow, the ratios of the inertia force to the buoyant and to the surface tension forces have been found to be most important. These dimensionless groups and their product are

$$\frac{\rho_L u_t^2}{dg(\rho_L - \rho_G)} \qquad \frac{\rho_L u_t^2 d}{\sigma} \qquad \frac{\rho_L^2 u_t^4}{g\sigma(\rho_L - \rho_G)} \tag{4.3.5}$$

It is interesting to note that Eqs. (4.3.1) and (4.3.2) employ the nondimensional groupings of Eqs. (4.3.4) and (4.3.5). Analytical solutions for a single bubble rising in an infinite medium can be developed only when the bubble geometry remains constant and can be specified. For example, in the case of a spherical bubble, one can use results available for a solid sphere. If the flow Reynolds number, Re, is low enough, Stokes law applies and the corresponding drag coefficient, C_D, is obtained from

$$C_D = \frac{24}{\text{Re}} = \frac{24}{u_t d \rho_L / \mu_L} \tag{4.3.6}$$

The drag force, on the projected bubble area, can then be set equal to the buoyancy force so that

$$\left(\frac{1}{4}\pi d^2\right) \frac{1}{2} C_D \rho_L u_t^2 = \pi \frac{d^3}{6} (\rho_L - \rho_G) g$$

and there results

$$u_t = \frac{d^2 g(\rho_L - \rho_G)}{18 \mu_L} \tag{4.3.7}$$

which corresponds to the first of the terminal velocities listed in Eq. (4.3.1).

Similarly, Davidson and Harrison [4.3.5] developed an inviscid solution for flow over a spherical cap bubble of diameter d with a half subtended angle, θ:

$$u_t^2 = \frac{4}{9} gd \frac{\rho_L - \rho_G}{\rho_L} \frac{1 - \cos\theta}{\sin^2\theta} \tag{4.3.8}$$

By setting the term $(1 - \cos\theta)/\sin^2\theta$ equal to 0.5 and using an equivalent bubble diameter d_e, the following relation was proposed for bubbles with a spherical cap geometry:

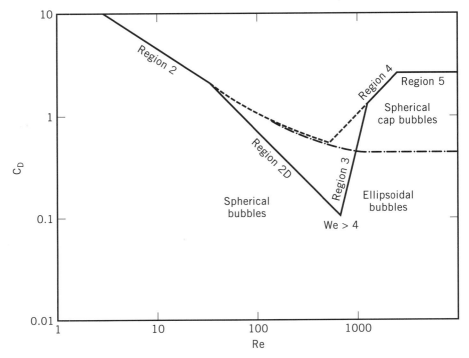

Figure 4.3.3 Drag correlation of air bubbles rising in water at 19°C: — distilled water, --- tap water. -·-·- solid sphere. (From Ref. 4.3.6.)

$$u_t = 0.71 \left(g d_e \frac{\rho_L - \rho_G}{\rho_L} \right)^{1/2} \quad (4.3.9)$$

which corresponds to the last of the terminal velocities listed in Eq. (4.3.1).

Kuo et al. [4.3.6] studied the trajectories of single bubbles flowing in suspension in water through various nozzles and evaluated them statistically to formulate correlations for the drag coefficient C_D. The correlations were recast in terms of the relative $\bar{u}_G - \bar{u}_L$, the corresponding relative Reynolds number, Re $= (\bar{u}_G - \bar{u}_L)\rho_L d/\mu_L$ and Weber number, We $= \rho_L(\bar{u}_G - \bar{u}_L)^2 d/\sigma$. The relations proposed by Kuo et al. are shown in Figure 4.3.3 and reproduced in Tables 4.3.1 and 4.3.2 for clean and tap water, but their fit of the data showed considerable scatter. Also, Kuo et al. observed a new effect, bubble shapes oscillating during their passage through some nozzles.

4.3.4 One-Dimensional Frictionless Bubble Flow Analysis

The first theoretical study of bubble flow was presented by Wallis [4.3.7]. He assumed that the flow was frictionless and that there is no variation in gas volume fraction or velocity across the channel. As the average gas volume fraction ap-

TABLE 4.3.1 Bubble Drag Correlations (Clean Water) in Terms of Reynolds and Weber Numbers[a]

Region	Drag Correlation C_D	Range
1	16/Re	Re < 0.49
2	20.68/Re$^{0.643}$	0.49 < Re < 33
2D	72/Re	Re > 33; We < 4
3		We = 4
4	We/3	We < 8
5	8/3	We > 8

Source: Ref. 4.3.6.

[a] $Re = (\bar{u}_G - \bar{u}_L)\rho_L d/\mu_L$; $We = \rho_L(\bar{u}_G - \bar{u}_L)^2 d/\sigma$.

proaches zero, the relative velocity of the gas with respect to the liquid, $\bar{u}_G - \bar{u}_L$, equals the terminal velocity u_t of a single bubble. Also, as the gas volume fraction $\bar{\alpha}$ approaches 1, Wallis took the relative velocity of the gas with respect to the liquid to be zero, and he wrote that

$$\bar{u}_G - \bar{u}_L = u_t(1 - \bar{\alpha}) \qquad (4.3.10)$$

For u_t, Wallis used expression (4.3.3), proposed by Peebles and Garber. The reason for this selection is that the equation proposed by Peebles and Garber does not contain the bubble diameter, and this eliminated the need to specify one more variable.

Later, Zuber [4.3.8] considered the case of laminar bubble flow and took into account the apparent viscosity of the continuous phase. If one neglects the acceleration and frictional losses, the momentum equation reduces to

$$-\frac{dp}{dz} = g[\rho_L(1 - \bar{\alpha}) + \rho_G\bar{\alpha}] = g\bar{\rho} \qquad (4.3.11)$$

The corresponding bubble drag equation is

$$\frac{\pi}{6}d^3\left(-\frac{dp}{dz} - g\rho_G\right) = \frac{1}{2}\rho_L C_D \frac{\pi d^2}{4}(\bar{u}_G - \bar{u}_L)^2 = 3\pi d \bar{\mu}_{TP}(\bar{u}_G - \bar{u}_L) \qquad (4.3.12)$$

where $\bar{\mu}_{TP}$ is the apparent viscosity of the continuous phase and Stokes law was

TABLE 4.3.2 Bubble Drag Correlations (Tap Water)

Region	Drag Correlation C_D	Range
1	16/Re	Re < 0.49
2	20.68/Re$^{0.643}$	0.49 < Re < 33
2B	6.297/Re$^{0.385}$	100 < Re
4	We/3	Re > 2065.1/We$^{2.68}$
5	8/3	We > 8

Source: Ref. 4.3.6.

used to obtain the last term of Eq. (4.3.12). If the viscosity $\overline{\mu}_{TP}$ is specified from Eq. (3.4.25) there results

$$\overline{u}_G - \overline{u}_L = u_t(1 - \overline{\alpha})^{3.5} \qquad (4.3.13)$$

A more general expression for bubble flow neglecting friction has been proposed by Zuber and Hench (4.3.1), who suggested that

$$\overline{u}_G - \overline{u}_L = u_t(1 - \overline{\alpha})^m \qquad (4.3.14)$$

Equation (4.3.14) was derived by solving for the relative velocity, $\overline{u}_G - \overline{u}_L$, from Eqs. (4.3.11) and (4.3.12) taking $\overline{\mu}_{TP} = \overline{\mu}_L$, and substituting for C_D the correlations proposed in References 4.3.2 to 4.3.4. The exponent m takes the value 1 for regions AB of Figure 4.3.2, $\frac{3}{4}$ and $\frac{1}{2}$ for regions BC and CE, respectively. [The $\frac{1}{2}$ exponent also applies for expression (4.3.3).]

It should be noted that Eq. (4.3.14) can be rewritten as

$$\frac{\overline{u}_{SG}}{\overline{\alpha}} - \frac{\overline{u}_{SL}}{1-\overline{\alpha}} = \frac{\overline{u}_{SG}}{\overline{\alpha}} + \frac{\overline{u}_{SG}}{1-\overline{\alpha}} - \frac{\overline{u}_{TP}}{1-\overline{\alpha}} = u_t(1-\overline{\alpha})^m \qquad (4.3.15)$$

where $\overline{u}_{SG}$ and $\overline{u}_{SL}$ are the average superficial gas and liquid velocities obtained by dividing the gas and liquid volumetric flow rates by the total channel area; $\overline{u}_{TP}$ is the average two-phase velocity and is equal to the sum of $\overline{u}_{SG}$ and $\overline{u}_{SL}$. Equation (4.3.15) is a general equation that describes frictionless bubbly flow. By assigning an appropriate expression to $\overline{u}_{TP}$, one can specify the value of $\overline{\alpha}$ for the following types of processes:

batch process: $\quad \overline{u}_{SL} = 0; \quad \overline{u}_{TP} = \overline{u}_{SG}$

co-current upward: $\quad \overline{u}_{TP} = \overline{u}_{SG} + \overline{u}_{SL} \qquad (4.3.16)$

countercurrent (liquid downward): $\quad \overline{u}_{TP} = \overline{u}_{SG} - \overline{u}_{SL}$

Another frequently used expression for frictionless bubble flow is the drift flux equation (3.5.36), which can be written

$$\overline{u}_G = C_0 \left(\frac{G_G}{\rho_G} + \frac{G_L}{\rho_L} \right) + 1.53 \left[\frac{\sigma g(\rho_L - \rho_G)}{\rho_L^2} \right]^{1/4} \qquad (4.3.17)$$

where G_L and G_G are the mass flow rates per channel flow area of the liquid and gas, respectively. Equation (4.3.17) attempts to recognize nonuniform profiles of gas volume fraction and velocity through the constant C_0.

Some investigators have employed the constant 1.18 of Peebles and Garber instead of the constant 1.53 proposed by Harmathy, which appears in the last term of Eq. (4.3.17). Others have adopted a modification of the constant C_0 suggested by Rouhani [4.3.9] for steam–water mixtures, or

$$C_0 = 1 + 0.21(1 - \bar{x}) \left(\frac{gD_H \rho_L^2}{G^2} \right)^{0.25} \qquad (4.3.18)$$

where D_H is the hydraulic diameter, G the total mass flow rate per unit area, and $\bar{x}$ the gas mass flow rate fraction. Many other variations of the value of C_0 have been reported, depending on the channel geometry, fluid properties, and means of introducing the gas into the stream. As pointed out in Sections 4.3.7 and 4.3.8, it is possible to have C_0 values below 1.

Equations (4.3.10), (4.3.11), (4.3.12), and (4.3.13) have been compared to available experimental data. Equation (4.3.13) was found particularly successful in dealing with problems of sedimentation and fluidization in solid–liquid systems. Equation (4.3.10) gives acceptable agreement with the liquid–gas data of various investigators even though its use of $(1 - \bar{\alpha})$ implies Stokes flow conditions. Similarly, Eq. (4.3.14) with $m = \frac{1}{2}$ and C_D specified from the Harmathy form of Eq. (4.3.3) gives good agreement with available test data even though it uses a turbulent expression exclusively for the terminal bubble velocity. The drift flux correlation of Eq. (4.3.17) tends to give the best overall agreement with available test data. This is illustrated in Figure 4.3.4 for air–water mixtures generated by Zuber and Findlay [4.3.10] to demonstrate the validity of Eq. (4.3.17). However, the differences between the various proposed frictionless relations for the average gas fraction $\bar{\alpha}$ are not so significant because they tend to be overwhelmed by the data scatter.

Up to this point, only the mass conservation or continuity equations have been utilized. However, it is possible to write a steady-state, frictionless, fully developed one-dimensional momentum equation for the gas and liquid in a channel of cross section A inclined at an angle θ from the horizontal (see Section 3.6.3.2):

$$-(1 - \bar{\alpha}) \frac{dp}{dz} + \frac{P_i \tau_i}{A} - g\rho_L (1 - \bar{\alpha}) \sin \theta = 0$$

$$-\bar{\alpha} \frac{dp}{dz} - \frac{P_i \tau_i}{A} - g\rho_G \bar{\alpha} \sin \theta = 0 \qquad (4.3.19)$$

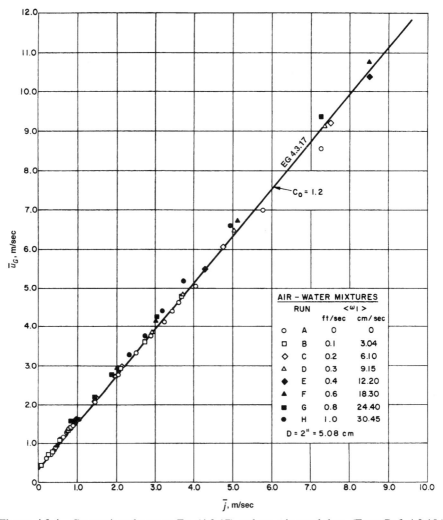

Figure 4.3.4 Comparison between Eq. (4.3.17) and experimental data. (From Ref. 4.3.10.)

where p represents the pressure, P_i the interfacial perimeter, and τ_i the corresponding interfacial shear stress. Eliminating dp/dz from Eqs. (4.3.19) yields

$$\frac{P_i \tau_i}{A} = g(\rho_L - \rho_G)(1 - \overline{\alpha})\overline{\alpha} \sin \theta \qquad (4.3.20)$$

The interfacial shear τ_i is usually defined as

$$\tau_i = \rho_L \frac{C_D}{8} |\overline{u}_r| \overline{u}_r \qquad (4.3.21)$$

where C_D is the interfacial drag coefficient and $\bar{u}_r$ is the void weighted relative gas/liquid velocity.

From the definition of the local drift flux U_{Gj}, the local relative velocity is

$$u_r = \frac{U_{Gj}}{1-\alpha} \qquad (4.3.22)$$

and the void weighted relative velocity $\bar{u}_r$ is, according to Andersen and Chu [4.3.11],

$$\bar{u}_r = \frac{\overline{U_{Gj}}}{\eta(1-\bar{\alpha})} \quad \text{with} \quad \eta = \frac{\overline{\alpha(1-\alpha)}}{\bar{\alpha}(1-\bar{\alpha})} \approx 1 \qquad (4.3.23)$$

where $\overline{U_{Gj}}$ is the void weighted drift flux velocity. By substituting Eq. (4.3.23) into Eqs. (4.3.20) and (4.3.21), there results

$$\frac{P_i C_D}{A} = \frac{8g(\rho_L - \rho_G)\bar{\alpha}(1-\bar{\alpha})^3 \sin\theta}{\rho_L \overline{U_{Gj}}^2} \qquad (4.3.24)$$

If the Zuber–Findlay drift flux model is utilized, $\overline{U_{Gj}}$ is equal to the terminal velocity specified by Harmathy for a single bubble under Eq. (4.3.3), or

$$\overline{U_{Gj}} = 1.53 \left[\frac{\sigma g(\rho_L - \rho_G)}{\rho_L^2}\right]^{1/4} \qquad (4.3.3)$$

which implies that, according to Eq. (4.3.24), the grouping $P_i C_D/A$ is exclusively a function of $\bar{\alpha}$ and the fluid properties. Also, *it is interesting that the grouping $P_i C_D/A$ maximizes with respect to $\bar{\alpha}$ when $\bar{\alpha} = 0.25$, a value often used for transition from bubble to slug flow*. Moreover, if we employ the Harmathy drag coefficient for a single bubble, $C_{D,SB}$, from Eq. (4.3.3), for vertical flow, and if we set P_i/A from the standard spherical bubble ratio of volume to interfacial area of $6\bar{\alpha}/d$, Eq. (4.3.24) requires that

$$C_D = 0.575 \left[\frac{g(\rho_L - \rho_G)d^2}{\sigma}\right]^{1/2} (1-\bar{\alpha})^3 \approx (1-\bar{\alpha})^3 C_{D,SB} \qquad (4.3.25)$$

In other words, once P_i/A is specified, the drag coefficient for one-dimensional frictionless bubble flow can be derived from Eq. (4.3.24) as well as the interfacial pressure drop, $P_i \tau_i / A$, from Eq. (4.3.20). Alternatively, if C_D is specified, P_i/A can be deduced from Eq. (4.3.24).

Finally, it is worth noting that many investigators have chosen to define the relative velocity between the gas and liquid as $\bar{u}_G - \bar{u}_L$, while in fact the gas volume–weighted relative velocity is

$$\bar{u}_G - \bar{u}_L = \frac{(1-\bar{\alpha}C_0)\bar{u}_G}{1-\bar{\alpha}} - C_0 \bar{u}_L \neq \bar{u}_G - \bar{u}_L \qquad (4.3.26)$$

174 FLOW PATTERNS

The two relative velocity expressions in Eq. (4.3.26) are different. They become equivalent only when $C_0 = 1$ or when the gas volume fraction and velocities are uniform across the flow cross section.

4.3.5 One-Dimensional Bubble Flow Analysis with Wall Friction

If friction at the channel wall is included, the steady-state, fully developed, one-dimensional gas and liquid momentum equations become Eqs. (3.6.44) and (3.6.45), or

$$-(1-\overline{\alpha})\frac{dp}{dz} - g\rho_L(1-\overline{\alpha})\sin\theta + \frac{P_i\tau_i}{A} - \frac{P_{WL}\tau_{WL}}{A} = 0 \quad (4.3.27)$$

$$-\overline{\alpha}\frac{dp}{dz} - g\rho_G\overline{\alpha}\sin\theta - \frac{P_i\tau_i}{A} - \frac{P_{WG}\tau_{WG}}{A} = 0$$

which can be combined to eliminate dp/dz and yield

$$\frac{P_i\tau_i}{A} = \overline{\alpha}\frac{P_{WL}\tau_{WL}}{A} - \frac{P_{WG}\tau_{WG}}{A}(1-\overline{\alpha}) + g(\rho_L - \rho_G)\overline{\alpha}(1-\overline{\alpha})\sin\theta \quad (4.3.28)$$

In Eqs. (4.3.27) and (4.3.28), P_{WL} and P_{WG} are the wall wetted perimeters by the liquid and gas and τ_{WL} and τ_{WG} the corresponding wall shear stresses.

For adiabatic bubble flow, one usually sets $P_{WG} = 0$ and $P_{WL} = P$, the channel wetted perimeter. Also, one can express the liquid wall shear stress from Eq. (3.5.21) in terms of the single-phase pressure drop, $(dp/dz)_{FL}$, or

$$\frac{P\tau_{WL}}{A} \approx \left(-\frac{dp}{dz}\right)_{FL}\frac{1}{(1-\overline{\alpha})^2} = \left(-\frac{dp}{dz}\right)_{FTP} \quad (4.3.29)$$

If we assume that $\overline{U}_{Gj}$ is not affected by wall shear, Eq. (4.3.28) becomes

$$\frac{P_iC_D}{A} = \frac{8}{\rho_L\overline{U}_{Gj}^2}\left[\left(-\frac{dp}{dz}\right)_{FL}\overline{\alpha} + g(\rho_L - \rho_G)\overline{\alpha}(1-\overline{\alpha})^3\sin\theta\right] \quad (4.3.30)$$

According to Eq. (4.3.30), the grouping P_iC_D/A is again a function of the fluid properties, the gas volume fraction $\overline{\alpha}$, and the single-phase liquid frictional pressure drop. If P_i is specified, the drag coefficient C_D can be established from Eq. (4.3.30).

Equation (4.3.30) can be rewritten for vertical flow and for $P_i A = 6\bar{\alpha}/d$ to get the drag coefficient C_D:

$$C_D = C_{D,SB} \left[(1 - \bar{\alpha})^3 + (1 - \bar{\alpha})^2 \frac{(-dp/dz)_{FTP}}{g(\rho_L - \rho_G)} \right] \qquad (4.3.31)$$

Equation (4.3.31) indicates that the bubble flow drag coefficient increases when wall shear is taken into account as long as the relation for $\overline{U}_{Gj}$ of Eq. (4.3.3) remains applicable. There is a good basis for that assumption since $\overline{U}_{Gj}$ was derived exclusively from the conservation equation. Equation (4.3.31) raises another important issue about the applicability of a single-bubble drag coefficient to multibubble flow and the need to modify it not only for gas volume fraction and wall friction but also for other potential influences, such as wall heat transfer or possibly accelerating flows. Finally, it should be noted that the single-bubble drag coefficient $C_{D,SB}$ involves the bubble diameter d and there is a need to deal with specifying d, particularly when faced with a bubble size distribution. This topic is covered in Section 4.3.7.

By employing Eq. (4.3.29) and taking the derivative of Eq. (4.3.30) with respect to $\bar{\alpha}$ and setting it equal to zero, the grouping $P_i C_D/A$ maximizes when the gas volume fraction reaches a transition value

$$\bar{\alpha}_{tr} = \frac{1}{4} \left[1 + \frac{(-dp/dz)_{FL}}{(1 - \bar{\alpha})^2 g(\rho_L - \rho_G) \sin \theta} \right] = \frac{1}{4} \left[1 + \frac{(-dp/dz)_{FTP}}{g(\rho_L - \rho_G) \sin \theta} \right] \qquad (4.3.32)$$

Equation (4.3.32) has been solved for $\bar{\alpha}_{tr}$ as a function of $(-dp/dz)_{FL}/g(\rho_L - \rho_G) \sin \theta$ and the results are shown in Figure 4.3.5. Equation (4.3.32) and Figure 4.3.5 show that the transition from bubble to slug flow occurs at increased gas volume fraction beyond the frictionless transition value $\bar{\alpha}_{tr} = 0.25$ as the liquid flow rate rises from zero. At $\bar{\alpha}_{tr} = 0.5$, the nondimensional liquid frictional pressure drop, $(-dp/dz)_{FL}/g(\rho_L - \rho_G)$, is equal to 0.25. Beyond that liquid flow rate the transition from bubble to slug flow is best kept at $\bar{\alpha} = 0.5$ because Eq. (4.3.32) breaks down at $\bar{\alpha} > 0.5$, indicating a transition to another flow pattern. The correlation employed by RELAP5/MOD3 (see Section 3.6.2) is plotted in Figure 4.3.5 for a tube diameter of 5 cm and saturated test water conditions at 74.5 bar. For simplification purposes, a constant friction factor of 0.020 was utilized in the calculations. The comparison is quite satisfactory, but the agreement can be expected to vary with channel size and fluid properties. It is also interesting to note the similar maximum value of $\bar{\alpha} = 0.5$ for bubble-to-slug flow transition employed by the computer system code RELAP5/MOD3.

Many investigators have chosen not to consider Eq. (4.3.28), which is a valid steady-state closure equation for $P_i \tau_i / A$. Instead, they have decided to rely on empirically developed drag coefficient correlations, combining them with a definition of the interfacial area in terms of $\bar{\alpha}$ and a bubble diameter d determined from a critical Weber number. This approach was illustrated in Section 3.6.3.1.1. Indirectly, it

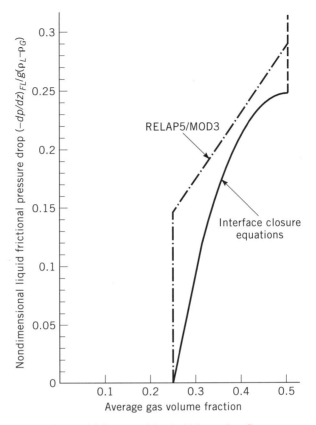

Figure 4.3.5 Transition bubble to slug flow.

specifies a drift flux or a relative gas-to-liquid velocity, and this fact may not be recognized.

4.3.6 Bubble Flow Friction Losses

In the presence of wall shear stress, frictional pressure drop data have been reported by Wallis [4.3.12] and by Rose and Griffith [4.3.13]. Wallis correlated his results by means of the expression

$$\phi_{TPL}^2 = 1 + 3 \frac{\bar{x}}{1 - \bar{x}} \frac{\rho_L}{\rho_G} (G \times 10^{-6})^{-1/3} \qquad (4.3.33)$$

where G is in lb/hr-ft^2 and ϕ_{TPL}^2 is the Martinelli multiplier. On the other hand, Rose and Griffith defined the following Reynolds number based on the liquid viscosity μ_L and the ratio of mean density $\bar{\rho}$ to the homogeneous density $\bar{\rho}_H$:

$$\text{Re}_{R,G} = \frac{G}{\bar{\rho}_H} \frac{D_H \bar{\rho}}{\mu_L} \tag{4.3.34}$$

and correlated their wall shear stress τ_w according to

$$\tau_w = f_L \frac{\bar{\rho}}{8} \left(\frac{G}{\bar{\rho}_H}\right)^2 \tag{4.3.35}$$

where f_L is a single-phase friction factor corresponding to $\text{Re}_{R,G}$.

In a subsequent study, Meyer and Wallis [4.3.14] have reevaluated all the available bubble flow data and found no satisfactory correlation for the frictional pressure drop. They concluded that a bubbly mixture behaves as a non-Newtonian fluid and that a more complex rheological description of the flow is needed. This conclusion needs to be tempered by the fact that in most bubble flows, the frictional losses can be neglected by comparison to the hydrostatic or head pressure drop. Also, it appears that relations of the type of Eq. (4.3.29) are able to describe the wall shear stress adequately because the data exhibit considerable scatter.

4.3.7 Bubble Distribution and Interfacial Area in Bubble Flow

There is general agreement that bubbly flow will exhibit a distribution of bubble sizes and shapes and that those bubble characteristics are variable and depend on flow conditions, fluid properties, and how the bubbles are generated or introduced into the liquid stream. Still, it has been customary to simplify the geometric structure of a bubbly two-phase stream by assuming that the bubbles are spherical and by defining a Sauter mean diameter d_{SM} as a function of time t:

$$d_{SM}(t) = \frac{\sum_{j=1}^{N} d_{V_j}^3(t)}{\sum_{j=1}^{N} d_{A_j}^2(t)} \tag{4.3.36}$$

where $d_{V_j}(t)$ and $d_{A_j}(t)$ are, respectively, the equivalent instantaneous volume and surface area of the bubbles of volume V_j and area A_j, or

$$d_{V_j}(t) = \left(\frac{6V_j}{\pi}\right)^{1/3} \quad \text{and} \quad d_{A_j}(t) = \left(\frac{A_j}{\pi}\right)^{1/2} \tag{4.3.37}$$

If the fluctuations in time of V_j and A_j are small, one can define a time-averaged

178 FLOW PATTERNS

Santer mean diameter, $\overline{d_{SM}}$, and utilize the classical expression for P_i/A such that

$$\frac{P_i}{A} = \frac{6\overline{\alpha}}{\overline{d_{SM}}} \tag{4.3.38}$$

If a bubble size diameter distribution function is available, it becomes possible to relate the Sauter mean diameter to other bubble diameters. For example, one of the earliest bubble size distribution functions was that of Nukiyama–Tanasawa [4.3.7]:

$$p^* = 4d^{*2}e^{-2d*} \tag{4.3.39}$$

where p^* is the probability of finding bubbles with a nondimensional diameter $d^* = d/d'$, where d' is the most probable bubble diameter. According to Eq. (4.3.39),

$$\frac{\overline{d_{SM}}}{d'} = \frac{\int_0^\infty d^{*3} p^* dd^*}{\int_0^\infty d^{*2} p^* dd^*} = \frac{5}{2} \tag{4.3.40}$$

One can similarly calculate an average volumetric bubble diameter $\overline{d_{V\,a}} = \sqrt[4]{7.5}d'$ or an average surface area bubble diameter $\overline{d_{Aa}} = \sqrt[3]{3}d'$. An important shortcoming of the distribution by Nukiyama–Tanasawa is that it does not specify a maximum bubble diameter $d_{\max}$ and it must be assumed as was done in RELAP5/MOD3, where it is presumed to be $d_{\max} = 3d'$.

Another important issue about bubble flow is the need to define a specific Sauter mean bubble diameter. Many investigators, particularly some of the current complex system computer codes, have chosen to employ a critical Weber number and to prescribe the critical Weber number and the bubble diameter to be used. As noted in Chapter 3, the maximum bubble diameter is utilized in RELAP5/MOD3 in the critical Weber number, and a value of 10 was chosen for the critical Weber number. If the critical Weber number approach is selected, it means that whatever bubble diameter appears in the Weber number, that diameter will decrease when the gas-to-liquid relative velocity increases. If a drift flux model applies, Eq. (4.3.21) states that as the gas volume fraction increases, the relative average velocity increases and the bubble diameter should decrease.

In a recent review of interfacial area in bubbly flow by Delhaye and Bricard (4.3.15), the exact opposite trend was reported. As shown in Table 4.3.3, the interface area and the Santer mean bubble diameter increase as the gas volumetric fraction increases. Also, it should be noted that both the interfacial area and the Santer mean diameter decrease as the liquid flow rate increases. All these trends, which are contrary to specifying a critical Weber number, can be deduced from the closure interface Eqs. (4.3.20), (4.3.24), (4.3.28), and (4.3.30).

TABLE 4.3.3 Bensler's Experimental Data[a]

Setup	D (m)	J_L (m/s)	J_G (m/s)	$\bar{\alpha}$	$\bar{d}_{SM}$ (mm)	$(P_i/A)_\text{ph}$ (m^{-1})	$(P_i/A)_\text{us}$ (m^{-1})	(dp/dz) (Pa/m^{-1})	ρ_L (kg/m^3)	ρ_G (kg/m)	μ_L (m · Pa · s)	μ_G (m · Pa · s)	σ (N/m)	(dp/dz) (Pa/m)
Bus	0.08	—	0.003	0.010	3.28	18.5	17.0	9682	997.1	1.24	0.889	0.018	0.0713	—
	0.08	—	0.016	0.050	3.57	84.8	75.4	9288	997.1	1.24	0.889	0.018	0.0713	—
	0.08	—	0.020	0.071	3.78	112.2	98.0	9095	997.5	1.24	0.926	0.018	0.0713	—
	0.08	—	0.035	0.101	4.03	150.6	136.9	8793	997.1	1.23	0.890	0.018	0.0713	—
	0.08	—	0.046	0.132	4.10	192.6	173.9	8499	997.5	1.23	0.924	0.018	0.0713	—
	0.08	—	0.059	0.150	4.26	211.7	228.3	8313	997.2	1.23	0.898	0.018	0.0713	—
	0.12	—	0.003	0.010	3.31	18.5	18.1	9685	997.4	1.23	0.915	0.018	0.0713	—
	0.12	—	0.010	0.040	3.56	68.1	73.5	9389	997.4	1.23	0.914	0.018	0.0713	—
	0.12	—	0.018	0.071	3.73	113.5	123.1	9095	997.4	1.23	0.916	0.018	0.0713	—
	0.12	—	0.027	0.101	3.96	153.1	170.8	8797	997.4	1.23	0.918	0.018	0.0713	—
	0.12	—	0.037	0.131	4.10	191.9	205.7	8503	997.4	1.23	0.917	0.018	0.0713	—
Torus	0.04	0	0.001	0.101	3.05	19.6	14.9	9696	998.3	1.68	1.011	0.018	0.0713	—
	0.04	0	0.005	0.040	3.64	66.7	55.9	9397	998.3	1.69	1.009	0.018	0.0713	—
	0.04	0	0.010	0.071	4.10	103.3	94.8	9103	998.3	1.69	1.012	0.018	0.0713	—
	0.04	0	0.014	0.101	4.38	138.9	132.5	8801	998.3	1.68	1.013	0.018	0.0713	—
	0.04	0	0.021	0.131	4.83	162.9	178.5	8511	998.4	1.68	1.017	0.018	0.0713	—
	0.04	0.503	0.036	0.060	3.96	90.9	74.2	9374	997.7	1.72	0.947	0.018	0.0713	172
	0.04	0.500	0.110	0.186	4.50	248.0	229.1	8369	997.7	1.71	0.949	0.018	0.0713	399
	0.04	0.500	0.180	0.265	4.89	325.2	377.0	7608	998.3	1.74	0.950	0.018	0.0713	409
	0.04	0.501	0.246	0.327	5.01	391.6	505.6	7080	997.8	1.77	0.952	0.018	0.0713	487
	0.04	1.002	0.036	0.047	3.63	77.7	43.7	9812	997.6	1.72	0.939	0.018	0.0713	485
	0.04	1.002	0.110	0.114	3.77	181.4	127.8	9282	997.7	1.72	0.937	0.018	0.0713	609
	0.04	1.004	0.183	0.165	4.05	244.4	220.1	8805	997.7	1.71	0.949	0.018	0.0713	630
	0.04	1.000	0.253	0.212	4.06	313.3	302.5	8411	997.8	1.73	0.955	0.018	0.0713	694
	0.04	1.997	0.035	0.030	2.34	76.9	45.8	10637	998.0	1.79	0.978	0.018	0.0713	1139
	0.04	1.998	0.105	0.061	2.77	132.1	103.3	10348	997.8	1.80	0.955	0.018	0.0713	1155
	0.04	1.997	0.171	0.089	3.40	157.1	155.6	10192	997.8	1.83	0.953	0.018	0.0713	1273
	0.04	2.005	0.234	0.123	3.80	194.2	201.0	10044	997.8	1.87	0.953	0.018	0.0713	1457
	0.04	3.001	0.033	0.021	1.88	67.0	43.7	11755	998.6	1.87	1.048	0.018	0.0713	2164
	0.04	3.000	0.100	0.045	2.08	129.8	91.6	11611	998.6	1.88	1.053	0.018	0.0713	2255
	0.04	3.001	0.163	0.060	2.35	153.2	140.0	11491	998.7	1.92	1.058	0.018	0.0713	2281
	0.04	3.001	0.224	0.085	2.79	182.8	178.2	11482	998.7	1.95	1.058	0.018	0.0713	2516

Source: From Ref. 3.2.15; data from Ref. 3.2.16.
[a]ph, photographic; us, ultrasonic.

In the case of bubble flow with no wall friction (i.e., no liquid flow), the interface shear stress τ_i can be rewritten from Eqs. (4.3.21), (4.3.22), (4.3.23), and (4.3.25):

$$\tau_i = \rho_L \frac{0.575}{8}\left[\frac{g(\rho_L - \rho_G)\overline{d}_{SM}^2}{\sigma}\right]^{1/2} (1 - \overline{\alpha})^3 \frac{\overline{U}_{Gj}^2}{(1 - \overline{\alpha})^2} = 0.167 g(\rho_L - \rho_G)\overline{d}_{SM}(1 - \overline{\alpha}) \quad (4.3.41)$$

An interesting finding of Delhaye and Bricard was that the interface shear remained relatively constant at 5.6 Pa for all air–water mixtures at room temperature. In fact, that is the value obtained from Eq. (4.3.41) for ($\overline{\alpha} = 0$) and $\overline{d}_{SM} = \overline{d}_{SM0}$, being the bubble diameter obtained from the Fritz [4.3.17] formula for no liquid flow:

$$\overline{d}_{SM0} = 0.0208\beta_0 \sqrt{\frac{\sigma}{g(\rho_L - \rho_G)}} \quad (4.3.42)$$

where β_0, the bubble contact angle in degrees, is set at about 61° for water. Under those conditions, the Sauter mean bubble diameter is obtained from

$$\frac{\overline{d}_{SM}}{\sqrt{\sigma/g(\rho_L - \rho_G)}} = \frac{0.0208\beta_0}{1 - \overline{\alpha}} \quad (4.3.43)$$

With wall friction, the drag coefficient of Eq. (4.3.31) should be utilized, giving

$$\overline{d}_{SM} = \frac{0.0208\beta_0}{1 - \overline{\alpha}} \frac{1}{1 + (-dp/dz)_{FL}/g(\rho_L - \rho_G)(1 - \overline{\alpha})} \quad (4.3.44)$$

and the interface term P_i/A is

$$\frac{P_i}{A}\sqrt{\frac{\sigma}{g(\rho_L - \rho_G)}} = \frac{6\overline{\alpha}(1 - \overline{\alpha})}{0.0208\beta_0}\left[1 + \frac{(-dp/dz)_{FTP}}{g(\rho_L - \rho_G)(1 - \overline{\alpha})}\right] \quad (4.3.45)$$

The corresponding nondimensional correlation proposed by Delhaye and Bricard was

$$\frac{P_i}{A}\left[\frac{\sigma}{g(\rho_L - \rho_G)}\right]^{1/2} = \left(7.23 - 6.82 \frac{Re_L}{Re_L + 3240}\right) 10^{-3} Re_G \quad (4.3.46)$$

where Re_L and Re_G are the liquid and gas Reynolds numbers based on the channel hydraulic diameter and the respective liquid and gas superficial velocity.

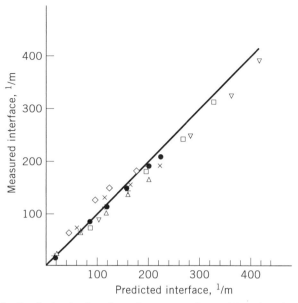

Figure 4.3.6 Prediction by interface closure equations of Bensler's interface data.

A comparison of Eq. (4.3.45) with the photographic interfacial data in Table 4.3.3 is shown in Figure 4.3.6, and the agreement is good but becomes less satisfactory as the water velocity increases. It should be mentioned that the measured values of $\bar{\alpha}$ and frictional pressure losses were employed to calculate the values in Figure 4.3.6. This was done because the bubble drift velocity decreased by a factor of about 2 from the bus to the torus tests. Furthermore, the constant C_0, which accounts for the gas volume fraction distribution across the flow cross section, decreased as the torus velocity increased, starting at a value of about 1 at a liquid velocity of 0.5 m/s, and dropping to about 0.8 at a liquid velocity of 3.0 m/s. These circumstances are probably due to the use of a small square channel of 0.04 m and high liquid velocities. However, they provide another illustration of the difficulties of modeling bubble flow distribution in all channel geometries and over a wide range of liquid velocities.

The observed experimental trends and the predictions from the closure interface equation make sense physically. In bubbly flow, as the gas volumetric fraction increases, the bubbles tend to group together and form bubble clusters. The rise velocity of the bubble clusters increases as the gas volumetric fraction increases, until transition to slug flow occurs, at which time the slug velocity remains relatively constant with further increases in gas volumetric fraction. This behavior has been observed in air–water mixtures and even more strikingly in laminar air–oil mixtures by Park [4.3.18], because there is an enhanced tendency with oil for the bubbles to group together and form clusters. According to Kalrach-Navarro et al. [4.3.19], this process occurs in three stages. The first stage corresponds to collision between two bubbles or between a bubble and a bubble cluster. The rate of bubble collision and the formation and growth of bubble clusters is encouraged by the increased velocity difference prevailing in the wake behind a large bubble or a bubble cluster. The second stage consists of draining the liquid film between the adjacent bubbles or within a bubble cluster. If the liquid film between bubbles reaches a critical thickness before being knocked apart by liquid-phase turbulence, the bubbles will coalesce and form larger bubbles or clusters. In an air–oil mixture the turbulent breakup of the bubble clusters is suppressed and the drainage time for the interstitial liquid is relatively long, thus encouraging the formation of bubble clusters and their coalescence. In turbulent air–water flows, the drainage time is much shorter and the bubbles do not coalesce or cluster as readily. This is particularly true as the liquid velocity increases. The physical picture supports the concept that the bubble mean Sauter diameter will tend to increase at the same time that the bubble rise velocity increases with gas volumetric fraction. Increasing the liquid velocity will diminish this trend and disrupt bubble clusters to produce smaller Sauter mean bubble diameters and delay their transition from bubble to slug flow.

The physical considerations above raise considerable doubt about the merit of employing a Weber critical number to specify the bubble diameter. The reason simply is that collision and coalescence of bubbles and liquid turbulence are the dominant mechanism in determining the appropriate Sauter mean diameter. Finally, it should be noted that if a critical Weber number is to be employed, the use of the maximum bubble diameter would most appropriate to fit some of the data of Table 4.3.3. However, the fit of test results is rather poor and does not exhibit the correct trend.

4.3.8 Multidimensional Bubble Flow and Analysis

In recent years, there has been increased interest in detailed mechanisms of bubble flow. Serizawa and Katoaka [4.3.20] were the first to measure the lateral gas volume fraction distribution and the corresponding turbulent fluctuations in a vertical pipe. They described four different major gas volume fraction patterns, which are plotted in Figure 4.3.7 in terms of the superficial liquid and gas velocities. Figure 4.3.7 shows that high liquid velocities and low gas flows tend to favor the presence of bubbles near the walls instead of the channel center. Similar results were obtained by Wang et al. [4.3.21], and some of their findings are reproduced in Figures 4.3.8 and 4.3.9. Wall peaking of the gas volume fraction is observed to occur for upflow, with the peak moving toward the center of the channel in downflow conditions. The liquid velocity in Figure 4.3.9 increases as the gas flow is raised, but the turbulent fluctuations remain relatively constant and are much higher than for liquid flow alone. Subsequently, Liu [4.3.22] found that the initial bubble size influences the gas volume fraction distribution and that this was particularly true at low liquid flows.

Lahey [4.3.23] has attempted to predict the multidimensional behavior of bubble flow and he has met with limited success after making several assumptions:

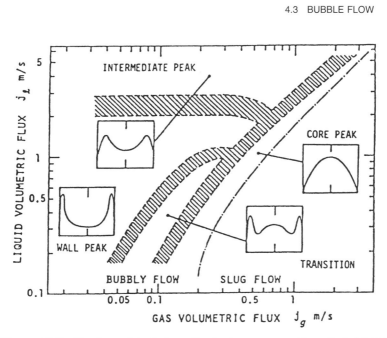

Figure 4.3.7 Simple model of phase distribution patterns. (From Ref. 4.3.20.)

- The gas bubble interfacial shear can be predicted from a drag coefficient C_D and the relative local gas to liquid velocity $u_G - u_L$. The drag coefficient employed was that of Kuo et al. [4.3.6] for tap water and for region 2B in Table 4.3.2. The bubbles are assumed to be spherical and to have a diameter d, so that the interfacial perimeter is $6\alpha/d$. It was not clear how the diameter d was specified, but it was probably obtained from a Weber number.
- A lateral lift was added to the momentum equation at the interface such that

$$M_{iG} = M_{iL} = C_L \rho_L \alpha (u_G - u_L) \frac{du_L}{dr} \quad (4.3.47)$$

where u_G and u_L are the local time average gas and liquid velocity and r is the radial distance measured from the pipe center. The coefficient C_L is taken to be 0.5 for inviscid flow of a single bubble, and it approaches zero for highly viscous flows. In the solution developed by Lahey, C_L was set at $C_L = 0.05$. Since $u_G - u_L$ is positive in upflow, Eq. (4.3.47) would lead to bubble concentration toward the wall. During downflow, the relative velocity, $u_G - u_L$, is negative and a bubble concentration toward the pipe center would result.
- Wall shear stress was assumed to be caused exclusively by the liquid flow and was calculated from single-phase relations using the liquid average velocity $\bar{u}_L$.
- The Reynolds stresses in the gas phase were neglected and the corresponding liquid stresses were specified from a single-phase liquid τ–ε model, where ε

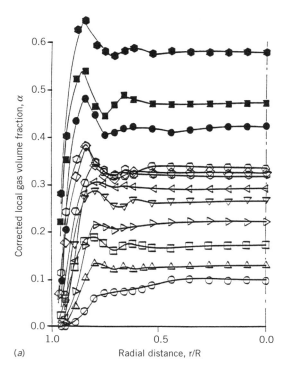

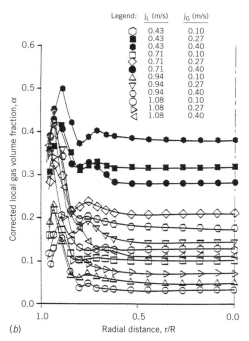

Figure 4.3.8 Void fraction profile: in (*a*) downward flow; (*b*) in upward flow. (Reprinted from Ref. 4.3.21 with permission from Elsevier Science.)

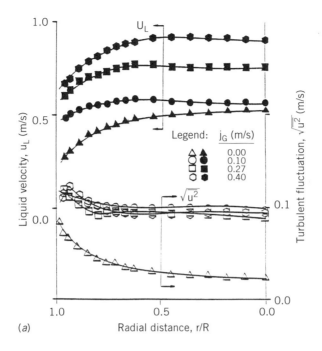

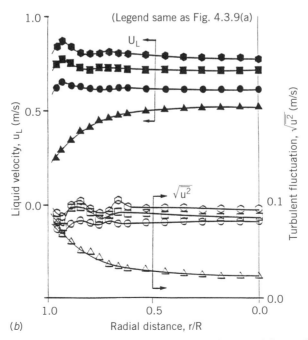

Figure 4.3.9 Liquid velocity and turbulent fluctuation: (*a*) in upward flows; (*b*) in downward flows. (Reprinted from Ref. 4.3.21 with permission from Elsevier Science.)

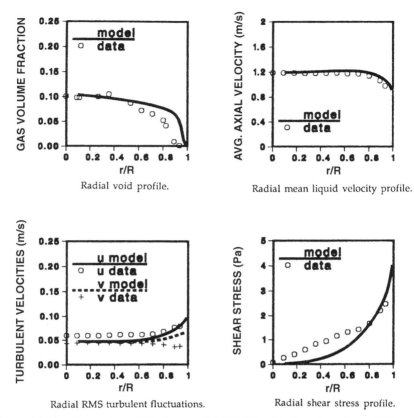

Figure 4.3.10 Comparison of Leahey model [4.3.23] with downflow data of Wang et al. [4.3.21]. (Reprinted from Ref. 4.3.23 with permission from Elsevier Science.)

is the eddy diffusivity. An additional source of turbulence was included due to the bubble-induced fluctuations in the liquid phase as they flow through it.

- The standard logarithmic liquid velocity profile was used next to the wall.

Although Leahey was able to get adequate predictions of Wang et al. test data as shown in Figure 4.3.10, he reiterated Liu's conclusion that bubble size and shape may play an important role in lateral gas volume distribution. In fact, as suggested by Serizawa et al. [4.3.24], there is evidence that larger bubbles will tend to migrate toward the center of the channel and that they behave differently from small spherical bubbles because they tend to deform in the flow field. Also, various other mechanisms have been proposed to account for the gas volume peaks near the walls [4.3.24]. For instance, according to Zun's bubble deposition model [4.3.25], the bubble penetrates laterally due to lift and diffusion with large-scale turbulent eddies acting to restrain this transverse penetration.

Most recently, Kocamustafaogullari and Huang [4.3.26] studied horizontal bubbly two-phase flow in detail and found that the local values of the gas volumetric fraction, interfacial area, and bubble passing frequency were nearly constant over the flow cross section. Slight local peaking close to the channels and a strong segregation due to buoyancy were reported as well as difficulties in reaching fully developed flow conditions with air–water mixtures at atmospheric pressure. Several two-dimensional measurements of gas volume fractions have been made in vertical geometries other than pipes. For instance, Furukawa et al. [4.3.27] have taken measurements in concentric annuli and found peaks in the gas volume fraction near the inner and outer walls. Phase distributions have been obtained in triangular and rectangular conduits by Sadatomu et al. [4.3.28], Sim and Lahey [4.3.29], and Jones and Zuber [4.3.30]. Wall and corner peaking was observed at low-gas-weight rate fractions, while at higher-gas-weight fractions the peak moved toward the interior of the channel. Furthermore, secondary flow patterns were noted to be present at the corners of the conduit and to contribute to increased gas volume fraction at those locations. Some of the same conclusions could be inferred from the parameter C_0 falling below 1.0 in the square torus channel of Table 4.3.3.

In summary, progress is being made on the subject of multidimensional bubble flow. However, the conditions in real bubble flows can be expected to be very complex, due to the presence of a spectrum of bubble sizes and shapes. Our understanding of the phenomena involved is far from complete, as attested to by the different models proposed for lateral distributions and their difficulty in coping with different flow geometries and directions (horizontal versus vertical). In other words, the available multidimensional models are not ready for introduction in complex system computer codes and it is doubtful that they will be ready in the near future.

4.3.9 Transition from Bubble Flow to Slug Flow

Studies of the bubbly slug transition have been carried out by Wallis [4.3.12], Radovcich and Moissis [4.3.31], and Moissis and Griffith [4.3.32]. They found that large initial bubble sizes and small pipes favored an early transition to slug flow. They also observed that for volume fraction of gas less than 10%, the rate of bubble agglomeration is usually slow, and bubble flow persists as long as the bubble diameter is not large. When the gas volume fraction is between 10 and 30%, agglomeration starts to occur more frequently and the amount of coalescence (i.e., transition to slug) increases with gas volume fraction, purity of the liquid, and its surface tension. When the gas volume fraction is greater than 30%, the collision between bubbles are so frequent that transition to slug flow always takes place.

As pointed out by Schwartzbeck and Kocamustafaogullari [4.3.33], a constant value of the gas volume fraction, $\bar{\alpha}$, may be the best way to describe the bubble-to-slug flow transition in a vertical channel. Taitel et al. [4.3.34] used $\bar{\alpha} = 0.25$, while Mishima and Ishii [4.3.35] employed simple geometric considerations to recommend a value of $\bar{\alpha} = 0.3$. Most recently, Lu and Zhang [4.3.36] suggested that the transition from bubble to intermittent flow will happen when the interfacial area A_i reaches a maximum. They used Eq. (4.3.17) with the Peebles and Garber constant of 1.18, and Eqs. (4.3.19), (4.3.21), and (4.3.26). In addition, they employed a bubble flow drag coefficient C_D from Lahey et al. [4.3.37]:

$$C_D = 26.34 \mathrm{Re}_b^m$$

$$m = -0.889 + 0.0034 \mathrm{Re}_b + 0.0014 (\ln \mathrm{Re}_b)^2 \quad (4.3.48)$$

$$\mathrm{Re}_b = \frac{\rho_L d_b (\bar{u}_G - \bar{u}_L)}{\mu_L}$$

The bubble diameter d_b was obtained from Kelly and Kazimi [4.3.38], which predicts an increased bubble diameter with $\bar{\alpha}$ and reduced liquid velocity:

$$d_b = \begin{cases} d_{b0} & \text{for } \bar{\alpha} < 0.1 \\ d_{b0} \left(\dfrac{9\bar{\alpha}}{1 - \bar{\alpha}} \right)^{1/3} & \text{for } \bar{\alpha} \geq 0.1 \end{cases} \quad (4.3.49)$$

$$d_{b0} = 0.9 \left[\frac{\sigma}{(\rho_L - \rho_G)g} \right]^{0.5} [1 + 1.34(\bar{u}_{SL})^{1/3}]^{-1}$$

The interfacial flow areas calculated are shown for refrigerant R-12 in Figure 4.3.11 for different liquid superficial velocities, $\bar{u}_{SL}$. The results tend to confirm a transition value of $\bar{\alpha} = 0.3$ and they show an increase in the transition gas volume fraction with liquid flow rate. The same method was applied to vertical flow with heat addition, which made it necessary to recognize vapor formation and acceleration terms in the steady-state momentum Eqs. (4.3.27). Also, the wall shear stress was subdivided in a liquid and gas component with $P_{WG}/A = 4\bar{\alpha}/D_H$ and $P_{WL} = 4(1 - \bar{\alpha})D_H$. Lu and Zhang [4.3.36] again obtained good comparison of their model to their test data taken with refrigerant R-12. The value of the Lu and Zhang work is that it confirmed that flow pattern transitions are dependent on wall friction and heat addition and that it supports a predictive methodology based on interface closure equations.

Another approach to predict the bubble-to-slug transition was developed by Pauchon and Banerjee [4.3.39]. They employed the two-fluid inviscid conservation and momentum equations, including a virtual mass term to determine the gaseous void propagation velocity and utilized the method of characteristics to determine when that pattern became unstable. Their model predicts that the transition from bubble to slug flow takes place when the gas volume fraction exceeds $\bar{\alpha} > 0.26$, which is in excellent agreement with the simplified frictionless interface closure prediction of 0.25 from Eq. (4.3.24).

Kalrach-Navarro et al. [4.3.19] applied the same approach to laminar air–oil bubble flow employing the experimentally determined and exponentially accelerat-

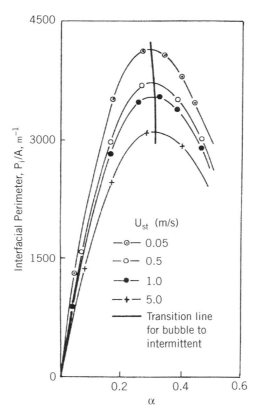

Figure 4.3.11 Calculated relation between P_i/A and $\bar{\alpha}$ for adiabatic flow condition (R-12, p = 10 bar, D = 0.02 m). (Reprinted from Ref. 4.3.34 with permission from Elsevier Science.)

ing cluster rise velocity from Park [4.3.18]. They found the transition to take place at $\bar{\alpha} > 0.16$, which was in agreement with Park's result for no oil flow. The corresponding transition values can be calculated from the closure equations (4.3.24) and from the laminar bubble terminal velocity and drag coefficient of Eq. (4.3.1). It is interesting to note that Eqs. (4.3.24) and (4.3.41) apply both to churn and laminar bubble flow with no liquid flow and that the laminar frictionless transition value is $\bar{\alpha} = 0.125$, which falls in the transition zone of 0.08 to 0.16 reported by Park [4.3.18].

In the case of wall friction, Eq. (4.3.31) is again applicable, except that now $(-dp/dz)_{FTP} = (-dp/dz)_{Fl}/(1 - \bar{\alpha})$ for laminar conditions and the grouping $P_i C_D/A$ becomes

$$\frac{P_i C_D}{A} = \frac{8(18\mu_L)^2 \bar{\alpha}(1-\bar{\alpha})^3}{g(\rho_L - \rho_G)\bar{d}_{SM}^4}\left[1 + \frac{(-dp/dz)_{FL}}{g(\rho_L - \rho_G)(1-\bar{\alpha})^2}\right] \quad (4.3.50)$$

With $\bar{d}_{SM}$ being proportional to

$$(1 - \bar{\alpha})^{-1}\left[1 + \frac{(-dp/dz)_{FL}}{g(\rho_L - \rho_G)(1 - \bar{\alpha})^2}\right]^{-1}$$

we can take the derivative of Eq. (4.3.50) with respect to $\bar{\alpha}$ and set it to zero to get the laminar transition value $\bar{\alpha}_{tr}$ as

$$\bar{\alpha}_{tr} = \left[8 - 2\frac{(-dp/dz)_{FTP}/g(\rho_L - \rho_G)(1 - \bar{\alpha})}{1 + (-dp/dz)_{FTP}/g(\rho_L - \rho_G)(1 - \bar{\alpha})}\right]^{-1} \quad (4.3.51)$$

Contrary to Park's findings, Eq. (4.3.51) continues to predict that $\bar{\alpha}_{tr}$ increases as the liquid flow increases. However, it is in excellent agreement with Park's measurement of $\bar{\alpha}_{tr} = 0.124$ with an oil velocity of 5.71 cm/s. This improved agreement with oil flow is attributed to the fact that it inhibits clustering and produces conditions much more like those described by Stokes law for spherical bubbles. Still, it points out that the simplified interface closure equations cannot take into account the impact of bubble clustering and their exponential increase in rise velocity with $\bar{\alpha}$.

In summary, the interface closure equations offer a simple and adequate means of predicting the transition from bubble to slug flow. They also confirm the approximate transition models employed in complex system computer codes.

4.3.10 Summary: Bubble Flow

1. Considerable progress has been in the understanding of gas–liquid bubble flow. There are good correlations for the drag coefficient and terminal velocity for single gas bubbles rising within a liquid. They employ appropriate nondimensional groupings and are based on test data that take into account changing bubble geometries and their deformation. The empirical correlations have been employed successfully in simplified analytical continuity models to approximately describe the behavior of a multitude of bubbles rising within a liquid. Two-dimensional measurements of bubble flow gas volume fraction and turbulent fluctuations have been obtained and are helping to formulate new mechanisms to explain the void distribution test data across the channel. The use of a maximum interfacial area or of the product P_iC_D/A to define the transition from bubble to intermittent slug flow holds considerable promise.

2. There remain several shortcomings in our understanding of bubble flow, including changes in bubble size and geometry with flow conditions which have not been described analytically to date. The same is true for bubble interactions, coalescence, and clustering. This complex part of flow bubble behavior is expected to depend on bubble size and their distribution and to influence the multidimensional gas volume fraction and turbulent fluctuations distributions. *It looks like (but it has*

not been proven) that this lack of microscopic details about bubble flow may not affect the overall (top-down) prediction of pressure and average phasic velocities and volume fractions by complex system computer codes. However, this microscopic behavior is expected to become more significant when dealing with such bottom-up phenomena as critical flow and critical heat flux, particularly if they are to be represented by definitive analytical models.

3. There are several differences and discrepancies in system computer codes in the application of results obtained from gas–liquid bubble flow pattern tests; for example:

- Many correlations have been proposed for the drag coefficient. They include those of Haberman and Morton [Eq. (4.3.1)], Peebles and Garber, Harmathy [Eqs. (4.3.3)], Kuo et al. (Tables 4.2.1 and 4.2.2), Lahey [Eq. (4.3.48)], Ishii and Chaula [Eq. (3.6.34)], and others. Furthermore, simplified and different versions of the correlations are employed in most system computer codes.
- Often empirical correlations for interfacial area, bubble diameter, and drag coefficient are chosen from different sources without checking their consistency. The interfacial closure laws derived from the steady-state momentum and energy conservation equations such as Eqs. (4.3.20) and (4.3.28) provide an excellent opportunity to check for consistency of the selected empirical sources and to avoid overspecifying the number of required correlations. Furthermore, most studies fail to recognize the potential influence of heat transfer or accelerating flows and even wall shear upon interfacial area and/or drag coefficient.
- There are many variations in the bubble diameter selected (maximum, average, Sauter mean, or most probable), in the critical Weber number, in the formulation of the relative velocity, and in the employed viscosity (liquid or two-phase). The use of a critical Weber number to specify a bubble diameter does not appear appropriate.
- In multidimensional bubble flow, various models have been postulated for the forces on the bubbles in a direction normal to the flow. The fact that the physical mechanisms in these models differ widely is an indication that our understanding is far from complete. The models still have to recognize a spectrum of bubble sizes and shapes and how they might distribute themselves across the cross section. Also, if a lateral lift force is introduced as in Eq. (4.3.47), it should appear in the corresponding local interfacial expression of the type of Eq. (4.3.28) and its influence on the grouping $P_i C_D/A$ considered.

In summary, the system computer codes have taken advantage of the bubble flow pattern data developed over the years, but they have taken significant liberty with the flow pattern transition and empirical correlations available. It is unfortunate that some of the changes or approximations adopted have not been tested for their impact. Also, there would be some merit to evaluating the variations employed in the various system computer codes and to identify a preferred strategy. Such efforts have not received priority to date because the codes have either yielded satisfactory predictions or may be relatively insensitive to the various forms of the bubble flow interfacial or wall shear terms employed. Still, the latter presumption, if true, needs validation.

4.4 SLUG FLOW

Slug flow is characterized by a succession of slugs of liquid separated by large gas bubbles whose diameter approaches that of the tube. Its behavior is illustrated schematically in Figure 4.4.1. It shows an "average" slug unit in a fully developed horizontal flow condition. The unit consists of a liquid slug region of length l_s and a liquid film and gas bubble region of length l_f. The liquid slug region propagates along the tube at a velocity $\bar{u}_s$, which is higher than the velocity of the liquid in the film region, $\bar{u}_{fs}$. Most of the liquid film is absorbed into the slug as it passes it by, and an equivalent amount of liquid is released downstream of the slug to regenerate the film. Gas from the film region tends to be entrained in the front of the slug as shown in Figure 4.4.1, and there is a mixing region of length l_m at that location. The picture shown can be further complicated by flow reversal in the liquid film and by a more complicated gas distribution in the slug body and liquid film. As illustrated in Figure 4.4.1, the pressure drop per slug unit is made up of three parts: an acceleration pressure drop to bring the liquid film content to the slug liquid velocity over the mixing length l_m, a frictional pressure drop over the slug length, and the pressure drop in the liquid film region, which tends to be small.

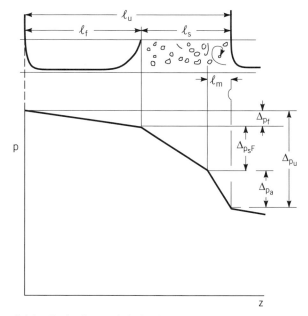

Figure 4.4.1 Basic flow unit in horizontal slug flow. (From Ref. 4.4.1.)

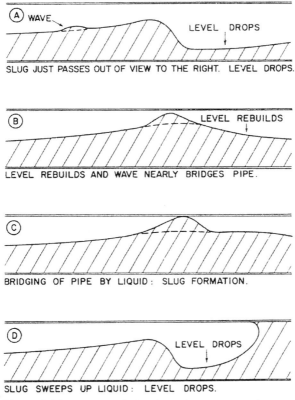

Figure 4.4.2 Process of slug formation. (From Ref. 4.4.1.)

Slug formation is an unsteady developing process illustrated in Figure 4.4.2, where in segment A, liquid and gas flow in a stratified pattern. The liquid layer decelerates and its level increases as it moves along the pipe. Waves also appear on the liquid surface. For appropriate liquid and gas velocities, the waves can grow enough for the rising liquid level and wave height to bridge the pipe as illustrated in segments B and C. When bridging occurs, the liquid in the bridge is accelerated to the gas velocity and it acts as a scoop picking up the slow-moving liquid in the film ahead of it. The fast-moving liquid builds up its volume and becomes a slug. The slug sheds liquid from its back to create a film with a free surface as depicted in Figure 4.4.1 and segment D of Figure 4.4.2. Since the slug sheds liquid at the same time that it picks up, its length stabilizes eventually.

This type of flow pattern inherently exhibits an intermittent behavior with large time variations of the mass flow rate, pressure, and velocity being present at any cross section normal to the flow pattern. Also, as anticipated, slug flow has been found to depend on the fluid properties, the channel geometry, and its direction (vertical versus horizontal). In the sections that follow, vertical slug flow is covered first and it is followed by a discussion of slug flow in horizontal channels.

4.4.1 Velocity of a Single Slug in Vertical Flow

The earliest solutions dealing with a large bubble rising in a channel were reported by Dumitrescu [4.4.2] and Davis and Taylor [4.4.3]. Both studies considered the problem of how fast a closed tube full of liquid will empty when the bottom is suddenly opened to the atmosphere. Both Dumitrescu and Davis and Taylor assumed that viscosity and surface tension effects could be neglected, and the asymptotic rise velocity of the bubble, u_t, was calculated from potential flow theory. They postulated that the velocity at the pipe wall was in the axial direction and that the pressure at the bubble boundary was constant. They found that the ratio $\rho_L u_t^2 / gD(\rho_L - \rho_G)$ was constant, so that

$$u_t = 0.35 \rho_L^{-1/2} [gD(\rho_L - \rho_G)]^{1/2} \qquad (4.4.1)$$

where D is the pipe diameter.

An investigation of the limits of this ideal fluid assumption was made by White and Beardmore [4.4.4], and they showed that the bubble rise velocity satisfies Eq. (4.4.1) when the viscosity and surface tension effects are negligible. When the role of surface tension is dominant, the grouping $(\rho_L - \rho_G)gD^2/\sigma$ approaches zero, and when it falls below 4, the bubbles cannot move at all. When viscous and surface tension forces are not negligible, the value of the nondimensional bubble velocity $\rho_L^{1/2} u_t / [gD(\rho_L - \rho_G)]^{1/2}$ varies from zero to a maximum of 0.35 depending upon the magnitude of the dimensionless groupings.

$$\frac{(\rho_L - \rho_G)gD^2}{\sigma} \quad \text{and} \quad \frac{g\mu_L^4(\rho_L - \rho_G)}{\rho_L^2 \sigma^3}$$

All of these trends have been confirmed experimentally by Wallis [4.4.5], who studied the rise of slug bubbles in a variety of fluids in vertical tubes of different diameters.

Wallis recommends the following expression for the slug velocity $\bar{u}_t$:

$$\frac{\bar{u}_t}{\sqrt{gD}} = 0.345[1 - \exp(0.0029 N_\mu)] \left(1 - \exp\frac{3.37 - N_E}{m}\right) \qquad (4.4.2)$$

where N_E, called the Eatvos number by Harmathy, is a nondimensional group previously mentioned which is obtained by forming the ratio of buoyant to surface tension forces:

$$N_E = \frac{gD^2(\rho_L - \rho_G)}{\sigma} \qquad (4.4.3)$$

The nondimensional grouping N_μ is

$$N_\mu = \frac{[D^3 g (\rho_L - \rho_G)\rho_L]^{1/2}}{\mu_L} \qquad (4.4.4)$$

with m in Eq. (4.4.2) depending on N_μ as follows:

$$m = \begin{cases} 10 & \text{for } N_\mu > 250 \\ 69 N_\mu^{-0.35} & \text{for } 18 < N_\mu < 250 \\ 29 & \text{for } N_\mu < 18 \end{cases}$$

Many years later, Dukler and Fabre [4.4.6] found that Eq. (4.4.2) continued to give as good a fit to the available test data as many subsequent models and correlations.

During his tests, Wallis observed three different types of bubble shapes. When viscosity dominated, the bubbles were rounded at both ends and there was no turbulence in the wake. When viscosity and surface tensions were negligible, the bubbles were rounded at the nose and flat at the tail. In this case, a strong toroidal vortex prevailed in the wake. Finally, when surface tension became significant, the bubble nose filled more and more of the tube, and the bubble exhibited an additional contraction expansion just prior to its tail. Subsequently, Runge and Wallis [4.4.7] attempted to extend Wallis's work to inclined circular tubes and were unable to correlate their data satisfactorily. They found that the bubble rise velocity peaked at an inclination of about 45°. It was generally greater than for a vertical tube except when the tube approached the horizontal, at which time the bubble velocity dropped rapidly to zero. The increased bubble velocity with inclination was related to changes in bubble geometry.

4.4.2 Vertical Slug Flow Analysis and Tests with No Wall Friction

The first theoretical analysis of slug flow was presented by Griffith and Wallis [4.4.8]. Starting from the rate of rise of a single bubble in a channel, they assumed the liquid inertia effects to be dominant, $\rho_L \gg \rho_G$, and they obtained for vertical tubes,

$$\bar{u}_G = \bar{u}_{TP} + 0.35 c \sqrt{gD} \qquad (4.4.5)$$

where $\bar{u}_{TP}$ is the total superficial velocity ($\bar{u}_{SG} \pm \bar{u}_{SL}$) or total volumetric flux $\bar{j}$ and c is an empirical constant that was specified in terms of a liquid Reynolds number from their tests.

Subsequently, Nicklin et al. [4.4.9] restudied slug flow in vertical tubes. They reported that the rising gas velocity is more appropriately equal to

$$\bar{u}_G = 1.2 \bar{u}_{TP} + 0.35 \sqrt{gD} \qquad (4.4.6)$$

where the constant 1.2 was inserted in front of the total superficial velocity $\bar{u}_{TP}$ to agree with the experimental results and to account for the nonuniform velocity profile ahead of the bubble. More generally, Eq. (4.4.6) can be written as follows:

$$\bar{u}_G = c_1 \bar{u}_{TP} + c_2 \sqrt{\frac{gD(\rho_L - \rho_G)}{\rho_L}} \qquad (4.4.7)$$

The constant c_2 takes into account the surface tension, viscosity, and geometry

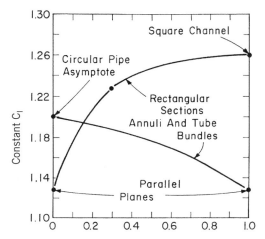

For Annuli, inside diameter/outside diameter
For Tube Bundles, 1 - Hydraulic diameter/rod diameter
For Rectangles, side/width

Figure 4.4.3 Slug flow in noncircular channels. (From Ref. 4.4.10.)

effects. The constant c_1 represents the degree to which the core of liquid in the center of the channel moves faster than the mean velocity.

Measurements of the slug bubble rise in noncircular channels have been reported by Griffith [4.4.10]. Practically all of his tests were in the region where inertia forces are dominant and the constant value c_1 assumed by the nondimensional bubble rise velocity is plotted in Figure 4.4.3 in terms of a geometric parameter describing the various test sections. It is observed that the constant c_1 in Eq. (4.4.7) varies with geometry and that the use of a hydraulic diameter in the relation (4.4.7) would have been in error. Griffith also reported that the observations of the rising bubbles were not always as expected. For instance, he noted that in rectangular channels the large dimension was the important one and that the liquid ran down the small sides. The distribution of bubbles was not symmetrical in multirod geometries, and a marked channeling prevailed in some of the multirod passages. Griffith also assumed fully developed turbulent liquid velocity profiles and calculated the values of the constant c_1 for various types of geometry. His predictions, which are valid for high liquid Reynolds numbers, ranged from 1.13 to 1.26.

It is interesting to note the similarity of Eq. (4.4.7) to the drift flux equation (4.3.17) [derived in Chapter 3, Eq. (3.5.36)]. It should also be realized that Eq. (4.4.7) can be rewritten as

$$\bar{u}_{SG} = c_1 \bar{\alpha}(\bar{u}_{SG} \pm \bar{u}_{SL}) + c_2 \bar{\alpha} \sqrt{\frac{gD(\rho_L - \rho_G)}{\rho_L}}$$

$$\bar{\alpha} = \frac{\bar{u}_{SG}}{c_1(\bar{u}_{SG} \pm \bar{u}_{SL}) + c_2 \sqrt{gD(\rho_L - \rho_G)/\rho_L}}$$

(4.4.8)

and it specifies the gas volume fraction $\bar{\alpha}$ in terms of the given superficial gas and liquid velocities, $\bar{u}_{SG}$ and $\bar{u}_{SL}$. As pointed out for Eq. (4.3.16), the plus sign in front of $\bar{u}_{SL}$ corresponds to co-current upward flow and the minus sign represents counter-current flow with the liquid traveling downward.

The expression in most favor currently is a drift flux model where

$$\bar{u}_G = 1.2\bar{j} + \bar{u}_t \quad (4.4.9)$$

where $\bar{j}$ is the volumetric flux and $\bar{u}_t$ is the velocity of a single large gas bubble rising in a static liquid which is obtained from Eq. (4.4.2). When the liquid flow between the gas bubbles is laminar, Frechon [4.4.11] found that the constant 1.2 assumes a higher value and increases as the laminar liquid Reynolds number decreases.

The validity of Eqs. (4.4.5) to (4.4.8) has been checked satisfactorily by the experiments of References 4.4.8 and 4.4.9 and by comparison of the predicted values of $\bar{\alpha}$ with a variety of gas volume fraction measurements over the expected range of application of vertical slug flow. In another investigation, Ellis and Jones [4.4.12] found that the dependence on the diameter D disappears as the tube size exceeds 4 to 6 in. In such large tubes one would not anticipate having a single slug occupying most of the channel area, and the resulting flow pattern would be more of the churn turbulent type as discussed under bubble flow. Equations (4.4.5) to (4.4.8) would no longer be applicable, and the maximum permissible bubble diameter based on Taylor hydrodynamic instability would become controlling.

4.4.3 Wall Friction in Vertical Slug Flow

As in bubble flow, the wall shear stress has been found negligible in several tests of vertical slug flow and it has often been neglected. As a matter of fact, if we examine closely the flow direction of the liquid film in co-current vertical slug flow, we find that it sometimes flows down along the walls and that the frictional pressure loss can be negative. However, at increased liquid velocities, the frictional pressure drop can be positive and important. It can be estimated by a method proposed by Griffith and Wallis [4.4.8]. First, they defined a liquid Reynolds number in terms of the total superficial velocity, $\bar{u}_{TP}$, or

$$(N_{Re})_L = \frac{G}{\bar{\rho}_H} \frac{D_H \rho_L}{\mu_L} = \frac{\bar{u}_{TP} D_H \rho_L}{\mu_L} \quad (4.4.10)$$

Next, they calculated the single-phase shear stress $(\tau_w)_L$ based on liquid flow in the pipe at the mixture velocity,

$$(\tau_w)_L = f_L \frac{\rho_L}{8} \left(\frac{G}{\bar{\rho}_H}\right)^2 \quad (4.4.11)$$

where f_L is the single-phase friction factor corresponding to $(N_{Re})_L$. The corresponding slug flow shear stress τ_w is finally obtained from

$$\tau_w = (\tau_w)_L \frac{(1-\bar{x})\bar{\rho}_H}{\rho_L} \qquad (4.4.12)$$

where the term $(1-\bar{x})\bar{\rho}_H/\rho_L$ represents the ratio of the liquid volumetric flow rate to the total volumetric flow. The success of Eq. (4.4.12) is predicated on the assumption that only the shear along the pipe area occupied by water slugs needs to be considered and the term $(1-\bar{x})\bar{\rho}_H/\rho_L$ is equal approximately to the fraction of pipe wall covered with liquid slugs. In other words, liquid film acceleration losses were neglected.

Moissis and Griffith [(4.3.32)] have studied developing slug flow in vertical tubes. In their tests, the slugs followed each other closely, and they found that when a slug is followed by a second slug at a distance of less than about six tube diameters, the following slug is faster, and eventually the two combine. To handle this type of flow, they proposed that the constant c_2 in Eq. (4.4.7) be taken as

$$c_2 = 0.35 + 2.8 e^{-1.06(l_s/D)} \qquad (4.4.13)$$

where l_s represents the liquid slug length. Equation (4.4.13) implies that fully developed vertical slug flow will exist rarely under practical conditions, which adds considerable complexity to the ability to model slug flow.

4.4.4 Interfacial Area and Shear in Frictionless Vertical Slug Flow

Simple geometric considerations have been proposed by Ishii and Mishima [4.4.13] to establish the interfacial area for fully developed slug flow as depicted in Figure 4.4.4. Let us first consider the case of a cylindrical Taylor bubble of diameter d_{TB} and length l_{TB} in a pipe of diameter D with no gas in the liquid film or in the liquid slug of length l_s. The gas volume fraction $\bar{\alpha}$ is

$$\bar{\alpha} = \left(\frac{d_{TB}}{D}\right)^2 \frac{l_{TB}}{l_{TB} + l_s} \qquad (4.4.14)$$

and the corresponding interfacial area P_i/A becomes

$$\frac{P_i}{A} = \frac{\pi d_{TB} l_{TB} + (\pi/2) d_{TB}^2}{(\pi/4) D^2 (l_{TB} + l_s)} = 4 \frac{\bar{\alpha}}{D} \frac{D}{d_{TB}} \left(1 + \frac{1}{2} \frac{d_{TB}}{l_{TB}}\right) \qquad (4.4.15)$$

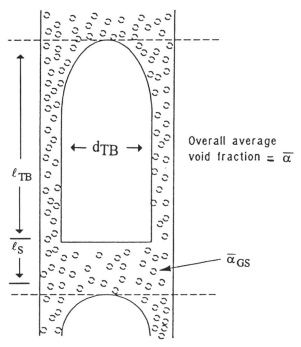

Figure 4.4.4 Slug flow patterns. (From Ref. 4.4.13.)

If we assume that the gas bubble has a spherical nose, the corresponding values are

$$\overline{\alpha} = \left(\frac{d_{TB}}{D}\right)^2 \frac{l_{TB}[1 - \frac{1}{6}(d_{TB}/l_{TB})]}{l_{TB} + l_s} \qquad \frac{P_i}{A} = 4 \frac{\overline{\alpha}}{D} \frac{D}{d_{TB}} \frac{1 + \frac{1}{4}(d_{TB}/l_{TB})}{1 - \frac{1}{6}(d_{TB}/l_{TB})} \qquad (4.4.16)$$

By recognizing that $d_{TB} \ll l_{TB}$, the preceding expressions can be simplified to Eq. (4.4.14) and

$$\frac{P_i}{A} \approx \frac{4\overline{\alpha}}{D} \frac{D}{d_{TB}} \qquad (4.4.17)$$

For the case of no wall friction, the closure equation (4.3.23) applies and with the drift flux velocity, $\overline{U}_{Gj}$, being derived from Eq. (4.4.1), there results

$$\frac{P_i C_D}{A} = 65.3 \frac{\overline{\alpha}(1 - \overline{\alpha})^3 \sin \theta}{D} \qquad (4.4.18)$$

With Eq. (4.4.17),

$$C_D = 16.3 \left(\frac{d_{TB}}{D}\right) (1 - \overline{\alpha})^3 \sin \theta \qquad (4.4.19)$$

In Reference P-8 it is reported that a slug is first formed when $d_{TB}/D = 0.6$, and at the transition location from bubble to slug flow, the constant in Eq. (4.4.19) becomes 9.8, which reproduces the value proposed by Ishii and Mishima [4.4.13] and utilized in Eq. (3.6.36). Beyond the transition point the ratio d_{TB}/D is usually set at 0.88 [e.g., in Eq. (3.6.28)], and the constant in Eq. (4.4.19) would be equal to 14.4. In any case, the important result is that the interfacial drag coefficient, C_D, is a constant dependent on the gas bubble geometry, which is multiplied by $(1 - \bar{\alpha})^3$. Furthermore, it is seen that the drag coefficient can be determined if P_i/A is specified, and vice versa, as long as the steady-state frictionless interfacial closure law is employed.

Let us next assume as shown in Figure 4.4.4 that part of the gas is in the form of small bubbles present in the liquid film and slug and that they occupy a volume fraction $\bar{\alpha}_{GS}$ of that volume, so that

$$\bar{\alpha} = \bar{\alpha}_{TB} + \bar{\alpha}_B = \bar{\alpha}_{TB} + \bar{\alpha}_{GS}(1 - \bar{\alpha}_{TB}) \quad \text{or} \quad \bar{\alpha}_{TB} = \frac{\bar{\alpha} - \bar{\alpha}_{GS}}{1 - \bar{\alpha}_{GS}} \quad (4.4.20)$$

where $\bar{\alpha}_{TB}$ and $\bar{\alpha}_B$ are, respectively, the volume fraction occupied by the Taylor gas bubble and the small gas bubbles. The interfacial area for the long slug bubble and for the small gas bubbles, which are presumed to be spherical and to have a diameter d_B are

$$\left(\frac{P_i}{A}\right)_{TB} = \frac{4}{D}\frac{D}{d_{TB}}\frac{\bar{\alpha} - \bar{\alpha}_{GS}}{1 - \bar{\alpha}_{GS}} \quad \left(\frac{P_i}{A}\right)_B = \frac{6\bar{\alpha}_{GS}}{d_B}\frac{1 - \bar{\alpha}}{1 - \bar{\alpha}_{GS}} \quad (4.4.21)$$

If the flow remains frictionless, Eq. (4.3.19) can be written as

$$\left(\frac{P_i\tau_i}{A}\right)_{TB} + \left(\frac{P_i\tau_i}{A}\right)_B = g(\rho_L - \rho_G)\bar{\alpha}(1 - \bar{\alpha})\sin\theta \quad (4.4.22)$$

If one introduces Eqs. (4.3.24), (4.4.19), (4.4.20), and (4.4.21), Eq. (4.4.22) becomes

$$\bar{\alpha}_{TB}(1 - \bar{\alpha}_{TB}) + \bar{\alpha}_B(1 - \bar{\alpha}_B) = \bar{\alpha}(1 - \bar{\alpha}) \quad (4.4.23)$$

Equation (4.4.23) can be satisfied only when $\bar{\alpha} = \bar{\alpha}_B$ or $\bar{\alpha} = \bar{\alpha}_{TB}$. This means that if small gas bubbles are present in the liquid at the same time as a Taylor bubble in slug flow, their total interfacial shear would exceed the value prescribed by the interfacial closure law. While this increase can be only as high as 16% when $\bar{\alpha} = 0.25$, it can reach 50% when $\bar{\alpha} = 0.5$. The implications of Eq. (4.4.23) are that when a Taylor bubble and small bubbles in the rest of the liquid are present simultaneously in slug flow, they interact with each other to reduce their respective drag coefficients. It is interesting to note that if for $\bar{\alpha} = 0.5$, we reduce the constant in Eq. (4.4.19) from 14.4 for $d_{TB}/D = 0.88$ by 50%, the drag coefficient constant would decrease to match the value of 9.8 derived for $d_{TB}/D = 0.6$. In other words, the product $16.3(d_{TB}/D)$ could be approximated by a constant of 9.8, as suggested in Reference 4.4.13 over a wide range of ratio d_{TB}/D. More specifically, if the split in volume fraction between the Taylor bubble and the small gas bubbles is known, one can use Eq. (4.4.22) to establish the decrease in their respective drag coefficients to satisfy it. Alternatively, if the respective drag coefficients are specified, the split in volume fraction needs to be adjusted to match Eq. (4.4.22).

In RELAP5/MOD3, an exponential function equation (3.6.27) was proposed to calculate the gas volume fraction $\bar{\alpha}_{GS}$. It was presumed at the transition from bubble to slug flow that $\bar{\alpha}_{GS} = \bar{\alpha}_{AB}$ or it is equal to the transition gas volume fraction. Beyond that value, the small gas bubble volume fraction $\bar{\alpha}_{GS}$ decreases and it is assumed to go to zero, or $\bar{\alpha}_{GS} \to 0$ at the transition from slug to annular flow. Physically, as noted in Section 4.3.7, for bubbly flow, the distribution of bubbles is not uniform in the axial direction, and bubbles locally will tend to group together, form clusters, and eventually become a Taylor bubble. At the transition from bubble to slug flow, there will be a sharp drop or discontinuity in the volume fraction occupied by the gas bubbles. Beyond that point, if the total gas volume fraction $\bar{\alpha}$ continues to increase, it is not clear that $\bar{\alpha}_{GS}$ would decrease continually as prescribed in RELAP5/MOD3. In fact, according to Ishii and Mishima [4.4.13], there will next be a transition from slug to churn annular flow, and it will occur when the gas volumetric fraction of the small bubbles in the liquid-slug section, $\bar{\alpha}_{GS}$, reaches the volumetric fraction of the slug bubble section, $\bar{\alpha}_{TB}$. This would require $\bar{\alpha}_{GS}$ to grow again beyond the transition point from bubble to slug flow. However, after the slug to churn annular flow transition, $\bar{\alpha}_{GS}$ is expected to decrease and eventually approach zero when a full annular flow pattern of a liquid film with a central gaseous core is established.

According to Eqs. (4.4.14) and (4.4.20), we can write

$$\bar{\alpha}_{GS} = \frac{\bar{\alpha} - (d_{TB}/D)^2[l_{TB}/(l_{TB} + l_s)]}{1 - (d_{TB}/D)^2[l_{TB}/(l_{TB} + l_s)]} \quad (4.4.24)$$

At the transition from bubble to slug flow, $\bar{\alpha} = \bar{\alpha}_{AB}$ and, according to Eq. (4.3.13), $\bar{\alpha}_{GS}$ approaches $\bar{\alpha}_{AB}$ only if $l_{TB} \to 0$ or $l_{TB}/(l_{TB} + l_s) \to 0$. Also, to get a meaningful value of $\bar{\alpha}_{GS}$, the ratio $l_{TB}/(l_{TB} + l_s)$ must be below $\bar{\alpha}_{AB}/(d_{TB}/D)^2$. If we employ the frictionless value of $\bar{\alpha}_{AB} = 0.25$ from Section 4.3.4 and at the same time set $(d_{TB}/D) \approx 0.6$ as suggested in Reference P-8, one finds that $l_{TB}/(l_{TB} + l_s)$ must be below 0.69 or $l_s \geq 0.46 l_{TB}$.

Let us arbitrarily assume that $l_s = l_{TB}$, $d_{TB}/D = 0.6$, and $\bar{\alpha} = \bar{\alpha}_{AB} = 0.25$; the value of $\bar{\alpha}_{GS}$ would then have to drop from 0.25 to 0.085 at the transition from bubble to slug flow. Let us next assume that as the gas volumetric fraction increases, the length of the liquid slug l_s remains equal to l_{TB} and that the Taylor bubble first increases primarily in diameter. The ratio d_{TB}/D would start at 0.6 and reach 0.88 when the transition to annular churn flow takes place. Also, it makes sense to assume that churn annular flow will occur at or before $\bar{\alpha}_{GS} = 0.25$, or the same value employed for the transition from bubble to slug flow. The transition from slug to churn annular for this illustration case would then occur at $\bar{\alpha} = 0.54$. Beyond that value of gas volumetric fraction, the Taylor bubble will presumably grow in length until $l_s \to 0$ and $\bar{\alpha}_{TB} = (d_{TB}/D)^2$. With $\bar{\alpha}_{GS} = 0$ and $d_{TB}/D = 0.88$ the churn annular to annular transition would happen when $\overline{\alpha} = 0.77$. Although it is recognized that there is no basis for the preceding evolution of slug to annular flow, it is still quite useful in demonstrating the shortcomings and constraints of slug flow models employed in complex system computer codes. The computer codes include constant formulations for interfacial areas, an oversimplified and most likely incorrect prediction of the gas volumetric fraction in the liquid slug and film section, and they fail to take into account the interaction between the Taylor bubble and the other small gas bubbles present in the liquid. In fairness to system computer codes, it should be mentioned that test data and modeling continue to be lacking for developing slug flow. Another important restraint not often recognized in slug flow modeling is that volumetric averages of gas and liquid fractions over a slug cell are being employed instead of flow cross-sectional averages.

4.4.5 Interfacial Area and Shear in Vertical Slug Flow with Wall Friction

The same basic approach can be applied to vertical slug flow with wall friction. First, Eq. (4.3.28) would replace Eq. (4.3.19). Second, the transition from bubble to slug flow would be influenced by wall friction and the transition gas volumetric fraction could grow from $\overline{\alpha}_{AB} = 0.25$ to $\overline{\alpha}_{AB} = 0.5$, depending on the wall friction (i.e., increasing liquid flow). As the bubble-to-slug transition volumetric fraction $\overline{\alpha}_{AB}$ goes up, there is less and less opportunity for churn annular flow to occur, and one may transition directly from slug to annular flow. Third, it is necessary to specify the wall interface and wall shear forces in slug flow. One could then proceed to assess the interface areas and shears and to recognize their interaction and that of the wall shear by utilizing Eq. (4.3.28). Finally, an expression for $\overline{\alpha}_{GS}$ or a simplified approach as put forward in Section 4.4.3 must be developed to take into account the impact of overall gas volumetric fraction, $\overline{\alpha}$. If all these steps are followed, a much more consistent treatment of slug flow with wall friction would result, and it may be preferable to purely empirical formulations and closure laws presently in use.

4.4.6 Horizontal Slug Flow Analysis and Tests

Slug flow in horizontal piping has received increased attention in recent years because it occurs in large and long oil pipelines, where it can produce "severe slugging" during flow over undulating terrain. The earliest horizontal models were rather simplified and similar to the vertical slug approach suggested by Griffith and Wallis. For example, Vermeulen and Ryan [4.4.14] assume that the liquid slug velocity was equal to the mixture velocity $\overline{u}_{TP} = G/\overline{\rho}_H$ and they obtained the same frictional wall shear as Griffith and Wallis, or

$$\tau_w = f_L \left(\frac{G}{\overline{\rho}_H}\right)^2 \frac{(1-\overline{x})\overline{\rho}_H}{8} \quad (4.4.25)$$

An additional pressure loss was considered due to the liquid film at the bottom of the pipe being scooped by the liquid slug and being accelerated to the liquid slug velocity. Simultaneously, to conserve mass, a liquid film is ejected from the rear of the slug, and its momentum is gradually lost until it settles at the bottom of the pipe. Vermeulen and Ryan assumed that the liquid film velocity is small when it is at the bottom of the pipe and that the area fraction of the pipe occupied by the liquid film is $\overline{\alpha}_{fL}$. The resulting acceleration pressure loss is therefore

$$\Delta p_{acc} = \rho_L \left(\frac{G}{\overline{\rho}_H}\right)^2 \frac{\pi D^2}{4} \overline{\alpha}_{fL} \qquad (4.4.26)$$

To calculate the pressure drop per unit length, the frequency of slugs per unit length, ω_s, must be known so that the number of slugs, N_s, per unit length becomes

$$N_s = \frac{\omega_s}{G/\overline{\rho}_H}$$

and the corresponding acceleration loss per unit length becomes

$$\frac{dp_{acc}}{dz} = \rho_L \left(\frac{G}{\overline{\rho}_H}\right)^2 \frac{\pi D^2}{4} \overline{\alpha}_{fL} \omega_s \qquad (4.4.27)$$

Vermeulen and Ryan employed their experimentally measured slug frequency and the empirical correlation by Lockhart and Martinelli [3.5.1] for the total liquid volume fraction to deduce the film volumetric fraction $\overline{\alpha}_{fL}$. The evaluation of the liquid-film acceleration is, therefore, not only empirically based, but also, it is inconsistent because of its simultaneous use of homogeneous flow in defining $\overline{u}_{TP}$ and slip flow in inferring $\overline{\alpha}_{fL}$ from Lockhart and Martinelli's work.

Several more detailed and accurate models have been proposed to describe the complex conditions prevalent in horizontal slug flow. The work of Dukler and Hubbard [4.4.1] is particularly noteworthy and it will be summarized next. Dukler and Hubbard concluded that in order to calculate the pressure drop across a slug, they needed to know the liquid film pickup and shedding rate per unit area, G_f, the liquid-film velocity where it is picked by the slug, $G_f/\overline{\alpha}_{fLe}\rho_L$, the average slug velocity $\overline{u}_s$, the liquid volume fraction in the slug, $\overline{\alpha}_{SL}$, the slug length l_s, and the length of the mixing zone, l_m. The parameter $\overline{\alpha}_{fLe}$ represents the fraction of the pipe occupied by the liquid film before its pickup by the slug.

Dukler and Hubbard assumed that there is no gas entrainment in the liquid film and no liquid droplet entrainment in the gas bubble. The observed rate of advance of the slug, $\overline{u}_s$, is the sum of the mean fluid velocity in the slug, $G/\overline{\rho}_H$, and of the rate of buildup at the front of the slug due to film pickup, or

$$\overline{u}_s = \frac{G}{\overline{\rho}_H} + \frac{G_f}{\rho_L \overline{\alpha}_{SL}} \qquad (4.4.28)$$

Equation (4.4.28) was rewritten as

$$\bar{u}_s = (1 + C)\frac{G}{\bar{\rho}_H} \quad \text{or} \quad \frac{G_f}{\rho_L \bar{\alpha}_{SL}} = C\frac{G}{\bar{\rho}_H} \quad (4.4.29)$$

By using standard turbulent velocity distribution in the slug, it was found that

$$C = 0.021 \ln(\text{Re}_s) + 0.022$$

$$\text{Re}_s = \frac{GD}{\bar{\rho}_H} \frac{\rho_L(\bar{\alpha}_{SL}) + \rho_G(1 - \bar{\alpha}_{SL})}{\mu_L(\bar{\alpha}_{SL}) + \mu_G(1 - \bar{\alpha}_{SL})} \quad (4.4.30)$$

The constant C has a value in the range 0.2 to 0.3 for typical gas and liquid flows. At a low mixture velocity, Dukler and Fabre [4.4.15] showed that when

$$\frac{G}{\bar{\rho}_H}\left[\frac{gD(\rho_L - \rho_G)}{\rho_L}\right]^{-0.5} < 3.5$$

the slug velocity becomes

$$\bar{u}_s = \frac{G}{\bar{\rho}_H} + \bar{u}_{s\infty} \quad (4.4.31)$$

where $\bar{u}_{s\infty}$ is the velocity at the bubble front in the case of drainage from a horizontal tube. The velocity $\bar{u}_{s\infty}$, according to Benjamin [4.4.16] and Webber [4.4.17] is

$$\bar{u}_{s\infty} = \left(0.54 - 1.76 N_E^{-0.56}\right)\sqrt{\frac{gD(\rho_L - \rho_G)}{\rho_L}} \quad (4.4.32)$$

The mass rate of pickup by the slug is

$$G_f = \left(\bar{u}_s - \frac{G_f}{\rho_L \bar{\alpha}_{fLe}}\right)\rho_L \bar{\alpha}_{fLe} \quad (4.4.33)$$

The liquid film velocity, $G_f/\rho_L\bar{\alpha}_{fL}$, was obtained by writing a mass balance across the plane where the liquid film is leaving the slug and across any other plane drawn normal to the film, where it occupies a volume fraction $\bar{\alpha}_{fL}$, so

$$\left(\bar{u}_s - \frac{G_f}{\rho_L \bar{\alpha}_{fL}}\right)\bar{\alpha}_{fL} = \left(\bar{u}_s - \frac{G}{\bar{\rho}_H}\right)\bar{\alpha}_{SL} \quad \text{or} \quad \frac{G_f}{\rho_L \bar{\alpha}_{fL}} = \frac{G}{\bar{\rho}_H}\left(1 - C\frac{\bar{\alpha}_{SL} - \bar{\alpha}_{fL}}{\bar{\alpha}_{fL}}\right) \quad (4.4.34)$$

In the Dukler and Hubbard model, the liquid-film thickness varies along the film, starting at $\bar{\alpha}_{fL} = \bar{\alpha}_{SL}$ right against the back of the slug and decreasing to

$\overline{\alpha}_{fL} = \overline{\alpha}_{fLe}$, where the liquid film is picked up by the liquid slug or a liquid-film length l_f away from the back of the downstream slug.

Dukler and Hubbard utilized Eq. [4.4.34] and a momentum equation for the stratified film to determine $\overline{\alpha}_{fLe}$ as a function of l_f. Finally, Dukler and Taitel developed the following expressions for the mixing length l_m and the slug length l_s shown in Figure 4.4.1 and for their pressure drops:

$$l_m = \frac{0.15}{g}\left(\frac{G}{\rho_L} - \frac{G_f}{\rho_L \overline{\alpha}_{fLe}}\right)$$

$$l_s = \frac{G/\overline{\rho}_H}{\omega_s(\overline{\alpha}_{SL} - \overline{\alpha}_{fLe})}\left[\frac{(1-\overline{x})\overline{\rho}_H}{\rho_L} + C(\overline{\alpha}_{SL} - \overline{\alpha}_{fLe})\right]$$

$$\Delta P_{SF} = 2f_s \left(\frac{G}{\overline{\rho}_H}\right)^2 \frac{(l_s - l_m)[\rho_L(\overline{\alpha}_{SL}) + \rho_G(1 - \overline{\alpha}_{SL})]}{D}$$

$$\Delta P_{acc} = C\rho_L \overline{\alpha}_{SL} \frac{G}{\overline{\rho}_H}\left(\frac{G}{\overline{\rho}_H} - \frac{G_f}{\rho_L \overline{\alpha}_{fLe}}\right)$$

(4.4.35)

where the friction factor f_s is calculated from Re_s as specified in Eq. (4.4.30). Dukler and Taitel obtained good agreement with their test results but only after they introduced their experimentally measured values of slug liquid volumetric fraction $\overline{\alpha}_{SL}$ and slug frequency ω_s. In other words, even in the case of Dukler and Hubbard's work, a complete model for horizontal slug flow is not yet available from first principles. Empirical correlations are usually required for the slug frequency ω_s and the liquid holdup $\overline{\alpha}_{SL}$ in the slug body.

The slug frequency has often been calculated from the Gregory and Scott [4.4.18] correlation

$$\omega_s = 0.0226\left[\frac{G(1-\overline{x})}{gD\rho_L}\left(\frac{19.75\overline{\rho}_H}{G} + \frac{G}{\overline{\rho}_H}\right)\right]^{1.2}$$

(4.4.36)

The slug frequency can also be predicted from a semitheoretical model developed by Tronconi [4.4.19], where $\Omega_s = \rho_L \omega_s D/G\overline{x}$ is a function of the Martinelli parameter ϕ_{LG}, as shown in Figure 4.4.5. The liquid holdup in the slug body has been measured by Gregory et al. [4.4.20] and their proposed correlation was

$$\overline{\alpha}_{SL} = \frac{1}{1 + (G/8.66\overline{\rho}_H)^{1.39}}$$

(4.4.37)

where $G/\overline{\rho}_H$ is in meters per second. A more mechanistic model for predicting $\overline{\alpha}_{SL}$ has been proposed by Barnea and Brauner [4.4.21].

Considerable horizontal slug flow data have been accumulated over the years, and a comparison of some of the available slug flow models to the data has been

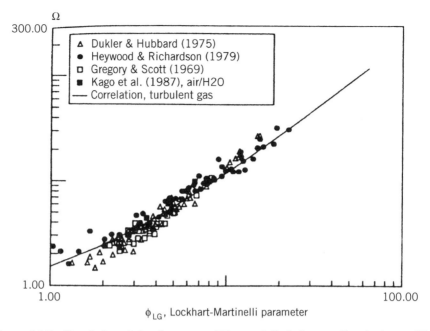

Figure 4.4.5 Correlation of slug frequency of Tronconi. Turbulent gas flow in the stratified flow at the pipe entry. (From Ref. 4.4.19.)

reported by Azzopardi et al. [4.4.22]. Three simplified versions of the Dukler and Hubbard model were employed in that study: (1) $\bar{\alpha}_{SL} = 1$ and $l_m = 0$, (2) $\bar{\alpha}_{SL} = 1$, and (3) $\bar{\alpha}_{SL}$ from Reference (4.4.20). The best comparison to the data was obtained for the most simplified case 1. This is not too surprising since Dukler and Hubbard pointed out that their model is not especially sensitive to the slug liquid volumetric fraction $\bar{\alpha}_{SL}$. Another striking finding of Azzopardi et al. comparison is that the pressure drop data spread is at least minus 60% and plus 30% around all the models and correlations tested.

At the present time, the disagreement of the models with test results appears large enough to justify the application of simplified slug flow models. For example, it makes sense to set $\bar{\alpha}_{SL} = 1$ in the Dukler and Hubbard model. This means that

$$\frac{G_f}{\rho_L} = C\frac{G}{\bar{\rho}_H} \qquad C = 0.021 \ln\left(\frac{GD\rho_L}{\mu_L \bar{\rho}_H}\right) \qquad \bar{\alpha}_{fLe} = \frac{2C}{1+C}$$

$$l_m = \frac{0.15}{g}\left[\frac{G}{\rho_L} - \frac{G}{\bar{\rho}_H}(1+C)\frac{1}{2}\right] \qquad l_s = \frac{G}{\bar{\rho}_H}\frac{1}{\omega_s}\frac{1+C}{1-C}\left(\frac{1-\bar{x}}{\rho_L}\bar{\rho}_H + C\frac{1-C}{1+C}\right) \quad (4.4.38)$$

$$\Delta P_{SF} = 2f_s\left(\frac{G}{\bar{\rho}_H}\right)^2 \frac{(l_s - l_m)\rho_L}{D} \qquad \Delta P_{acc} = C\rho_L\left(\frac{G}{\bar{\rho}_H}\right)^2\left(1 - \frac{1-C}{2}\right)$$

Equations (4.4.38) provide an adequate solution to horizontal slug flow if they are coupled with a slug frequency calculated, for example, from the Tronconi curve of Figure 4.4.4.

In summary, our physical understanding of horizontal slug flow has improved significantly over the years. A rather complex gas–liquid configuration has emerged that is intermittent over time. Good attempts have been made to describe it analytically, but their predictive capability does not support their added complexity or their inclusion in complex system computer codes.

4.4.7 Transition from Vertical Slug to Annular Flow

Several approaches have been suggested to describe the slug-to-annular transition in vertical flow. The most accepted approach was developed by Wallis [4.4.23]. He assumed that the transition to annular flow occurs when the vapor shear stress is sufficient to carry up the liquid film (i.e., stop downward liquid flow). This phenomenon, referred to as *flooding*, is of great importance in the design of bubble towers and other process equipment. It is covered in detail in Chapter 5.

"Flooding" in a tube with counterflow of gas is illustrated in Figure 4.4.6. On the left is the extreme case of practically no water flow; as the gas velocity increases, flooding occurs when a liquid drop attached to the tube by surface tension begins to slide upward and eventually is entrained by the gas. (This transition from A to B does not necessarily lead to slug flow.) The case of a thin liquid film is depicted in the center of Figure 4.4.6; the waves on the surface of the film increase in amplitude suddenly as a critical gas velocity is reached; at that time, the large liquid drops break away from the crests of the waves and are entrained by the upward gas flow. At low water flows, the bridging between waves is intermittent, as depicted in the center sketch. At high liquid flows, the bridging is complete, and a transition to slug flow occurs as shown at the extreme right of Figure 4.4.6. The transition

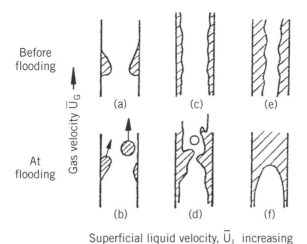

Figure 4.4.6 Flooding regimes in countercurrent gas–liquid flows.

from E to F is also representative of the case of practically no gas flow; the maximum water flow under condition F is then the rate that will just prevent a slug of gas from rising.

As discussed in Chapter 5, flooding has been shown to depend on the following nondimensional superficial gas and liquid velocities:

$$u_G^* = \bar{u}_{SG} \sqrt{\rho_G} \, [gD(\rho_L - \rho_G)]^{-1/2}$$
$$u_L^* = \bar{u}_{SL} \sqrt{\rho_L} \, [gD(\rho_L - \rho_G)]^{-1/2}$$
(4.4.39)

Its occurrence can be predicted from an expression of the form

$$\sqrt{u_G^*} + c' \sqrt{u_L^*} = c \tag{4.4.40}$$

where c' and c are constants depending upon fluid properties and channel geometry.

One of the difficulties in establishing the transition from slug to annular flow is the ability to measure it accurately. The most common technique has relied on visual observations. This approach is subjective, and the defined transition point depends on the observer. More quantitative evaluations can be obtained by using an electrical probe that determines the electrical resistance between the tip of the probe and the channel wall as a function of time. The recorded signal can then be used to determine when the tube is no longer bridged with slugs of water. This definition of the transition point is still arbitrary since there is a rather broad region in which slugs are detected few and far between but do not disappear completely. Another method suggested by Wallis et al. [4.4.24] consists of detecting a liquid bridge across the gas core by sampling the flow with a liquid entrainment probe at the tube axis. In slug flow, the percent of liquid entrained is high, and it decreases rapidly during the transition to annular flow. A critical lower gas velocity can then be defined for the transition condition.

Wallis compared all of the foregoing methods and found that the entrainment probe gave the best agreement with the condition where the slug flow correlation for gas volume fraction stopped being valid. The electrical probe tended to overestimate the gas velocity at transition, and the visual observations were too subjective. Based on these findings, Wallis recommended the empirical equation

$$\sqrt{u_G^*} - 0.6 \sqrt{u_L^*} = 0.4 \tag{4.4.41}$$

to predict the transition from slug to annular flow. Equation (4.4.41) gives a much lower gas flow rate than the following equation, determined by Griffith and Haberstroth [4.4.25] with a conducting probe:

$$\sqrt{u_L^*} - 0.6 \sqrt{u_L^*} = 0.9 \tag{4.4.42}$$

It is interesting to note that Eq. (4.4.41) was found by Wallis to correspond to a constant gas volume fraction between 0.75 and 0.8. The possibility of using such a constant gas volume fraction to predict the transition condition can also be inferred from the measurements reported by Griffith [4.4.26] for steam–water mixtures. The data he obtained with a conducting probe are reproduced in Figure 4.4.7, and it is observed that the transition line can be represented at each pressure by a constant steam quality value corresponding to a gas volume fraction on the order of 0.8.

Some investigators have chosen to define an intermediate transition point between slug and annular flow. It has been referred to as intermittent or churn annular flow. It comes about from acceleration of the large gas bubbles so that they are lined up next to each other and the tail of one large gas bubble touches the nose of the following bubble. The intermediate liquid slugs become unstable and are destroyed by the gas bubble trailing wake effect. A simplified model for predicting the transition to churn annular can be constructed by assuming that all the gas bubbles are absorbed into a Taylor bubble at the transition from bubble to slug flow, or that $\overline{\alpha}_{GS} = 0$ at $\overline{\alpha} = \overline{\alpha}_{AB}$. Churn annular flow is presumed to take place when $\overline{\alpha}_{GS}$ rebuilds to $\overline{\alpha}_{AB}$ with no further growth in the Taylor bubble. By using Eq. (4.4.24), there results

$$\left(\frac{d_{TB}}{D}\right)^2 \frac{l_{TB}}{l_{TB} + l_s} = \overline{\alpha}_{AB} \qquad (4.4.43)$$

$$\overline{\alpha}_{DE} = 2\overline{\alpha}_{AB} - \overline{\alpha}_{AB}^2$$

where $\overline{\alpha}_{DE}$ corresponds to the transition from slug to churn annular flow. Equation (4.4.43) predicts that $\overline{\alpha}_{DE} = 0.44$ when $\overline{\alpha}_{AB} = 0.25$ and $\overline{\alpha}_{DE} = 0.75$ when $\overline{\alpha}_{AB} = 0.5$. It also means that the churn annular transition gas volumetric fraction would go up with increased liquid velocity and reduced pipe diameter.

Taitel and Dukler [4.4.27] have suggested that the transition to churn annular flow occurs where

$$\frac{\overline{u}_{SL}}{\sqrt{gD}} = 0.245 \left(\frac{L_e}{D}\right) - 0.22 \frac{\overline{u}_{SG}}{\sqrt{gD}} \qquad (4.4.44)$$

where $\overline{u}_{SL}$ and $\overline{u}_{SG}$ are the respective liquid and gas superficial velocity and L_e is the equivalent pipe length to its exit.

They also suggested that the full transition to annular flow takes place when liquid entrainment reduces the presence of large liquid waves. This entrainment-induced transition criterion is defined as

$$\frac{\overline{u}_{SG}}{[\sigma g(\rho_L - \rho_G)/\rho_G^2]^{1/4}} = 3.1 \qquad (4.4.45)$$

4.4.8 Transition from Horizontal Slug to Intermittent or Annular Flow

For horizontal flow, the flow pattern is stratified at low gas flow rates. As the gas velocity increases, waves appear on the surface of the stratified liquid and they can

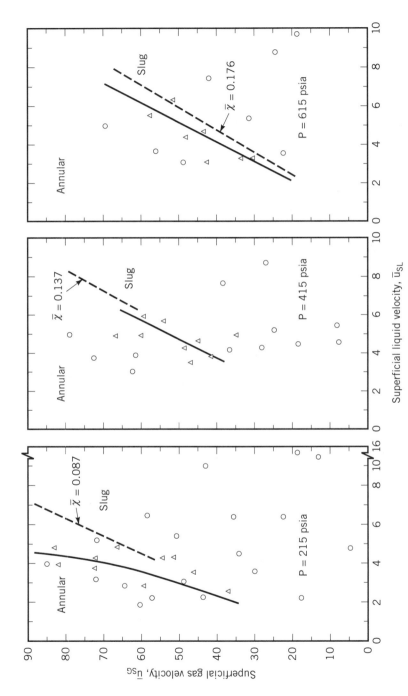

Figure 4.4.7 Transition from slug to annular flow for steam flow in a 0.375-in. pipe. (From Ref. 4.4.25.)

grow to form a stable slug when the liquid volume fraction is large enough to maintain such a slug. This requires the gas volume fraction, $\bar{\alpha}$, to be below 0.5 or the liquid volume fraction to be above 0.5. If the gas volume fraction is above 0.5, the waves are swept up around the wall and the annular flow transition takes place directly from the stratified wavy flow pattern. If the gas volume fraction is below 0.5, intermittent slug flow will result. This intermittent flow will transition to dispersed bubble flow when the turbulent fluctuations in the liquid film exceed the buoyant forces which make the gas rise to the top of the pipe.

According to Taitel and Dukler [4.4.27], the transition from stratified to intermittent or annular flow will occur when

$$\frac{\rho_G \bar{u}_{S,G}^2}{g(\rho_L - \rho_G)D} > \left\{ \frac{(\pi/4)(1 - H_l/D)^2 \bar{\alpha}^2}{[1 - (2(H_l/D) - 1)^2]^{1/2}} \right\}^{1/2} \tag{4.4.46}$$

where D is the pipe diameter and H_L the depth of liquid in the pipe. Dispersion of the gas is visualized to take place when

$$\bar{u}_L \geq \left[\frac{4\bar{\alpha}}{S_i} \frac{g}{f_L} A \left(1 - \frac{\rho_G}{\rho_L}\right) \right]^{1/2} \tag{4.4.47}$$

where A is the pipe area, S_i the interface perimeter, $\bar{u}_L$ the liquid average velocity, and f_L the corresponding friction factor.

4.4.9 Severe or Terrain Slugging

When a line with a downward inclination ends up with a vertical riser, or when a pipe is laid in a hilly terrain, the lower section of the pipe can accumulate liquid and prevent the gas passage. The gas is compressed until it overcomes the riser liquid head. A long liquid slug is thus formed which is pushed ahead of the expanding gas and which can produce severe slugging. The process of severe slugging formation is illustrated in Figure 4.4.8. In Figure 4.4.8a, the slug is formed; in Figure 4.4.8b, the riser fills up with water, permitting the gas to reach the bottom of the riser and producing a blowout as shown in Figure 4.4.8c. In Figure 4.4.8d, the remaining liquid in the riser falls back to the bottom of the riser and the process of severe slugging starts all over again.

According to Taitel[4.4.28], severe slugging will not occur when the pressure at the top of the riser, p_s, exceeds the following criterion:

$$\frac{p_s}{p_0} > \frac{(\bar{\alpha}/\bar{\alpha}')l - h}{p_0/\rho_L g} \tag{4.4.48}$$

where p_0 is the atmospheric pressure, $\bar{\alpha}$ the gas volumetric fraction in the holdup

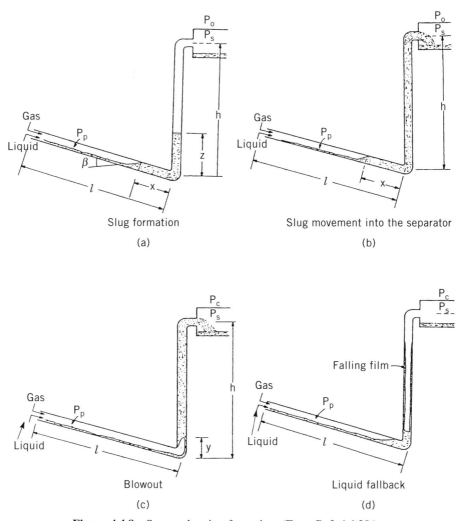

Figure 4.4.8 Severe slugging formation. (From Ref. 4.4.28.)

line of length l, $\overline{\alpha}'$ the gas holdup in the gas cap penetrating the liquid column, and h is the riser length. The value of the gas holdup $\overline{\alpha}$ can be calculated from the stratified model discussed in Section 4.5.1; the value of the gas holdup $\overline{\alpha}'$ here was obtained from a simplified model described in Reference 4.4.28, based on the work of Fernandez et al.[4.4.29]. This vertical slug model starts from Eq. (4.4.6) for the gas velocity. A liquid mass balance relative to a coordinate system that moves with the transitional Taylor bubble velocity $\overline{u}_G$ is written to yield

$$\overline{\alpha}_{fL}(\overline{u}_G + \overline{u}_f) = \overline{\alpha}_{SL}(\overline{u}_G - \overline{u}_L) \qquad (4.4.49)$$

where $\overline{\alpha}_{fL}$ and $\overline{\alpha}_{SL}$ are the liquid volume fraction in the liquid moving downward and in the slug, $\bar{u}_f$ is the liquid film velocity around the Taylor bubble (taken positive for downward flow), and $\bar{u}_L$ is the liquid velocity in the slug. In the liquid slug, Fernandez et al. assumed that gas bubbles were present and traveling at Harmathy terminal rise velocity, Eq. (4.3.3), so that

$$\bar{u}_L = \bar{u}_{TP} - 1.53 \left[\frac{g(\rho_L - \rho_G)\sigma}{\rho_L^2}\right]^{1/4} (1 - \bar{\alpha}_{SL}) \qquad (4.4.50)$$

The liquid film around the Taylor bubble was considered to be a free-falling film for which the film velocity $\bar{u}_f$ can be expressed in terms of its thickness, δ:

$$\bar{u}_f = \left\{\frac{(\delta/D)^{1-m}}{k[\mu_L^2/D^3 g(\rho_L - \rho_G)\rho_L]^{1/3}(4\rho_L D/\mu_L)^m}\right\}^{1/m} \qquad (4.4.51)$$

In Eq. (4.5.51) k and m for laminar flow are 0.909 and $\frac{1}{3}$, respectively. For turbulent flow [i.e., $(4\rho_L \bar{u}_f \delta/\mu_L) > 1000$], Fernandez et al. suggested that $k = 0.0682$ and $m = \frac{2}{3}$.

The film liquid holdup can be expressed in terms of the film thickness δ:

$$\bar{\alpha}_{fL} = 4\frac{\delta}{D} - 4\left(\frac{\delta}{D}\right)^2 \qquad (4.4.52)$$

For a prescribed value of $\bar{\alpha}_{SL}$, it is possible to solve Eqs. (4.4.6) and (4.4.49) to (4.4.52) for $\bar{u}_f$, δ, and $\bar{u}_L$ for prescribed values of $\bar{u}_{TP} = G/\bar{\rho}_H$. In Reference 4.4.28, the slug liquid volumetric fraction $\bar{\alpha}_{SL}$ was taken at 0.7 (a higher value is suggested in Section 4.4.4) as being indicative of bubble-to-slug flow transition and it was found that $\bar{\alpha}_{fL} = 0.11$ for a wide range of liquid and gas superficial velocities, ranging from 0.01 to 10 m/s in a 5-cm pipe. This means that $\bar{\alpha}' = 0.89$ and that the Taylor bubble has a diameter $d_{TB}/D = 0.94$ and is relatively independent of the liquid and gas flow rates. It is also worth noting that

$$\bar{u}_{LS} = \bar{u}_L \bar{\alpha}_{SL} \frac{l_s}{l_u} - \bar{u}_f \bar{\alpha}_{fL} \frac{l_u - l_s}{l_u} \qquad (4.4.53)$$

$$\frac{l_s}{l_u} = \frac{\bar{u}_{LS} + \bar{u}_f \bar{\alpha}_{fL}}{\left\{\bar{u}_{TP} - 1.53[g(\rho_L - \rho_G)\sigma/\rho_L^2]^{1/4}(1 - \bar{\alpha}_{SL})\right\}\bar{\alpha}_{SL} - \bar{u}_f \bar{\alpha}_{fL}}$$

The total liquid holdup in the slug cell is

$$1 - \overline{\alpha} = \frac{l_s}{l_u} \overline{\alpha}_{SL} + \left(1 - \frac{l_s}{l_u}\right) \overline{\alpha}_{fL} \qquad (4.4.54)$$

where l_s is the length of the slug and l_u is the length of the total slug cell as shown in Figure 4.3.1. Finally, by employing Eq. (4.4.50), one can determine whether severe slugging will occur or how to choose p_s to avoid it. It is also worth noting that in a transient mode, very large liquid slugs can be formed if the piping arrangement allows or encourages it.

4.4.10 Summary: Slug Flow

1. Considerable progress has been made in an understanding of gas–liquid slug flow. There are good correlations for the terminal velocity of a single large gas bubble rising in a vertical and horizontal pipe and for the average gas volumetric fraction. Simplified analytical models have been developed to describe the behavior and the pressure drop of a multitude of liquid slugs separated by gas bubbles as well as the transition from slug flow to other flow patterns.

2. There remain several shortcomings in our understanding of slug flow, including detailed knowledge about the form and amount of gas present in the liquid film and slug, the periodic changes in liquid-film flow direction, the frequency of slugs, the developing nature of slug flow, the exact impact of channel geometry, and the lack of data beyond air–water mixtures at atmospheric pressure. As in the case of bubble flow, this lack of microscopic details does not appear to significantly affect the overall (top-down) prediction of average phasic velocities, volume fraction, and wall friction in complex systems.

3. The system computer codes have taken considerable liberty with the results obtained from gas–liquid slug flow pattern tests:

- There is a tendency to avoid this type of flow pattern. For example, in the TRAC-BF1MOD1 system computer code, the bubble flow region is just expanded up to the annular flow pattern. In the TRAC-PF1/MOD1 system computer code (3.5.13), an empirical interpolation is employed from the effective bubble size d_b at the end of bubble flow at $\overline{\alpha}_{tr} = 0.3$ and $G < 2000$ kg/m²·s, to a Taylor bubble diameter, d_{TB}, which reaches the hydraulic diameter D_H at $\overline{\alpha} = 0.5$, or

$$d_{TB} = \frac{D_H - d_b}{0.2} \overline{\alpha} + 2.5 d_b - 1.5 D_H \qquad (4.4.55)$$

No systematic tests or evaluation of this recipe have been carried out. Furthermore, the bubbly flow interfacial drag coefficient is employed. In the RELAP system computer code, an effort is made to recognize the presence of gas

bubbles in the liquid film and slug, but again an arbitrarily developed exponential function is employed to predict its value [see Eq. (3.6.27)]. Also, as already noted, the same interfacial drag coefficient tends to be employed for the various gas bubbles.
- There is no recognition of the impact of channel geometry, which can be significant, as shown in Figure 4.4.3. There is no effort to deal with the developing nature of slug flow, and the wall shear stresses are specified from empirical correlations.
- There is no recognition of the interfacial closure equation that can be derived from steady-state slug flow momentum equations, and it is presumed that the interfacial closure laws are not dependent on wall shear, heat transfer, and flow acceleration.
- The flow pattern transition expressions have been simplified considerably.
- Simplified models such as the severe slugging model tend to be as accurate as complex models, which try to simulate some of the flow microscopic details.

In summary, system computer codes have made limited use of the slug flow pattern information developed over the years. However, the models employed may be adequate for the purpose they have to serve in system computer codes. Here again, this presumption needs to be tested and validated.

4.5 HORIZONTAL STRATIFIED FLOW

A stratified flow pattern is characterized by separation of the liquid and gas flows, with the liquid moving downstream along the bottom of the horizontal pipe or channel and the gas flowing concurrently above it. When the gas velocity is low, the gas–liquid interface appears undisturbed. However, as the gas velocity is increased, waves appear on the gas–liquid interface and the flow pattern is referred to as stratified wavy. With additional increase in gas velocity, large-amplitude waves of long wavelength result and the interface surface becomes highly irregular. This can lead to slug formation when the liquid volume fraction is large. At very high gas velocities, atomization of the waves by the shearing gas flow occurs and an asymmetrical annular flow pattern develops when the gas volume fraction exceeds 50%.

4.5.1 Analysis of Smooth Stratified Flow

This particular flow pattern was analyzed by Taitel and Dukler [4.4.27], who applied the Lockhart–Martinelli methodology to the inclined channel configuration shown

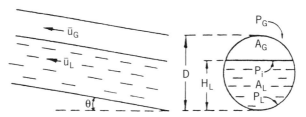

Figure 4.5.1 Parameters in equilibrium stratified flow.

in Figure 4.5.1. By assuming a fully developed flow with no significant acceleration or change in properties, the momentum equations for the liquid and gas phases are Eqs. (3.6.44) and (3.6.45) or Eqs. (4.3.27), which can be combined to eliminate dp/dz and yield the interfacial shear closure Eq. (3.6.45) or (4.3.28).

$$\frac{P_i \tau_i}{A} = \overline{\alpha}\frac{P_{WL}\tau_{WL}}{A} - \frac{P_{WG}\tau_{WG}}{A}(1-\overline{\alpha}) + g(\rho_L - \rho_G)\overline{\alpha}(1-\overline{\alpha})\sin\theta \quad (4.3.28)$$

Smooth stratified flow has an inherent advantage over the Lockhart–Martinelli analysis because the interfacial perimeter P_i and the liquid and gas wetted perimeters P_{WL} and P_{WG} can be specified in terms of the pipe diameter and the gas volume fraction $\overline{\alpha}$ or the ratio H_L/D, where H_L is the height of the liquid layer in the pipe and D the pipe diameter. For example,

$$\overline{\alpha} = \frac{1}{\pi}\left[\cos^{-1}\left(2\frac{H_L}{D}-1\right) - \left(2\frac{H_L}{D}-1\right)\sqrt{1-\left(2\frac{H_L}{D}-1\right)^2}\right]$$

$$\frac{P_i}{D} = \sqrt{1-\left(2\frac{H_L}{D}-1\right)^2} \quad (4.5.1)$$

$$\frac{P_{WL}}{D} = \pi - \cos^{-1}\left(2\frac{H_L}{D}-1\right) \quad \frac{P_{WG}}{D} = \cos^{-1}\left(2\frac{H_L}{D}-1\right)$$

The interfacial and wall shear stresses were specified from

$$\tau_{WL} = \frac{f_L \rho_L \overline{u}_L^2}{8} \quad \tau_{WG} = \frac{f_G \rho_G \overline{u}_G^2}{8} \quad \tau_i = \frac{f_i \rho_G (\overline{u}_G - \overline{u}_L)^2}{8} \quad (4.5.2)$$

4.5 HORIZONTAL STRATIFIED FLOW

with the wall friction factors f_L and f_G being defined in terms of the corresponding Reynolds numbers,

$$f_L = C_L \left(\frac{D_L \rho_L \bar{u}_L}{\mu_L}\right)^{-n} \quad f_G = C_G \left(\frac{D_G \rho_G \bar{u}_G}{\mu_G}\right)^{-m} \quad (4.5.3)$$

Because the interface is smooth, Taitel and Dukler assumed that $f_i = f_G$ and that $\bar{u}_G \gg \bar{u}_L$, so that

$$\tau_i \approx \frac{f_G \rho_G \bar{u}_G^2}{8} \quad (4.5.4)$$

The hydraulic diameters D_L and D_G were obtained from

$$\frac{D_L}{D} = \pi \frac{D}{P_{WL}}(1 - \bar{\alpha}) \quad \frac{D_G}{D} = \pi \bar{\alpha} \frac{D}{P_{WG} + P_i} = \pi \bar{\alpha} \left(\frac{P_{WG}}{D} + \frac{P_i}{D}\right)^{-1} \quad (4.5.5)$$

By substituting all the preceding relations into the interfacial shear Eq. (4.3.28), there results

$$X^2 \left(\frac{\bar{u}_L}{\bar{u}_{SL}} \frac{D_L}{D}\right)^{-n} \left(\frac{\bar{u}_L}{\bar{u}_{SL}}\right)^2 \frac{P_{WL}}{(1-\bar{\alpha})P}$$

$$- \left(\frac{\bar{u}_G}{\bar{u}_{SG}} \frac{D_G}{D}\right)^{-m} \left(\frac{\bar{u}_G}{\bar{u}_{SG}}\right)^2 \left[\frac{P_{WG}}{P\bar{\alpha}} + \frac{P_i}{P(1-\bar{\alpha})} + \frac{P_i}{P\bar{\alpha}}\right]$$

$$- \frac{(\rho_L - \rho_G) g \sin \theta}{(dp/dz)_{FG}} = 0 \quad (4.5.6)$$

where X^2 is the Martinelli parameter,

$$X^2 = \frac{(dp/dz)_{FL}}{(dp/dz)_{FG}} \quad (4.5.7)$$

and $\bar{u}_{SL}$ and $\bar{u}_{SG}$ are the superficial liquid and gas velocity and $(dp/dz)_{FL}$ and $(dp/dz)_{FG}$ are the corresponding single-phase frictional losses. The total wetted perimeter is P, with $P = \pi D = P_{WL} + P_{WG}$.

With the exception of X^2 and $(\rho_L - \rho_G)g \sin \theta/(dp/dz)_{FG}$ all the other variables in Eq. (4.3.29) are related to H_L/D, and the solution of Eq. (4.3.29) yields the curves

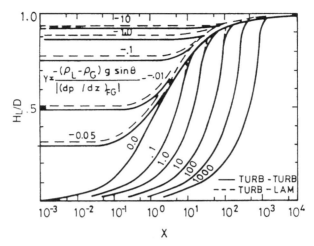

Figure 4.5.2 Equilibrium liquid level for stratified flow. (From Ref. 4.3.27.)

of Figure 4.5.2, where H_L/D is plotted versus the Martinelli parameter X at constant values of $(\rho_L - \rho_G) g \sin \theta /(dp/dz)_{FG}$. Solutions are given for turbulent liquid flow and laminar and turbulent gas conditions. Available stratified test data show that the Taitel and Dukler model underpredicts the friction pressure drop and overpredicts the liquid holdup or height. This is not unexpected because waves are practically always present at the gas–liquid interface, and the interface shear will be higher than presumed under Eq. (4.5.4).

4.5.2 Interfacial Waves and Friction during Stratified Flow

The best studies of interfacial waves and friction during stratified flow have been done in transparent rectangular channels because observations and measurements are much easier to perform. The work of Andritsos and Hanratty [4.5.1] and Shi and Kocamustafaogullari [4.5.2] were carried out with air and water and they are in excellent agreement. For a fixed water flow rate, the first waves to appear are two-dimensional and of small amplitude. The initiation of these regular two-dimensional waves occurs (4.5.1) when

$$\bar{u}_{SG} \geq 2\bar{\alpha} \left[\frac{(1 - \bar{\alpha})\mu_L(\rho_L - \rho_G)g}{\rho_L \rho_G \bar{u}_{SL} S} \right]^{1/2} \tag{4.5.8}$$

where $\bar{u}_{SG}$ and $\bar{u}_{SL}$ are the superficial gas and liquid velocities and S is Jeffrey's sheltering coefficient, which was suggested to be taken as $S = 0.06$ in Reference 4.5.1.

As the gas flow increases, the length of the waves decreases and the wave structure takes the form of irregular large-amplitude waves. The large-amplitude waves are formed [4.5.1] when

$$\bar{u}_{SG} = \bar{u}_{SKH} \left[\tanh\left(\frac{k_m \bar{H}_L}{10}\right) \right]^{-0.1} \frac{1}{\alpha^2} \left(\frac{\theta_w}{\theta}\right)^{0.025} \quad (4.5.9)$$

where $\bar{u}_{SKH}$ and k_m are, respectively, the gas superficial velocity and the characteristic wave number predicted by the Kelvin–Helmholtz (KH) theory:

$$\bar{u}_{SKH} = \frac{\bar{\alpha}}{1-\bar{\alpha}} \bar{u}_{SL} + \frac{1}{\bar{\alpha}} \left[\frac{k_m \sigma}{\rho_G} + \frac{(\rho_L - \rho_G)g}{k_m \rho_G} \right]^{1/2} \tanh[k_m(D - \bar{H}_L)]^{1/2} \quad (4.5.10)$$

$$k_m = \left(\frac{\rho_L g}{\sigma}\right)^{1/2}$$

D is the pipe diameter, $\bar{H}_L$ is the time-averaged liquid height, and

$$\theta = \frac{\rho_L}{\rho_G} \left(\frac{\bar{\alpha}\sigma}{\mu_L \bar{u}_{SG}}\right)^2 \quad (4.5.11)$$

with θ_w corresponding to the value of θ for water.

Above a certain gas velocity, droplets or liquid filaments are torn from the interface by the flowing gas, and water reaches the top surface of the channel for the first time. This atomization transition can be predicted from

$$\bar{u}_{SG} = 1.8 \, [\bar{u}_{SG} \text{ of Eq. } (4.5.9)] \quad (4.5.12)$$

Wave frequencies and wave propagation velocities are also provided in References 4.5.1 and 4.5.2 as well as the interfacial friction factor, f_i, which is the most relevant parameter to predicting two-phase flow in complex systems. The interfacial shear was calculated from the steady-state momentum equations for the gas and liquid phases:

$$-\frac{dp}{dz} = \frac{P_{WL}\tau_{WL}}{A} + \frac{P_{WG}\tau_{WG}}{A}$$

$$\frac{P_i \tau_i}{A} = \bar{\alpha} \frac{P_{WL}\tau_{WL}}{A} - \frac{P_{WG}\tau_{WG}}{A}(1 - \bar{\alpha}) \quad (4.5.13)$$

By using the measured values of $(-dp/dz)$, P_{WL}/A, P_{WG}/A, and P_i/A, the liquid wall shear stress τ_{WL} is calculated from the first of Eqs. (4.5.13), with the gas wall shear τ_{WG} being obtained from equations of the type (4.5.2) and (4.5.3).

Next, the value of τ_i is deduced and the interfacial friction data are presented in terms of the ratio f_i/f_G. Andritsos and Hanratty suggested that

$$\frac{f_i}{f_G} = \begin{cases} 1 & \text{for } \overline{u}_{SG} \leq \left(\frac{\rho_{G0}}{\rho_G}\right)^{1/2} \text{ (5 m/s)} \\ 1 + 15 \left(\frac{\overline{H}_L}{D}\right)^{0.5} \left[\frac{\overline{u}_{SG}}{5}\left(\frac{\rho_G}{\rho_{G0}}\right)^{1/2} - 1\right] & \text{for higher } \overline{u}_{SG} \end{cases} \quad (4.5.14)$$

In Eqs. (4.5.14), $\overline{u}_{SG}$ is in meters per second and ρ_{G0} is the value of gas density at atmospheric pressure. The comparison of Shi and Kocamustafaogullari data to Andritsos and Hanratty correlations is shown in Figure 4.5.3, and it is very good except at relatively low gas flow rates where the measurements of pressure drop are questionable.

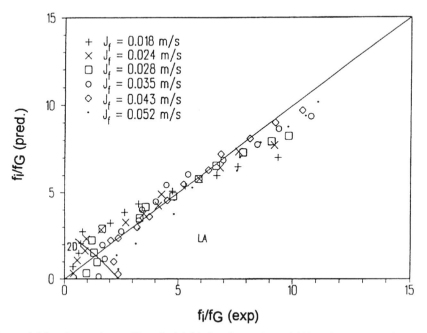

Figure 4.5.3 Comparison of interfacial friction factor data of Shi and Kocamustafaogullari [4.5.2] with Andritos and Hanratty's [4.5.1] correlation. (Reprinted from Ref. 4.5.2 with permission from Elsevier Science.)

4.5.3 Summary: Horizontal Stratified Flow

1. Good information has been generated about gas–liquid stratified flow. There is a very good understanding of the interfacial behavior, including formation and types of traveling waves. The interfacial friction factor is significantly in excess of that obtained for a smooth interface as the gas velocity becomes high. Both the data reduction and analyses of stratified flow have made full use of the interfacial closure equations. The only significant remaining shortcoming is the overemphasis on water–air studies at low pressure. Additional tests at different pressures and with other fluids could help validate the proposed empirical correlations for wave transitions and the interfacial friction factor.

2. Few system computer codes recognize the available stratified flow pattern information, and when they do, they tend to use a smooth interface or $f_i = f_G$. This is acceptable as long as the gas velocity is low enough not to generate large-amplitude waves and to increase the interface friction factor. Under most practical circumstances, interfacial friction factors patterned after Eq. (4.5.14) would be much more appropriate.

3. Considerable basic work has been undertaken in measuring stratified flow wave frequency, amplitude, and propagation velocity. It has not found its way into computer system codes because it is probably difficult, if not impossible, to incorporate it in present two-fluid models.

4. First-principle analytical models describing the interfacial behavior, the wave size, and their growth are not yet available.

4.6 ANNULAR FLOW

Annular flow is characterized by the presence of a continuous liquid film on the channel wall that surrounds a central core of gas laden with a varying amount of entrained liquid droplets. Annular flow is found in many practical applications, such as liquid film evaporators and most high-pressure steam–water systems. It is particularly difficult to analyze because of the large number of different and comparable dynamic forces that must be taken into account. Both viscous and inertial effects are important inside the liquid film; surface tension forces control the stability of the liquid film at the interface, and the drag of the gas along the film determines its motion; finally, near the slug-annular transition, gravity governs whether the film flows up or down. There is a strong interaction between the foregoing forces, and their respective role determines the type of two-phase annular flow pattern that will occur.

4.6.1 Types of Annular Flow Patterns

There are three principal types of annular flow patterns. The first one is found near the annular-to-slug flow transition; it consists of a liquid film with waves of rather large and irregular amplitude. Referred to here as *unsteady or flooding waves*, they are a product of the degeneration of slug flow to annular flow; their amplitude decreases as the gas flow rate is increased. At low liquid flow rates and beyond a minimum critical gas velocity, the interface waviness is small, and a minimum of liquid droplets is carried in the core. The flow pattern here resembles the *"ideal" annular pattern*, where the gas and liquid are assumed to be separated and their *interface smooth*. As the liquid flow is increased and an upper critical gas velocity is reached, *roll waves* appear on the surface of the liquid film. They are periodic in nature and travel on top of the liquid film at a rate several times that of the liquid surface. The front of the roll waves rises sharply from the liquid film while their rear portion tapers off gradually. The appearance of the roll waves coincides with increased entrainment as drops are being sheared away from the top of the waves by the gas stream. A further increase in gas flow raises the presence of liquid droplets in the gas core and eventually leads to atomization of the liquid film (i.e., *fog* or *mist* flow).

With the exception of the ideal annular flow, adequate analytical solutions have not been developed to describe the more complex types of annular flow. In fact, very few fundamental and meaningful predictions have been published for a wavy interface and liquid entrainment in the gas core. Instead, reliance has to be placed on empirical correlations for the interface friction factor and for the liquid entrainment and deposition rates.

4.6.2 Two-Dimensional Analysis of Ideal Annular Flow

In practice, ideal annular flow is found rarely because the gas flows are high enough to produce waves at the liquid interface and liquid droplets of different sizes which are entrained in the gas core. Levy [4.6.1] presented the first solution of ideal laminar and turbulent annular flow. He utilized one-seventh power laws for the turbulent liquid and gas velocity profiles and obtained a theoretical prediction of the two-phase pressure drop and gas volume fraction. Similar analyses have been published by Calvert and Williams [4.6.2] and Anderson and Mantzouranis [4.6.3] for logarithmic distribution. Dukler [4.6.4, 4.6.5] has dealt with the case of a vertical falling film, and his results have been extended to upward flow by Hewitt [4.6.6].

A simplified two-dimensional solution of the ideal annular flow will be presented here primarily to provide some perspective about its limitation. It assumes that gravity effects can be neglected with respect to frictional pressure losses. It consists of a gas core of density ρ_G and radius a and a liquid film of density ρ_L and thickness $(b - a)$ flowing in a vertical circular pipe of radius b. If we assume that the static pressure p is constant over the entire cross section, the shear stress τ at any radial position r and axial distance z is equal to

$$\tau = \left[\left(-\frac{dp}{dz}\right) - \rho'\right]\frac{r}{2} \qquad (4.6.1)$$

4.6 ANNULAR FLOW

where dp/dz is the pressure loss in the flow direction z and ρ' is the density averaged over the cylinder of radius r.

$$\rho' = \frac{2}{r^2} \int_0^r \rho r \, dr \tag{4.6.2}$$

The shear stress at the wall τ_w can be written as

$$\tau_w = \left[\left(-\frac{dp}{dz}\right) - \bar{\rho}\right]\frac{b}{2} \tag{4.6.3}$$

where $\bar{\rho}$ is the average fluid density over the pipe of radius b. If we neglect head losses with respect to frictional pressure, Eqs. (4.6.1) and (4.6.3) yield

$$\frac{\tau}{\tau_w} = \frac{r}{b} \tag{4.6.4}$$

In turbulent flow, the shear stress τ can be written as

$$\tau = \rho l^2 \left(\frac{du}{dy}\right)^2 \tag{4.6.5}$$

where y is the distance measured from the wall, and the mixing length l is assumed to be the same for the liquid and gas and is given by the accepted single-phase relation

$$l = Ky\sqrt{1 - \frac{y}{b}} \tag{4.6.6}$$

K is the mixing-length constant and is usually taken as equal to 0.4.

Substitution of Eqs. (4.6.5) and (4.6.6) into Eq. (4.6.4) and integration with respect to y gives the velocity distribution in the liquid film and gas core:

$$u_L - u_i = \frac{1}{K}\sqrt{\frac{\tau_w}{\rho_L}} \ln \frac{y}{b-a}$$

$$u_G - u_i = \frac{1}{K}\sqrt{\frac{\tau_w}{\rho_G}} \ln \frac{y}{b-a} \tag{4.6.7}$$

where u_i represents the interface velocity, which is presumed to be the same for the gas and liquid. If we let, as in single-phase flow, the liquid velocity goes to zero at $y = y_o$, where

$$\frac{y_0 \rho_L}{\mu_L} \sqrt{\frac{\tau_w}{\rho_L}} = 0.111 \qquad (4.6.8)$$

The interface velocity u_i can be solved for from Eqs. (4.6.7), and the velocity profiles can be integrated to obtain the liquid and gas mass flow rate per unit channel area G_L and G_G. There results for $1 - y_0/b \approx 1$,

$$G_L = \frac{\rho_L(1 - a^2/b^2)}{K} \sqrt{\frac{\tau_w}{\rho_L}} \left[2.2 + \ln \frac{(b-a)\sqrt{\tau_w/\rho_L}\,\rho_L}{\mu_L} - 0.5 \frac{3 + a/b}{1 + a/b} \right]$$

$$G_G = \frac{a^2}{b^2} \frac{\rho_G}{K} \sqrt{\frac{\tau_w}{\rho_L}} \left[\ln \frac{(b-a)\sqrt{\tau_w/\rho_L}\,\rho_L}{\mu_L} + 2.2 \right] \qquad (4.6.9)$$

$$+ \frac{1}{2} \frac{\rho_G}{K} \sqrt{\frac{\tau_w}{\rho_G}} \left[2 \ln \frac{b}{b-a} - \frac{a}{b}\left(\frac{a}{b} + 2\right) \right]$$

The corresponding expression for single-phase water flow in the pipe at the rate G_L is obtained by setting $a = 0$ in Eqs. (4.6.9), or

$$G_L = \frac{\rho_L}{K} \sqrt{\frac{\tau_L}{\rho_L}} \left[\ln\left(\frac{b\rho_L}{\mu_L}\sqrt{\frac{\tau_L}{\rho_L}}\right) + 0.7 \right] \qquad (4.6.10)$$

where τ_L is the wall shear stress for single-phase liquid flow at the rate G_L. By combining Eqs. (4.6.9) and (4.6.10), the following expression is obtained for the Martinelli frictional multiplier $\phi_{TPL}^2 = \tau_w/\tau_L$:

$$\phi_{TPL}\left(1 - \frac{a^2}{b^2}\right)$$

$$= \frac{0.7 + \ln\left(\frac{b\rho_L}{\mu_L}\sqrt{\frac{\tau_L}{\rho_L}}\right)}{\ln\left(\frac{b\rho_L}{\mu_L}\sqrt{\frac{\tau_L}{\rho_L}}\right) + 2.2 - \ln\left(1 + \frac{a}{b}\right) + \ln\left[\phi_{TPL}\left(1 - \frac{a^2}{b^2}\right)\right] - 0.5\frac{3 + a/b}{1 + a/b}}$$

$$(4.6.11)$$

The corresponding slip ratio, defined as the average gas velocity divided by the average liquid velocity, is equal to

$$\frac{\bar{u}_G}{\bar{u}_L} = 1 + \frac{\phi_{TPL}(1 - a^2/b^2)}{4\sqrt{2K}}\sqrt{f_L}\left\{\frac{3 + a/b}{1 + a/b} - \left(\frac{\rho_L}{\rho_G}\right)^{1/2}\left[2\frac{b}{a}\ln\left(1 - \frac{a}{b}\right) + 1 + \frac{2b}{a}\right]\right\} \quad (4.6.12)$$

where f_L is the liquid single-phase friction factor corresponding to τ_L.

It can be shown that for all practical purposes, Eq. (4.6.11) reduces to

$$\phi_{TPL}\left(1 - \frac{a^2}{b^2}\right) = \phi_{TPL}(1 - \bar{\alpha}) \approx 1 \quad (4.6.13)$$

In other words, under ideal annular conditions, the liquid film behaves as if the rest of the channel were filled with liquid, and the usefulness of Eq. (4.6.13) first derived in Reference 4.6.1 has been noted in Chapter 3. Similarly, Eq. (4.6.12) exhibits some of the correct characteristics for slip ratio. The slip ratio decreases as $(\rho_L/\rho_G)^{1/2}$ is reduced and as the flow rate goes up. It should be noted, however, that if one had chosen to form the multiplier ϕ^2_{TPG} instead of ϕ^2_{TPL}, one would obtain

$$\phi^2_{TPG} = \frac{\tau_w}{\tau_G} \approx \frac{1}{\bar{\alpha}^2} \quad (4.6.14)$$

where τ_G is the single-phase wall shear stress for gas flow at the rate G_G. Equation (4.6.14) implies that the gas–liquid interface is smooth and that the interfacial friction factor and gas pressure drop are the same as in single-phase gas flow except for the increased gas velocity due to the presence of a liquid film. Equation (4.6.14) can be shown to be unsatisfactory, because while in an annular pattern the liquid behaves as if the gas were not present, the gas core is greatly influenced by the presence of liquid and, in particular, by the presence of waves at the interface. The waves will increase the interfacial friction of the gas and will reduce the slip of the gas with respect to the liquid film.

The ideal two-dimensional annular flow solution is therefore correct at best only in predicting the liquid-film behavior and the gas flow if there are no waves at the gas–liquid interface. Even under those conditions, the proposed turbulent flow solution may be subject to a serious physical flaw. By allowing the liquid and gas velocity to be equal at the interface, the solution requires, as pointed out by Levy and Healzer [4.6.7], that the gas shear stress velocity at the interface, u^+_{iG}, be equal to

$$u^+_{iG} = \frac{u_{ig}}{\sqrt{\tau_w/\rho_L}} = u^+_{iL}\sqrt{\frac{\rho_G}{\rho_L}} \quad (4.6.15)$$

Since usually $\rho_G/\rho_L \ll 1$, the gas shear stress velocity u^+_{iG} will tend to fall within the laminar range. For example, for $\rho_G/\rho_L = 0.0025$ and $u_{iL}^+ = 15$, the result would be $u_{iG}^+ = 0.75$. This means that the interface on the gas side would be laminar, while it would be turbulent on the liquid side. This is not physically acceptable, and for that reason, changes have been proposed to the ideal annular flow model. In particular, the gas velocity at the interface must be larger than the liquid velocity and waves formed to accommodate that transition.

4.6.3 Modification of Ideal Two-Dimensional Annular Flow Model

One approach suggested by Gill and Hewitt [4.6.8] and Hewitt and Hall-Taylor [4.6.9] was to assume that the liquid film acts like sand roughness in single-phase pipe flow. For fully developed roughness flow, the gas logarithmic velocity distribution would become

$$u_G^+ = \frac{1}{K} \ln \frac{y}{b-a} + 8.5 \tag{4.6.16}$$

and the corresponding interfacial friction factor would be [4.6.10]

$$f_i = \left(4 \log \frac{b}{b-a} + 3.36\right)^{-2} \tag{4.6.17}$$

Equation (4.6.17) presumes a significant jump in gas velocity, u_{iG}, at the interface by comparison to the interfacial liquid velocity, u_{iL}, or

$$\frac{u_{iG}}{u_{iL}} = \sqrt{\frac{\rho_L}{\rho_G}} \frac{8.5}{5.5 + 2.5 \ln\left[(b-a)\sqrt{\tau_w/\rho_L}/\mu_L\right]} \tag{4.6.18}$$

This ratio u_{iG}/u_{iL} would be equal to 11.33 for the example of $\rho_G/\rho_L = 0.0025$ and $u_{iL}^+ = 15$ used previously. It now becomes excessive, which explains why the interfacial friction equation (4.6.17) is low by comparison to test data.

Another plausible approach is to employ the value $u_{iG}^+ = 5.5$ utilized in single-phase flow to define the zero-velocity location in the turbulent gas flow, so that

$$u_G^+ = 2.5 \ln \frac{y}{b-a} + 5.5 = 2.5 \ln \frac{y}{3.32(b-a)} + 8.5 \tag{4.6.19}$$

Equation (4.6.19) defines an equivalent sand roughness k_s such that $k_s = 3.32(b-a)$ and the corresponding interfacial friction factor becomes

$$f_i = \left(4 \log \frac{b}{b-a} + 1.28\right)^{-2} \tag{4.6.20}$$

Equation (4.6.20) is similar but still 20% low by comparison to an accepted Wallis formulation (4.6.11) for interfacial friction factor, discussed later, which employs four times the liquid thickness.

Other variations of the ideal gas annular flow have been evaluated by Wallis (4.6.11). They include the assumption of a homogeneous gas core containing liquid droplets. The previous relation would apply except for replacing the gas density ρ_G by the core density ρ_C and the gas viscosity μ_G by the core viscosity μ_C. Also, there would be a need to specify the degree of liquid entrained in the gas core. Another suggested correction by Wallis was to recognize that the interfacial friction factor f_i is more appropriately defined,

$$f_i = \frac{8\tau_i}{\rho_G(\bar{u}_C - u_{iL})^2} \quad (4.6.21)$$

instead of assuming that the interface liquid velocity u_{iL} is small and negligible with respect to the average gas, $\bar{u}_G$, or core velocity, $\bar{u}_C$. This neglect may be reasonable, particularly when $\rho_G \ll \rho_L$, but it becomes less justifiable when the core density increases due to pressure or when droplets are being present. Under those circumstances, the friction factor is best calculated by integrating the velocity difference, $u_C - u_{iL}$ over the core flow area with u_{iL} obtained from Eqs. (4.6.7) and (4.6.8).

All the preceding ideal annular flow models offer a simplified treatment of the liquid and gas interface. For that reason, they remain inadequate in their predictions of interface behavior, and attempts have been made to improve them by introducing a multilayer liquid film structure.

4.6.4 Two-Dimensional Analysis of Annular Flow with a Two-Layer Liquid Film

Observations of annular flow indicate that the liquid film can be divided into two layers: a continuous liquid layer of thickness δ_B and a disturbed, wavy layer of thickness, $\delta_{max} - \delta_B$, as shown in Figure 4.6.1. The continuous layer is in direct contact with the channel wall and it supports an intermittent layer where the wave crests can grow to a height δ_{max} measured from the wall. The mean thickness of the two-layer liquid film is represented by δ.

Many investigators [4.6.12–4.6.15] have found that the gas flow rate has the dominant influence on the base film thickness δ_B. Dobran [4.6.16] was able to correlate δ_B in upflow, downflow, and for different tube diameters by

$$\frac{\delta_B}{D} = 140 N_L^{-0.433} \mathrm{Re}_C^{-1.35} \quad (4.6.22)$$

$$N_L = \left[\frac{gD^3\rho_L(\rho_L - \rho_C)}{\mu_L^2}\right]^{1/2} \qquad \mathrm{Re}_C = \frac{\bar{u}_{SG}\rho_C D}{\mu_G}$$

N_L is a two-phase Grashoff number (where D is the pipe diameter), $\bar{u}_{SG}$ the superficial gas velocity, and ρ_C the homogeneous core density.

228 FLOW PATTERNS

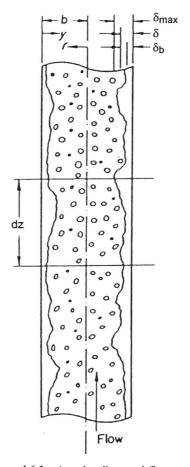

Figure 4.6.1 Annular-dispersed flow model.

Equation (4.6.22) is compared to the data of References 4.6.12 to 4.6.14 as shown in Figure 4.6.2. It is seen that only the gas velocity has a great influence on the thickness of the continuous liquid layer and that it decreases sharply with increased gas flow. On the other hand, the wavy layer thickness, $\delta_{max} - \delta_B$, depends on the gas as well as the liquid-film Reynolds number. It increases with the gas velocity and with the liquid-film Reynolds number as the wave structure changes from a smooth to a three-dimensional small amplitude, to a roll wave interfacial pattern. The thickness, $\delta_{max} - \delta_B$, reaches a maximum at the point of breakdown of the roll waves and at the onset of liquid entrainment. Beyond that point, the wavy layer thickness decreases until the onset of mist flow. At low gas velocities, the height of the wave crests is much larger in upflow than in downflow.

Moeck and Stachiewicz [4.6.15] were the first to suggest that beyond a purely liquid base layer of thickness δ_B, there exists an intermittent layer of decreasing density such that its radial gas volumetric fraction α is obtained from

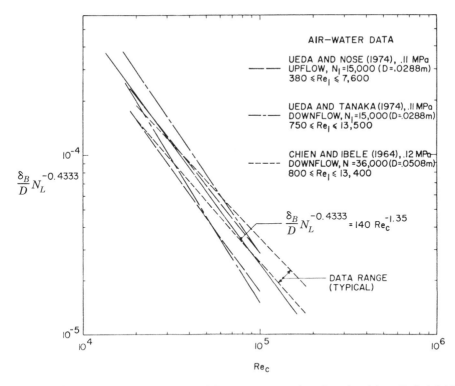

Figure 4.6.2 Continuous liquid layer thickness representation. (Reprinted from Ref. 4.6.16 with permission from Elsevier Science.)

$$1 - \alpha = (1 - \alpha_C)\frac{y - \delta_B}{\delta_{max} - \delta_B} \tag{4.6.23}$$

where α_C represents the constant gas volume fraction in the gas core beyond δ_{max}. They also defined a mean liquid-film thickness δ and employed the normal logarithmic liquid profile from 0 to δ and a linear profile beyond that point which is tangent to the logarithmic shape at that location. Finally, they correlated their data empirically for air–water flows to obtain the following expression for the interface friction factor:

$$f_i = 0.020\left[1 + 545\left(\frac{2\delta}{D}\right)^{1.42}\right]$$
$$\tau_i = \tfrac{1}{8}f_i\rho_G\,(\bar{u}_G - \bar{u}_{\delta_{max}})^2 \tag{4.6.24}$$

where $\bar{u}_G$ is the gas velocity in the core and $\bar{u}_{\delta_{max}}$ is the velocity at the crest of the waves. With appropriate treatment of the liquid droplets in the gaseous core, Moeck and Stakiewicz were able to offer an adequate but empirically based prediction of annular flow which took into account the presence of waves at the interface and liquid entrainment within the central core.

Levy and Healzer [4.6.7] introduced a similar transition layer where the density ρ varied exponentially as a function of the distance y from the wall, or

$$\frac{\rho}{\rho_L} = \left(\frac{y}{\delta}\right)^{-\beta} \tag{4.6.25}$$

with ρ approaching the gas (ρ_G) or the core (ρ_C) density at the end of the transition layer y_t. Levy and Healzer found that the only plausible mixing-length solution required a constant velocity in the transition layer which they connected to the wave velocity at the interface. In addition, they developed an empirical correlation for the exponent β in Eq. (4.6.25) to be able to predict two-phase annular flow.

Instead, Dobran [4.6.16] employed the test data of References 4.6.11 to 4.6.13 to develop an expression for the wavy layer momentum diffusivity, $(\epsilon_M)_{wl}$:

$$\left(\frac{\rho_L \epsilon_M}{\mu_L}\right)_{wl} = 1 + 1.6 \times 10^{-3} (\delta^+ - \delta_B^+)^{1.8} \tag{4.6.26}$$

$$\delta^+ = \frac{\delta \rho_L \sqrt{\tau_w/\rho_L}}{\mu_L} \qquad \delta_B^+ = \frac{\delta_B \rho_L \sqrt{\tau_w/\rho_L}}{\mu_L}$$

Equation (4.6.26) predicts reduced values of the momentum diffusivity within the wavy layer by comparison to those obtained from the Prandtl mixing length Eqs. (4.6.5) and (4.6.6). By employing Eq. (4.6.26) and a Wallis type of interface friction factor, Dobran was able to get acceptable agreement with the momentum and heat transfer test data of Chien and Ibele [4.6.12] and Ueda and Tanaka [4.6.13].

All three preceding two-layer models of the liquid film lead to improved representations of the prevailing physics at the interface in annular two-phase flow and to closer agreement with the test data. This improved performance, however, does not come about without increasing the number of required empirical correlations [e.g., Eq. (4.6.22) or Eq. (4.6.26)]. In other words, a complete first-principle description of the gas–liquid wavy interface is still not available.

4.6.5 Waves and Interfacial Friction Factor in Annular Flow

As discussed previously, the gas–liquid interface in annular flow is mostly covered with waves which play a significant role in determining the two-phase flow behavior. The occurrence of waves and their characteristics were studied starting in the early 1960s by Hewitt and Hall-Taylor [4.6.9]. Their results are reproduced in Figure 4.6.3, which shows that at very low liquid flows and high gas flow rates, the water is incapable of wetting the tube wall and a mist flow pattern exists. Above a critical gas velocity and at low liquid-flow rates, a liquid film is formed and, particularly for falling films, only *small surface ripples* exist at the interface. Beyond a critical liquid flow rate, *disturbance waves* begin to appear. Initially, they are comingled with ripple waves, but as the liquid flow is raised, disturbance waves are formed

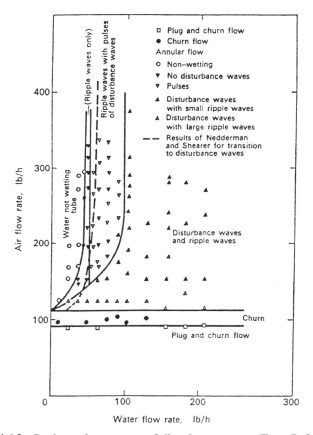

Figure 4.6.3 Regions of occurrence of disturbance waves. (From Ref. 4.6.20.)

primarily and they control the behavior at the interface. As illustrated in Figure 4.6.3, in annular flows below the critical gas velocity and close to the transition to slug flow, waves occur over the full range of liquid velocities. The film flow can then be intermittent and the waves are rather large and do not have a smooth, steady, or consistent profile. In between the waves, a liquid film exists that is decelerating and even reversing its direction before the arrival of the next wave. These waves are characteristic of *churn annular flow* and are referred to as flooding waves because they are associated with the flooding mechanisms in countercurrent vertical flow.

4.6.5.1 Ripple Waves
The ripple-wave regime exists below a critical liquid-film Reynolds number. According to Cousins et al. [4.6.17], the liquid-film Reynolds should be below 200, while Asali and Hanratty [4.6.18] found that ripple waves occur up to liquid-film Reynolds numbers of 330 for vertical flows. The gas velocity also must be above a critical velocity. For an air–water mixture, the gas speed must

exceed 20 m/s. In the ripple-wave region, the interfacial friction factor f_i is usually represented as an augmentation of the single-phase friction factor, f_{SG}, or

$$\frac{f_i}{f_{SG}} = 1 + C\,(\delta_G^+ - 4) \qquad (4.6.27)$$

where δ_G^+ is the nondimensional liquid-film thickness such that

$$\delta_G^+ = \frac{\delta \rho_G}{\mu_G}\sqrt{\frac{\tau_i}{\rho_G}} \qquad (4.6.28)$$

The constant C was found to be equal to 0.045 at gas velocities in excess of 25 m/s and to increase for gas velocities between 20 and 25 m/s. Equation (4.6.27) exhibits a cutoff point at $\delta_G^+ \leq 4$, implying that below or at this value the liquid film is thin enough that the waves are damped by the pipe wall and the interfacial friction factor reduces to that of a smooth tube, or $f_i/f_{SG} = 1$. Hewitt and Hall-Taylor [4.6.17] have suggested a cutoff value of 5. Under most practical circumstances, δ_G^+ is not much larger than 4, and the ideal annular flow model tends to be applicable. In fact, the test results of Nedderman and Shearer [4.6.19] showed no enhancement of interface friction factor for vertical flow in the small-ripple regime.

4.6.5.2 Disturbance Waves

An early and extensive investigation of disturbance or roll waves in vertical upward flow was carried out by Hall-Taylor et al. [4.6.20]. They found that the roll waves are thick compared to the liquid film and that they slide atop the film with a velocity an order of magnitude greater than that of the film. The velocity of the waves appears to be controlled entirely by the gas flow rate, while the number of waves is governed by the liquid rate. The roll waves have been identified as the source of entrainment by high-speed movies which show liquid ligaments torn from the crest of the disturbance waves and their breakup into clouds of liquid droplets in the gas core.

Disturbance waves have been found to have the following characteristics (References 4.6.21 and P-8):

- The ratio of the wave amplitude to the mean film thickness is large (on the order of 5).
- The wave surface is not smooth, and within the wave there are many subpeaks and troughs.
- The wave is rather long by comparison to its amplitude.
- The wave spacing is not uniform and can have a multipeak distribution function. The wave spacing decreases as the liquid flow is increased and the gas flow rate is increased.
- The wave frequency is initially high (near the liquid injection point) and drops as the waves coalesce. The wave frequency increases with liquid flow rate.

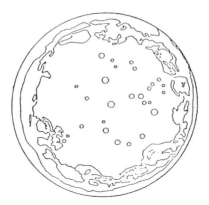

Figure 4.6.4 Cross-sectional view of annular flow behavior. (From Ref. 4.6.22.)

- Waves never break up or pass through each other. Sometimes, they merge to form one high-velocity wave which slows down thereafter.

Little hope exists at the present time for understanding the detailed characteristics of disturbance waves or the processes at work. The prevailing conditions are complex, inconsistent, and not susceptible to theoretical treatment. This is illustrated in Figure 4.6.4 by a cross-sectional view of annular flow sketched by Srivastoa [4.6.22]. Some efforts have been made to analyze periodic stable roll waves. For example, Chung and Murgatroyd [4.6.23] assumed that the size of the wave was determined by a balance between surface tension and static pressure differences across the waves. By postulating dynamic similarity in the pressure distribution, they get that the height h of the wave measured from the edge of the continuous film is given by

$$\frac{hp_d}{\sigma} = \text{constant} \quad (4.6.29)$$

where the constant has about the value 2 for a typical disturbance waveshape. The drag pressure p_d is obtained from

$$p_d = \frac{C_D}{h} \int_0^h \frac{1}{2}\rho_G (\bar{u}_G - \bar{u}_R)^2 \, dy \quad (4.6.30)$$

where C_D is the drag coefficient and is equal to

$$C_D = 15 \left(\frac{h}{B}\right)^2 \quad (4.6.31)$$

B is the length of the roll wave at its base. In Eq. (4.6.30), $\bar{u}_R$ is the mean roll wave velocity. It is obtained by using the accepted turbulent single-phase velocity profile in the liquid film. It is assumed to correspond to the value of $u_R^+ = \bar{u}_R/\sqrt{\tau_w/\rho_L}$ taken at half the total wave height or at

$$\left(\frac{\delta_B + h}{2}\right)^+ = \frac{\delta_B + h}{2}\frac{\rho_L}{\mu_L}\sqrt{\frac{\tau_w}{\rho_L}} \qquad (4.6.32)$$

where δ_B is the base liquid-film thickness.

Chung and Murgatroyd also showed that the ratio of undisturbed liquid-film thickness δ_B to the roll wave height h was equal to 0.25 for laminar flow and 0.12 for turbulent flow. They, finally, pointed out that the interfacial shear τ_i is made up of the shear on the flat part of the film plus the drag on the waves, so that

$$\tau_i = \tau_w + Nhp_d \qquad (4.6.33)$$

N is the number of waves per unit length of flow channel, and it can be calculated for an assumed waveshape and a value of B (found to be equal to about $7h$ for air–water mixtures).

Equation (4.6.33) states that the interface shear stress depends on the shape and number of interface waves. This fact has been recognized by many investigators, who have found that the frictional pressure drop in the gas core depends on the form and amplitude of the waves atop the liquid film. For instance, Chien and Ibele [4.6.12] have shown that the two-phase frictional pressure drop $(-dp/dz)_{FTP}$ in annular downward flow can be predicted from

$$\left(-\frac{dp}{dz}\right)_{FTP} = -f'_G \rho_G \frac{\bar{u}_{SG}^2}{2D} \qquad (4.6.34)$$

where the superficial friction factor f'_G associated with the mean gas superficial velocity $\bar{u}_{SG}$ is related to the waviness of the liquid film according to

$$f'_G = 1.2\left(\frac{\delta_M - \delta_B}{D}\right)^{0.485} \qquad (4.6.35)$$

where δ_M represents the mean film thickness and δ_B corresponds to the continuous undisturbed liquid layer, which can be calculated from Eq. (4.6.22) or Figure 4.6.2.

Other correlations of the interfacial friction factor have been developed in terms of the wave properties (i.e., frequency, spacing, and height), but such properties of wave behavior are not generic or validated enough to permit their application to system computer codes. The best results have been obtained for disturbance waves in horizontal stratified flow. It should be noted here that even though some of the disturbance waves in horizontal flow exhibit many of the same characteristics as reported for those in vertical flows, they are usually generated atop thick horizontal films. Their behavior is less complicated because the wave height is small compared to the film thickness and their interfacial stability can be predicted from conventional fluid mechanics, as discussed briefly in Section 4.5. However, empirical correlations are still being employed to predict their interfacial frictional losses and the same is true of vertical two-phase annular flow.

4.6.5.3 Empirical Correlations for Interfacial Friction Factor for Vertical Annular Flow

An initial semiempirical attempt to predict the interfacial friction factor for vertical annular two-phase flow was made by Levy [4.6.24]. He presumed that gravitational forces could be neglected with respect to shear forces so that the liquid-film flow rate and its thickness could be coupled by relations of the type derived for the ideal annular flow model equation (4.6.9). Also, he determined that the shear stress on the core side of the interface was dominated by the sharp density gradient existing at that location and could be expressed as

$$\frac{\rho_G}{G_G}\left(\frac{\rho_L}{\rho_G}\right)^{1/3}\left(\frac{g\rho_L}{-dp/dz}\right)^{-n}\sqrt{\frac{(-dp/dz)(b/2)}{\rho_L}} = F\left(\frac{b-a}{b}\right) \quad (4.6.36)$$

where $n = 0$ for $(-dp/dz) > g\rho_L$ and $n = \frac{1}{3}$ for $(-dp/dz) \leq g\rho_L$. In Eq. (4.6.36), the function F of the liquid-film thickness divided by the pipe radius b was obtained by comparison with test data, as illustrated in Figure 4.6.5. This method can be used to predict the film thickness for a given pressure drop $(-dp/dz)$ or the pressure drop from the liquid-film thickness. It was found to give an approximate, yet acceptable agreement with the test results as long as the liquid film was turbulent ($y^+ > 30$), the head losses were not excessive in vertical flow, and the liquid film was relatively uniform around the perimeter of a horizontal channel.

The pressure drop subdivision by Levy into two regions of $(-dp/dz) > g\rho_L$ or $< g\rho_L$ came about from the realization that the frictional pressure drop in the liquid film $(-dp/dz)_{LTP}$ for vertical two-phase annular flow was equal to

$$\left(-\frac{dp}{dz}\right)_{LTP} = \frac{2\tau_w b - 2\tau_i a}{b^2 - a^2} = -\frac{dp}{dz} - g\rho_L \quad (4.6.37)$$

It was postulated that unsteady waves could start to appear as soon as $(-dp/dz)_{LTP}$ became negative or $(-dp/dz) < g\rho_L$. This meant that the shear transmitted to the liquid at the interface exceeded the shear the liquid could dissipate at the channel wall and the interfacial wave characteristics would change.

Wallis [4.6.11] derived a much simpler expression by plotting the interfacial friction factor f_i versus the film thickness divided by the pipe diameter D. He found that

$$f_i = f_{SG}\left(1 + 300\frac{\delta}{D}\right) \quad (4.6.38a)$$

236 FLOW PATTERNS

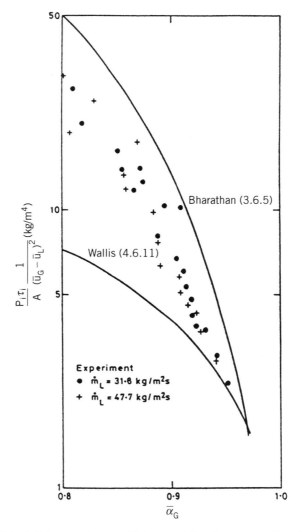

Figure 4.6.5 Interfacial shear stress coefficient with flooding waves. (Reprinted from Ref. 4.6.30 with permission from Elsevier Science.)

where f_{SG} represents the constant single-phase gas turbulent friction factor. Equation (4.6.38a) has sometimes been rewritten as

$$f_i = f_{SG}\left(1 + 360\frac{\delta}{D}\right) \quad (4.6.38b)$$

where f_{SG} is the Blasius friction factor for single-phase gas flow, or $f_{SG} = 0.316/\mathrm{Re}_G^{0.25}$.

Wallis pointed out that Eq. (4.6.38a) applies to a wavy annular film that is about equivalent to a sand roughness of four times the film thickness. Wallis also showed that Levy's function F could be broken down into two terms: The first was a property function represented by $\frac{1}{2}\sqrt{\rho_L/\rho_G}$, and the second term is a function of δ/D, which agreed with Wallis's relation as long as $(-dp/dz) > g\rho_L$ and $2\delta/D \leq 0.1$. Beyond that film thickness [i.e., in the case of $(-dp/dz) < g\rho_L$], Levy's model would predict an increased interfacial friction factor.

Expressions of the type of Eq. (4.6.38a) have been proposed subsequently by many other investigators. The Moeck and Stackiewicz correlation Eq. (4.6.24), was covered already. Similarly, good correlation of high-pressure steam–water upward vertical flow data was obtained by Nigmatulin et al. [4.6.25] and Subbotin et al. [4.6.26] by using

$$f_i = 0.032\left[1 + 210\left(\frac{2\delta}{D}\right)^{1.3}\right] \qquad (4.6.39)$$

All the preceding correlations of interfacial friction factor are in approximate agreement and they provide a powerful method of predicting this key variable as long as intermittent or flooding waves are not present.

Some doubt was raised by Whalley and Hewitt [4.6.27] about their universality and their ability to cope with disturbance waves with high liquid entrainment in the gas core or high-pressure systems where the ratio of liquid to gas density is reduced. Subsequent studies tend to indicate that the difficulties are more likely due to an excessive amount of entrained liquid drops and the turbulence suppression they might cause in the gas core. These preliminary findings are covered in Section 4.6.6.

4.6.5.4. Flooding Waves and Their Interfacial Friction Factor

Because of their irregular and unsteady behavior, the flooding waves near the slug-to-annular flow pattern transition are even more difficult to study and predict. Different regions of the liquid film have been observed simultaneously to be flowing upward or downward. The waves are rather large and there are large filamentlike liquid discontinuities at the interface which are not easily representable by some kind of velocity or shear distribution. In between the waves, a liquid film exists that is decelerating behind the wave, and its deceleration can lead to flow reversal before the next wave arrives. Measurements of the interfacial friction factor were carried out by Bharathan and Wallis [4.6.28] and Abe [4.6.29]. They found that the interfacial friction factor is much higher than the values predicted from correlations for disturbance waves.

The correlation proposed by Bharathan [4.6.28] was covered in Chapter 3 and provided by Eqs. (3.6.37) to (3.6.39). Its comparison to Wallis's equation (4.6.38a) is shown in Figure 4.6.5, where the grouping

$$\frac{P_i \tau_i}{A} \frac{1}{(\bar{u}_G - \bar{u}_L)^2}$$

is plotted against the gas volume fraction $\bar{\alpha}$. There is a significant difference between flooding and disturbance waves. It starts at $\bar{\alpha} = 0.95$ and reaches a factor in excess of 7 at $\bar{\alpha} = 0.8$. In Figure 4.6.5, recent data obtained by Govan et al. [4.6.30] are plotted, and they exhibit considerably increased interfacial friction factors by comparison to Wallis's correlation for disturbance waves. They also show a high scatter which is not unexpected, due to the intermittent nature of the flow.

The same finding has been reported for recent vertical test data taken at low-gas-vapor Reynolds numbers. These tests were performed with steam and water to simulate the water reflooding phase during a loss-of-coolant accident in a pressurized water reactor. The liquid-film Reynolds numbers were in the range 330 to 7800, and the vapor Reynolds numbers were between 300 and 13,000. Kelly and Freitas [4.6.31] reported that the currently available correlations for interfacial friction factor were inadequate for low-gas-vapor Reynolds numbers (i.e., below 10^4). They were able to correlate their data with

$$\frac{f_i}{f_{SG}} = 1 + \frac{4.22 \times 10^5}{\text{Re}_G^{1.5}} (\delta_G^+ - 5) \tag{4.6.40}$$

where δ_G^+ is obtained from Eq. (4.6.28). At $\text{Re}_G < 10^4$, Eq. (4.6.40) yields higher interfacial friction factors than the correlations developed by Asali and Hanratty [4.6.18] for upflow in the roll-wave regime:

$$\frac{f_i}{f_{SG}} = 1 + \frac{0.45}{\text{Re}_G^{0.2}} (\delta_G^+ - 4) \tag{4.6.41}$$

It is apparent that there are significant uncertainties in our knowledge and in our formulations for the interfacial friction factor for flooding waves.

4.6.6 Drift Flux Model for Annular Flow

Once it is recognized that the annular flow interfacial factor is best established from an empirical correlation, the interfacial closure equation (4.3.28) can be utilized to obtain the drift flux velocity $\overline{U}_{Gj}$ in terms of the gas volume fraction $\bar{\alpha}$. It is assumed that the wetted gas perimeter P_{WG} is equal to zero and that the wetted liquid perimeter is the pipe wetted perimeter where $P = \pi D$ where D is the pipe diameter. Also, to assure consistency with the Wallis interfacial friction factor equation (4.6.38a), no liquid entrainment is considered in the gas core, so that utilizing Eqs. (4.3.28) and (4.3.29), there results

$$\frac{P_i \tau_i}{A} = \frac{P_i \rho_G}{A} \frac{\overline{U}_{Gj}^2 f_i}{8 (1 - \bar{\alpha})^2} = \left[\left(-\frac{dp}{dz} \right)_{FL} \frac{\bar{\alpha}}{(1 - \bar{\alpha})^2} + g(\rho_L - \rho_G)\bar{\alpha}(1 - \bar{\alpha}) \right] \tag{4.6.42}$$

In Eq. (4.6.42), the term $(-dp/dz)_{FL}$ represents the frictional pressure drop if the liquid was only flowing in the pipe. With $P_i/A = 4\sqrt{\bar{\alpha}}/D$ and for a liquid-film thickness δ small compared to the pipe diameter, so that $4\delta/D = 1 - \bar{\alpha}$, Eq. (4.6.42) yields

$$\overline{U}_{Gj}^2 = \frac{2D}{\sqrt{\bar{\alpha}}\,\rho_G\,f_{SG}[1 + 75(1 - \bar{\alpha})]} \left[\left(-\frac{dp}{dz}\right)_{FL} \bar{\alpha} + g(\rho_L - \rho_G)\bar{\alpha}(1 - \bar{\alpha})^3 \right] \quad (4.6.43)$$

In the case where gravity can be neglected with respect to wall shear or $-(dp/dz)_{FL} \gg g(\rho_L - \rho_G)(1 - \bar{\alpha})^3$, Eq. (4.6.43), with the additional assumption $f_{SL} \approx f_{SG}$, simplifies to

$$\left(\frac{\overline{U}_{Gj}}{\overline{u}_{SL}}\right)^2 = \frac{\sqrt{\bar{\alpha}}}{1 + 75(1 - \bar{\alpha})} \frac{\rho_L}{\rho_G} \quad (4.6.44)$$

where $\bar{u}_{SL}$ is the superficial liquid velocity. Equation (4.6.44) can be combined with the standard expression for drift flux to determine the actual gas velocity $\bar{u}_G$ and the corresponding actual liquid velocity $\bar{u}_L$ in terms of $\bar{\alpha}$ if the superficial liquid and gas velocities are specified. Where gravity forces are important, the same result can be derived from Eq. (4.6.43).

4.6.7 Gas Core with Entrained Liquid in Annular Flow

In annular two-phase flow, liquid droplets are usually present in the gas core. The liquid drops are not uniform and they vary both in shape and size. Their distribution across the channel cross section is not constant, with most of the measurements showing a minimum of entrained liquid in the center of the channel. The behavior of the liquid drops and the prediction of the liquid flow that is entrained have an impact on the properties of annular flow for several reasons. There is a drag force exerted by the gas flow on the cloud of droplets present in the stream. For a single drop of diameter D_p, the drag force

$$F_D = \frac{1}{2}\left(\frac{\pi D_p^2}{4}\right) C_D \rho_G (\bar{u}_G - \bar{u}_{D_p})^2 \quad (4.6.45)$$

where C_D is the drag coefficient, $\bar{u}_G$ the actual average gas velocity, and u_{D_p} the drop velocity.

If we assume the droplets to be spherical in shape, the number of droplets dN per unit core volume dv is given by

$$dN = \frac{6(1 - \alpha_c)}{\pi D_p^3} dv \quad (4.6.46)$$

where α_c is the volume fraction occupied by the gas core. The drag force F_D per unit core volume becomes

$$F_D = \frac{3}{4}(1 - \alpha_c)\rho_G \int_0^1 \frac{C_D(\bar{u}_G - u_{D_p})^2}{D_p} dv \qquad (4.6.47)$$

In Eqs. (4.6.45) to (4.6.47), the gas velocity in the core is assumed to be constant at its average value, while the liquid drop velocity is allowed to vary with its size and will tend to decrease as the drop diameter increases. Furthermore, as implied by the preceding formulation, the liquid droplets will travel at a reduced velocity by comparison to the gas.

Also, drops will be entrained from the top of the waves traveling along the liquid film and will be returned from the gas core to the liquid film. Under steady-state conditions the entrainment rate will equal the deposition rate, but there will be a net exchange of momentum from the core to the liquid film because the drops leaving the peak of the waves will have a velocity $\bar{u}_{\delta_{max}}$, while the drops returning to the film possess an increased velocity, u_{D_p}. The entrainment rate, E (i.e., the deposition rate) is normally related to the concentration C of liquid droplets in the gas core, or

$$E = \bar{k}_d C \qquad (4.6.48)$$

where $\bar{k}_d$ is an average droplet mass transfer coefficient.

It can be seen that the presence of liquid drops in the gas core requires:

- Recognition of a gas drag force on the liquid droplets. This implies that we know the drag coefficient, C_D, the size of the drops, and the relative velocity of the gas with respect to that of the droplets. This drag force must be added to the core momentum equation.
- Addition of a momentum exchange term between liquid film and gas core, $E(u_{D_p} - \bar{u}_{\delta_{max}})$. This term is added to the liquid momentum equation and subtracted from the core equation. The impact on the interfacial closure equation is, however, zero. To calculate this momentum exchange term, it is necessary to know the droplet velocity u_{D_p}, the velocity of the peak of the waves $\bar{u}_{\delta_{max}}$, the mass transfer coefficient $\bar{k}_d$, and the concentration C of liquid droplets in the gas core.
- Specification of size distribution if the liquid drops are not assumed to be uniform, and integrals across the core cross section must be employed instead of simply average values. This adds further complexity to the analysis.

It is not surprising, therefore, that most system computer codes tend to fall back upon a simplified representation of the gas core where the gas and liquid drops are presumed to be mixed homogeneously. In the sections that follow, it will be shown that considerable information is now available on liquid droplet size, their drag coefficient, the amount of liquid entrained, and the entrainment/deposition rate to generate improved models. However, our present knowledge is still primarily in the form of empirical correlations and primarily for air–water mixtures at low pressure, which may explain the reluctance to incorporate them in complex system computer codes.

4.6.7.1 Liquid Droplet Characteristics Wicks [4.6.32] was the first to perform extensive measurement of drop sizes and their distribution as a function of air and water velocity. He found that the drop size decreases with increased gas flow rate and become dependent on liquid flow rate when it becomes large. Large apparent drop sizes are observed at large liquid rates because of liquid bridging and the tearing of liquid filaments. Maximum drop sizes reported by Wicks are reproduced in Figure 4.6.6. Wicks also found that the drop size distribution fit an upper-limit lognormal distribution, with the dominant number of drops having a diameter between 0.1 and 0.2 of the maximum drop size. Photographic studies by James et al. [4.6.33] have shown that the drops move in straight lines from the point of entrainment to the point of deposition; the drops tend to deposit on the liquid film

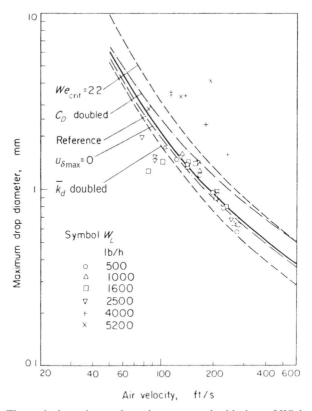

Figure 4.6.6 Theoretical maximum drop size compared with data of Wicks [4.6.32]. (Reprinted from Ref. 4.6.15 with permission from Elsevier Science.)

rather than bounce off it. The initial transverse velocity of drops is about equal to the shear stress velocity. The mean drop size is around 100 μm and the very small drops (diameter < 10 μm) tend to interact with the gas-phase turbulence and follow the eddies in the gas core.

Additional measurement of liquid particles entrained from liquid waves have been made by Azzopardi et al. [4.6.34], who recommended the following correlation for the average liquid droplet diameter $\overline{D}_P$:

$$\frac{\overline{D}_P}{D_H} = 1.91 \frac{\text{Re}_{\overline{D}_P}^{0.11}}{\text{We}_{\overline{D}_P}^{0.6}} \left(\frac{\rho_G}{\rho_L}\right)^{0.6} + 0.4 \frac{\overline{u}_L}{\overline{u}_G}(1 - \overline{\alpha}) \qquad (4.6.49)$$

where D_H is the channel hydraulic diameter and $\overline{u}_L$ and $\overline{u}_G$ are the actual average liquid and gas velocities. The droplet Reynolds number $\text{Re}_{\overline{D}_P}$ and the Weber number $\text{We}_{\overline{D}_P}$ are obtained from the standard definitions and by assuming that the droplet velocity u_{D_P} is negligible with respect to the gas velocity in the Weber number:

$$\text{Re}_{\overline{D}_P} = \frac{\overline{u}_G \overline{D}_P \rho_G}{\mu_G} \qquad \text{We}_{\overline{D}_P} = \frac{\rho_G \overline{u}_G^2 \overline{D}_P}{\sigma} \qquad (4.6.50)$$

In many practical applications, axial spray injectors are used to form liquid drops, and their diameter $\overline{D}_{P0}$ must be specified. It is expressed in terms of the gas and liquid superficial velocities $\overline{u}_{SG}$ and $\overline{u}_{SL}$, as, for example, in Reference 4.6.35, where

$$\overline{D}_{P0} = \frac{0.585 \sqrt{\sigma/\rho_L}}{\overline{u}_G - \overline{u}_L} + 53.2 \left(\frac{\mu_L}{\sqrt{\sigma/\rho_L}}\right)^{0.45} (\overline{u}_{SG} - \overline{u}_{SL})^{1.5} \qquad (4.6.51)$$

In those cases where spray systems are employed, the droplet diameter will change from its initial value at the spray location, and it needs to be evaluated along the flow stream, taking into account the fluxes of deposited and entrained droplets and their breakup as the Weber number reaches a critical value.

A typical set of annular flow drop size data obtained by Azzopardi et al. [4.6.34] is shown in Figure 4.6.7, where the Sauter mean droplet diameter is plotted against liquid and gas mass flow rates. Also, simultaneously, measurements of drop velocity and size have been made by Azzopardi et al., and the results show that the drop velocities are much lower than that of the gas.

4.6.7.2 Liquid Droplet Drag Coefficient
The droplet drag coefficient C_D in Eq. (4.6.45) was assumed by Moeck and Stachiewicz [4.6.15] to behave as if the drop were a solid particle. Moeck and Stachiewicz defined C_D from

$$C_D = 0.4(1 + 63.5\text{Re}_{D_P}^{-0.8})$$
$$\text{Re}_{D_P} = \frac{\rho_G(\overline{u}_G - u_{D_P})D_P}{\mu_G} \qquad (4.6.52)$$

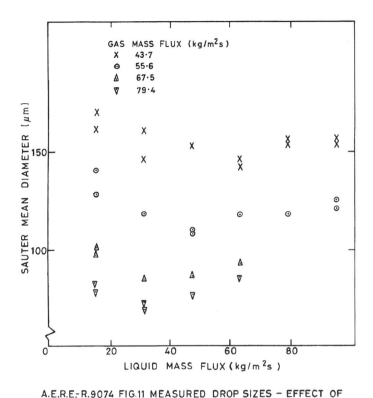

Figure 4.6.7 Measured drop sizes: effect of liquid flow rate. (From Ref. 4.6.34.)

for $0.5 < \text{Re}_{D_p} < 10^5$. Ishii and Chawla [4.6.36] recommended instead for a solid particle or an undistorted liquid droplet

$$C_D = \begin{cases} \dfrac{24}{\text{Re}_{D_p}} (1 + 0.1 \text{Re}_{D_p}^{0.75}) & \text{for } \text{Re}_{D_p} < 1000 \\ 0.45 & \text{for } 1000 < \text{Re}_{D_p} < 10^5 \end{cases} \quad (4.6.53)$$

Equations (4.6.52) and (4.6.53) apply to a single particle or an undistorted liquid drop in an infinite gas medium. Ishii and Chawla recognized that the viscosity of a multiparticle–gas system will increase in proportion to the presence of particles. They recommended a dependence of $(1 - \alpha_d)^{-2.5}$, where α_d is the core volume fraction occupied by the particles or undistorted liquid droplets. Also, they employed a similarity approach to derive the following drag coefficient for a cloud of undistorted droplets in an infinite gas medium:

$$C_D = C_D(SD_P) \frac{1 + 17.67(1 - \overline{\alpha}_d)^{18/7}}{18.67(1 - \overline{\alpha}_d)^3} \quad (4.6.54)$$

Ishii and Chawla also defined a drag coefficient for distorted liquid drops, which is covered under mist flow in Section 4.7. In Eq. (4.6.54), C_D (SD_P) stands for the drag coefficient for a single undistorted drop and is given by Eq. (4.6.53). Equation (4.6.54) states that the drag coefficient for a cloud of undistorted droplets in an infinite gas medium rises as the volume fraction occupied by the drops increases. Once C_D is specified, it becomes possible to calculate the relative velocity of the drop with respect to the gas. This is best done by writing a separate field momentum equation for the liquid droplets.

4.6.7.3 Separate Field Momentum Equation for the Liquid Droplets

As was done by Moeck and Stachiewicz, a separate momentum balance is written for the cloud of droplets except that it is generated here in terms of the average droplet diameter $\overline{D}_P$ and its corresponding velocity $\overline{u}_{\overline{D}_P}$. For an elemental length dz in the flow direction, the various forces and momentum exchanges are

$$\text{buoyant force} = -g\overline{\alpha}_d(\rho_L - \rho_G)A_c\,dz$$

$$\text{gas drag force} = F_D A_c\,dz = \frac{3}{4}\overline{\alpha}_d\rho_G C_D \frac{(\overline{u}_c - \overline{u}_{\overline{D}_P})^2}{\overline{D}_P}$$

$$\text{entrainment momentum} = E\overline{u}_{\delta_{\max}} P_c\,dz$$

$$\text{deposition momentum} = -E\overline{u}_{\overline{D}_P} P_c\,dz$$

(4.6.55)

In Eqs. (4.6.55), $\overline{\alpha}_d$ is the volume fraction occupied by the drops in the core and A_c is the core area, which is measured from the tip of the waves. The core interfacial perimeter is $P_c = \pi D_c$, where D_c is the diameter of the core of area A_c. Also, it was assumed that the drag coefficient C_D is relatively constant ($\text{Re}_{\overline{D}_P} > 1000$) and that the deposition rate is equal to the entrainment rate E. The entrainment rate is obtained from Eq. (4.6.48) which can be expressed in terms of the mass flow of liquid drops $W_{\overline{D}_P}$ and the gas mass flow rate W_G:

$$E = \overline{k}_d\rho_G \frac{W_{\overline{D}_P}}{W_G} = \overline{k}_d\rho_G \frac{\overline{\alpha}_d\overline{u}_{\overline{D}_P}\rho_L A_c}{\rho_G\overline{u}_G\overline{\alpha}A} \approx \overline{k}_d\rho_L\overline{\alpha}_d \frac{\overline{u}_{\overline{D}_P}\overline{\alpha}_c}{\overline{u}_G\overline{\alpha}}$$ (4.6.56)

The momentum balance now becomes

$$\frac{3}{4}\rho_G C_D \frac{(\overline{u}_G - \overline{u}_{\overline{D}_P})^2}{\overline{D}_P} = g(\rho_L - \rho_G) + \overline{k}_d\rho_L \frac{4}{D_c} \frac{\overline{\alpha}_c\overline{u}_{\overline{D}_P}}{\overline{\alpha}\overline{u}_G}(\overline{u}_{\overline{D}_P} - \overline{u}_{\delta_{\max}})$$ (4.6.57)

For simplification purposes, it is often presumed that $A_c \approx \overline{\alpha}A$, which eliminates the need to know $\overline{\alpha}_c$ and $\overline{\alpha}$. It is interesting to note that Moeck and Stachiewicz derived an expression identical to Eq. (4.6.57) except that it was applicable to any droplet size by replacing $\overline{u}_{\overline{D}_P}$ with $u_{\overline{D}_P}$ and $\overline{D}_P$ with D_P as well as with $u_{\overline{D}_{P,\max}}$ and $D_{p,\max}$, respectively, for the maximum size droplet. Equation (4.6.57) is easier to derive and is much more consistent with the usual cross-section averaging employed in all two-fluid models. Equation (4.6.57) indicates that the relative velocity, $\overline{u}_G - \overline{u}_{\overline{D}_p}$, increases with average drop diameter. This will lead to droplet breakup at a critical Weber number, We$_{crit}$, which is usually defined in terms of the maximum droplet diameter $D_{P,max}$:

$$\text{We}_{crit} = \rho_G \frac{D_{P,max}(\overline{u}_G - u_{\overline{D}_{P,max}})^2}{\sigma} \approx 12 \qquad (4.6.58)$$

Moeck and Stachiewicz solved Eqs. (4.6.57) and (4.6.58) for $u_{\overline{D}_{P,max}}$ and $D_{P,max}$ to obtain the values plotted in Figure 4.6.6. They found that the maximum droplet diameter $D_{P,max}$ is dependent primarily on the air velocity. The value of the core diameter, D_c, was assumed to be the pipe diameter. The relative influence of $\overline{u}_{\delta,max}$ and We$_{crit}$ have a minimal impact, according to Figure 4.6.6. Similarly, significant variations ($\pm 50\%$) in the parameter $\overline{k}_d$ can be tolerated as long as its absolute value is about right. Moeck and Stachiewicz used $\overline{k}_d = 0.8$ ft/sec, based on air–water tests. Also, the reference case plotted in Figure 4.6.6 assumes that the drag coefficient C_D is obtained from Eq. (4.6.52), We$_{crit} = 13$, and $\overline{u}_{\delta,max} = 0.07\overline{u}_G$.

With a specified liquid droplet distribution, it now becomes possible from $D_{P,max}$ to calculate the average drop diameter, $\overline{D}_P$ and its corresponding velocity, $\overline{u}_{\overline{D}_p}$, from Eq. (4.6.57). It is interesting to note that if we neglect the last term in Eq. (4.6.57), which deals with the momentum exchange between the liquid film and the liquid droplets, Eq. (4.6.58) can be combined with the simplified Eq. (4.6.57) to get the following approximation for the maximum drop diameter:

$$\overline{u}_c - u_{\overline{D}_{P,max}} \approx \left[\frac{4}{3} \frac{g(\rho_L - \rho_G)}{\rho_G^2} \frac{\sigma \text{We}_{crit}}{C_{D_{P,max}}} \right]^{1/4}$$

A more accurate treatment of the gas core and its liquid droplets can be derived if the deposition/entrainment rate can be established as discussed next.

4.6.7.4 Entrained Liquid Rate and Entrainment and Deposition Rates

Entrainment will occur in practically all annular flows. We shall here distinguish two properties of the liquid being transported in the gas core. The first refers to the *entrained* liquid rate, which describes the rate of liquid droplet flow W_{D_p}, or the fraction of liquid droplet flow in the core to the total liquid flow, W_{D_p}/W_L; the second is the *entrainment rate*, which determines the rate of exchange of liquid from the film to the core, which has been previously represented by the symbol E. Entrained liquid rates have been obtained either by measuring the liquid-film flow rate and subtracting it from the total liquid flow, or by using one or several extraction probes in the central part of the channel and assuming an entrained liquid distribution in the gas core. The entrained liquid rates have been found to increase with both gas and liquid rates. The entrained liquid flow decreases with system pressure and is

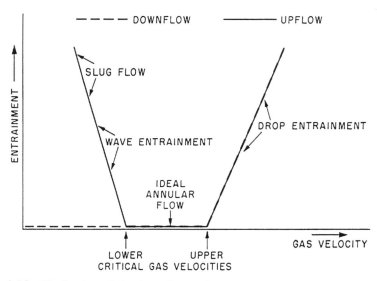

Figure 4.6.8 Idealization of the dependence of entrainment on gas velocity. (From Ref. 4.6.37.)

highly dependent on the liquid injection method employed at the entrance of the test section.

In vertical upward flow, entrainment occurs when the gas velocity is high enough to produce disturbance waves and to strip liquid from the top of the waves or when the gas velocity is low enough and is unable to support a continuous liquid film at the wall against gravity. An idealization of those possibilities [4.6.37] is shown in Figure 4.6.8, including the potential presence of a rather small zone of ideal annular flow where there is negligible entrainment in the gas core. In the case of downflow, a continuous liquid film can exist down to zero gas flow, and this is illustrated by a dashed line in Figure 4.6.8.

Predictions of the fraction of liquid being entrained and of the entrainment and deposition rates have relied to-date primarily on experimental measurements and empirical correlations. One of the correlations often employed for entrainment liquid rate (W_{D_p}/W_L) was developed by Ishii and Mishima [4.6.38]. It was specified previously in Eqs. (3.6.42) and (3.6.43). Its match against experiments is illustrated in Figure 4.6.9 for steam–water mixtures. A slight correction by General Electric is also shown. Another approach, developed by Hewitt and Govan [4.6.39] was first to define a critical liquid film velocity $\bar{u}_{Lfc}$ in a pipe of diameter D below which there is no entrainment:

$$\bar{u}_{Lfc} = \frac{\mu_L}{D\rho_L}\exp\left(5.85 + 0.425\frac{\mu_G}{\mu_L}\sqrt{\frac{\rho_G}{\rho_L}}\right) \qquad (4.6.59)$$

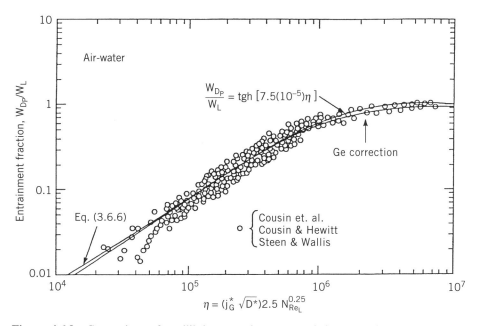

Figure 4.6.9 Comparison of equilibrium entrainment correlation to various data. (From Ref. 4.6.38.)

Beyond that critical film velocity, the corresponding mass entrained rate per unit perimeter, M_E, is obtained from

$$\frac{M_E}{\overline{u}_{SG}\rho_G \pi D} = 5.75 \times 10^{-5} \left[\rho_L (\overline{u}_{LF} - \overline{u}_{Lfc})^2 \frac{D\rho_L}{\sigma \rho_G^2} \right]^{0.316} \quad (4.6.60)$$

which applies only when $\overline{u}_{LF} > \overline{u}_{Lfc}$.

The deposition or entrainment rate E is related normally to a mass transfer coefficient $\overline{k}_d$ and to the concentration C of liquid droplets in the gas core as proposed in Eqs. (4.6.48) and (4.6.56). For many years, the deposition or entrainment rate mass transfer coefficient was based on the work of McCoy and Hanratty [4.6.40], which is reproduced in Figure 4.6.10. The nondimensional mass transfer coefficient $\overline{k}_d / \sqrt{\tau_i / \rho_G}$ is plotted in Figure 4.6.10 against a dimensionless relaxation time:

$$\tau^+ = \frac{\overline{D}_P^2 \tau_i}{18 \mu_G^2} \quad (4.6.61)$$

where τ_i is the interfacial shear stress and $\overline{D}_P$ is the average droplet diameter.

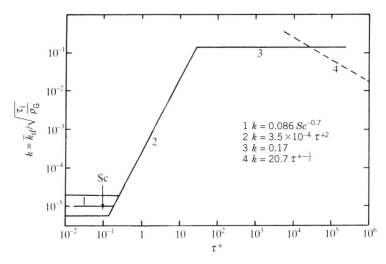

Figure 4.6.10 Deposition of liquid drops. (Reprinted from Ref. 4.6.40 with permission from Elsevier Science.)

Different lines are plotted in regions of Figure 4.6.10 depending on the value of the Schmidt number, Sc, defined from

$$\text{Sc} = \frac{\mu_G}{D_f \rho_G} \tag{4.6.62}$$

where D_f is the diffusion coefficient. More recently, Hewitt and Govan [4.6.39] have recommended that

$$\bar{k}_d \sqrt{\frac{\rho_G D}{\sigma}} = \begin{cases} 0.18 & \text{if } \frac{C}{\rho_G} < 0.3 \\ 0.083 \left(\frac{C}{\rho_G}\right)^{-0.65} & \text{if } \frac{C}{\rho_G} > 0.3 \end{cases} \tag{4.6.63}$$

where C is the liquid droplet concentration. Equations (4.6.63) provide for the role of the surface tension σ, which has been a missing link for several years.

It should be noted that the preceding empirical recipes developed for entrained liquid rate and entrainment have not been tested over a wide range of fluid conditions and that they exhibit considerable scatter with respect to the available test data. The reliance upon such correlations is also indicative of our lack of full understanding of the entrainment process.

4.6.7.5 Closure Equations for Annular Flow with Entrainment

For annular flow with liquid entrainment, the volume fraction occupied by the core is denoted by $\bar{\alpha}_c$ and the droplets occupy a fraction $\bar{\alpha}_d$ of the core or a volume fraction

$\overline{\alpha}_c \overline{\alpha}_d$. The liquid film occupies a volume fraction $(1 - \overline{\alpha}_c)$ and the gas continues to occupy a volume fraction labeled $\overline{\alpha}$. If we assume that only the liquid wets the vertical pipe wall, one can write three closure equations:

Liquid film: $(1 - \overline{\alpha}_c)\left(-\dfrac{dp}{dz}\right) - g\rho_L(1 - \overline{\alpha}_c)$

$$+ \frac{P_i \tau_i}{A} - \frac{P_{WL}\tau_{WL}}{A} + \frac{P_i}{A} E(\overline{u}_{\overline{D}_P} - \overline{u}_{\delta\max}) = 0 \quad (4.6.64)$$

Gas: $\overline{\alpha}_c \left(-\dfrac{dp}{dz}\right) - g\rho_G \overline{\alpha}_c - \dfrac{P_i \tau_i}{A} - F_D \overline{\alpha}_c = 0$

Droplets: $-g(\rho_L - \rho_G)\overline{\alpha}_c \overline{\alpha}_d + F_D \overline{\alpha}_c - \dfrac{P_i}{A} E(\overline{u}_{\overline{D}_P} - \overline{u}_{\delta\max}) = 0$

One can add the preceding three equations and obtain the standard steady-state relation for the pressure drop. As an alternative, one can add the second and third equations and obtain a momentum balance for the core. In so doing, the drag force F_D given by Eq. (4.6.47) or (4.6.55) stops having a role in the core conservation of momentum. With the total liquid and gas flow specified, one can use an empirical correlation for entrained liquid, such as Eqs. (3.6.42) and (3.6.43), and determine the amount of liquid flowing in the film and the amount being entrained. With an empirically specified value of the mass transfer coefficient $\overline{k}_d$, the entrainment rate E can be calculated from Eq. (4.6.56). If we assume $\overline{\alpha}_c$, we can calculate the interface perimeter P_i, and the interface friction factor fi from a Wallis type of relation. We can also determine the average film velocity $\overline{u}_{fL}$ and the term $P_{WL}\tau_{WL}/A$. As a first approximation, one sets $\overline{\alpha} = \overline{\alpha}_c$; the average gas velocity $\overline{u}_G$ can then be calculated. With $D_{P,\max}$ expressed in terms of the critical Weber number, and with $\overline{u}_{\delta\max} = 2\overline{u}_{fL}$, Eqs. (4.6.57) and (4.6.58) can be solved for $u_{D_{P,\max}}$ and $D_{P,\max}$ assuming an appropriate drag coefficient. With a specified droplet distribution, the average droplet size $\overline{D}_P$ can be calculated from the maximum droplet diameter and the corresponding velocity $\overline{u}_{\overline{D}_P}$ obtained from Eq. (4.6.57). From the entrained liquid flow and the calculated values of $\overline{u}_{\overline{D}_P}$ and $\overline{D}_P$, one can determine $\overline{\alpha}_c \overline{\alpha}_d$ and a new value of $\overline{\alpha}$. This iteration is repeated until $\overline{u}_G$ and $\overline{u}_{\overline{D}_P}$ and are stabilized. One can now calculate τ_i and combine the top two equations of Eq. (4.6.64) to eliminate $-dp/dz$ and to obtain

$$\frac{P_i \tau_i}{A} = g(\rho_L - \rho_G)\overline{\alpha}_c(1 - \overline{\alpha}_c)$$

$$+ \frac{P_{WL}\tau_{WL}}{A}\overline{\alpha}_c - \frac{P_i}{A}E(\overline{U}_{D_P} - \overline{U}_{\delta\max})\overline{\alpha}_c - F_D \overline{\alpha}_c(1 - \overline{\alpha}_c) = 0 \quad (4.6.65)$$

The value of $\overline{\alpha}_c$ is reiterated until Eq. (4.6.65) is satisfied. Finally, the total pressure $-dp/dz$ is obtained by adding the head loss $\rho_L(1 - \overline{\alpha}) + \rho_G \overline{\alpha}$ to the wall frictional pressure $P_{WL}\tau_{WL}/A$. A steady-state solution to annular flow with liquid entrainment is therefore available through a three-field formulation. It requires the assumption that the interfacial friction factor correlation applies to annular flow both without and with entrainment and that $\bar{u}_{\delta\max}$ can be estimated as a multiple of the mean liquid film velocity. It requires the specification of a liquid droplet distribution. While Moeck and Stachiewicz used the drag coefficient C_D for a single solid particle, the multiparticle correlation of Ishii and Chawla would have been preferable. In fact, the correlation for distorted particles covered in Section 6.10 would be the best choice.

Physically, as part of the liquid is entrained, it travels at a higher velocity than it would have if it were not entrained. The total liquid volume fraction is thus reduced and the corresponding gas volume fraction increases, and at the same time the gas average velocity is lowered. Also, because the liquid film is less thick, the corresponding interfacial friction factor is reduced because the ratio δ/D decreases in the Wallis formulation. The interface perimeter P_i goes up slightly, but the net impact is a decrease in the interfacial pressure drop. The same trend can be derived from Eq. (4.6.65) because of the need to subtract the drag force of gas upon the liquid droplets and the momentum transferred from the liquid droplets to the liquid film. This drag force increases as the amount of entrained liquid rises because as noted in Eq. (4.6.54), the drag coefficient increases as $(1 - \bar{\alpha}_d)$ goes up.

This possibility was recognized by Owen [4.6.41] when he performed pressure drop tests with high gas and liquid flow rates and noticed that the pressure gradient went through a maximum as the gas flow rate was increased while keeping the liquid flow rate constant. One possible explanation Owen provided for this behavior was that the entrained droplet concentration increased to the point where it was suppressing the gas core turbulence and hence reducing the interfacial friction factor. Owen correlated the mixing length K as a function of the ratio $\rho_G \bar{u}_G^2 / \rho_H \bar{u}_H^2$, where the subscript H represents homogeneous conditions. Owen found that the mixing length decreased as the amount of entrained droplets rose, as shown in Figure 4.6.11. This is contrary to the ideal annular model, which presumes homogeneous flow in the core and the single-phase value of the mixing length. This suppression of turbulence by entrained liquid droplets has not yet found its way into system computer codes because it requires further validation.

In summary, annular two-phase flow with entrained liquid is much more complicated and much more difficult to predict than the idealized flow pattern of a liquid film and a gas core, but it can be done if the addition of a third field to describe the cloud of droplets is used to obtain a more representative solution. It becomes necessary to recognize the gas drag force upon the liquid drops and the transfer of momentum from the entrained droplets to the liquid film. Even then, empirical correlations for the amount of liquid being entrained, for the interfacial friction factor f_i for the droplet distribution and for the mass transfer coefficient $\bar{k}_d$, are necessary. There is preliminary experimental evidence that the presence of droplets in the core reduces the interfacial shear and suppresses turbulence in the gas core, and the model for annular flow with entrainment supports it.

It should also be noted that the use of a drag coefficient for solid spheres must be questioned and that a distorted particle drag coefficient would be preferable. Few

4.6 ANNULAR FLOW 251

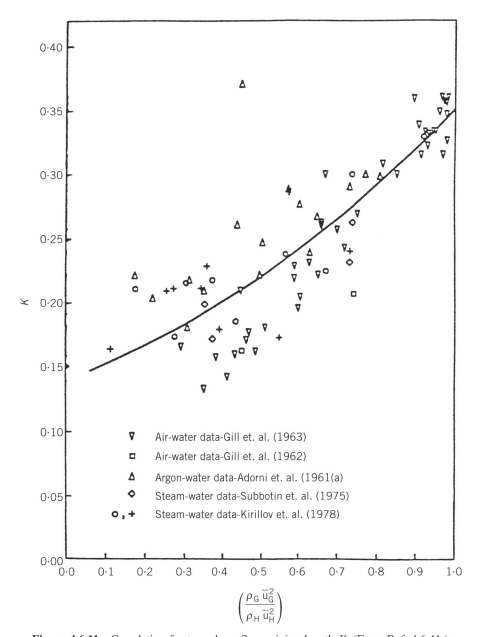

Figure 4.6.11 Correlation for two-phase-flow mixing length K. (From Ref. 4.6.41.)

complex system computer codes to date have taken advantage of the recent increase of knowledge about annular two-phase flow with liquid entrainment. The reason may be that the added modeling complexity may not be justified by the improvements in the accuracy of the predictions. However, this conjecture has not been checked adequately.

4.6.8 Heat Transfer in Annular Two-Phase Flow

Heat transfer tests have been performed with annular patterns because they are important to several industrial applications, particularly where they involve boiling or condensation. They are also important because they provide additional insight into the behavior of annular two-phase flow. Most of the solutions have relied upon single-phase treatment for heat transfer within the liquid film, and the Reynolds analogy between momentum and heat transfer has generally been used to predict the heat transfer coefficient between the liquid film and the gas stream. For example, Petukhov's [4.6.42] correlation for rough pipes has been applied, where

$$\mathrm{Nu}_f = \frac{(f_i/2)\mathrm{Re}_f \mathrm{Pr}_f}{1.07 + 12.7(f_i/2)^{0.5}(\mathrm{Pr}_f^{2/3} - 1)} \qquad (4.6.66)$$

where f_i is the Wallis interfacial factor and Nu_f, Re_f, and Pr_f are the Nusselt number, Reynolds number, and Prandtl number for the liquid film based on the pipe diameter.

A potentially superior approach has been suggested by Dobran [4.6.16], who employed Eq. (4.6.26) for the wavy layer momentum diffusivity to evaluate the temperature distribution beyond the base layer. Dobran defined a turbulent Prandtl number that represents the normal single-phase ratio of turbulent momentum to heat transfer eddy diffusivity and an effective Prandtl number that represents the ratio of the wavy layer momentum to the corresponding heat transfer diffusivity. He got best agreement with available air–water data, where the turbulent Prandtl number was set at the same value of 0.8 as the normal single-phase fluid Prandtl number but with the effective Prandtl number taken at 0.5. By setting all three Prandtl numbers at 0.8, the heat transfer coefficient decreased slightly.

Of special interest is the fact that the universal logarithmic velocity profiles of the ideal annular flow pattern overpredict the heat transfer by about 50%. This is attributed to overestimating the turbulence of the liquid film in the wavy layer and Dobran's equation (4.6.26) supports that conclusion. In other words, the waves are not as effective in transporting heat as has often been assumed in the past. This finding is supported by recent multidimensional studies of disturbance waves carried out by Jayanti and Hewitt [4.6.43] using computational fluid dynamics (CFD)

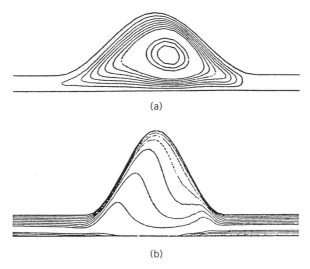

Figure 4.6.12 Turbulent viscosity distribution in disturbance wave calculated using low-Reynolds-number κ–ε model: (*a*) turbulent viscosity distribution; (*b*) temperature contours. (From Ref. 4.6.43).

methods. The results are illustrated in Figure 4.6.12, showing the turbulent viscosity distribution and the temperature contours calculated using the low-Reynolds-number κ–ε model. A typical disturbance wave with a sinusoidal shape and prescribed amplitude and wavelength dimensions was evaluated. It was found that the disturbance waves were circulating zones of turbulence traveling over a laminar substrate film. For a fixed heat flux and a fixed interface temperature, the isotherms are distorted by the recirculating flow in the wave. This leads to an increased heat transfer coefficient, but it is reduced from that obtained from one-dimensional turbulent analysis. Hewitt [4.6.21] points out that such CFD calculations are helping to elucidate the prevailing mechanisms in wavy interfaces but are far from generating definitive answers. Furthermore, to date they can be carried out only for presumed and idealized wave geometrical configurations.

4.6.9 Transition from Annular to Mist Flow

The start of liquid entrainment has been related to when a liquid drop can be supported by the gas flow. This same criterion has been employed to establish when flooding occurs in large pipes, and the most commonly quoted correlation is that of Pushkina and Sorokin [4.6.44], which requires that

$$\bar{u}_{SG} \geq 3.1 \left[\frac{\sigma g (\rho_L - \rho_G)}{\rho_G^2} \right]^{1/4} \tag{4.6.67}$$

which is identical to Eq. (4.4.45).

The transition to complete mist flow depends on whether or not the channel surface is heated. With sufficient heat addition, the transition to mist flow will occur much earlier. In the case of adiabatic flow, there is a minimum film thickness beyond which the liquid will not cover the entire channel surface. This minimum film thickness, $\delta_{L,\min}$, is established from a force balance on the creeping film, which yields

$$\delta_{L,\min} = \left(\frac{18\sigma\mu_L^2}{g^2\rho_L^3}\right)^{0.2} \qquad (4.6.68)$$

The corresponding core volume fraction $\overline{\alpha}_c$ is

$$\overline{\alpha}_c = \left(1 - \frac{2\delta_{L,\min}}{D_H}\right)^2 \qquad (4.6.69)$$

Equation (4.6.69) yields very small values of $\overline{\alpha}_c$ (e.g., in excess of 0.999 for water at atmospheric pressure).

4.6.10 Summary: Annular Flow

1. There exists a good physical understanding of the phenomena involved in annular two-phase flow, including an adequate prediction of the velocity profile in the base liquid film; identification of the types of waves (ripple, disturbance, and flooding) formed beyond the base liquid film and the boundaries over which they occur; empirically based correlations for the corresponding interfacial friction factor; and approximate prescriptions for the amount of liquid entrained in the gas core, the size of the liquid droplets, their distribution, and their deposition and entrainment rate. A full fundamental understanding and a generic model of annular flow is, however, still lacking because of the inability to analytically describe the formation and properties of waves at the gas–liquid interface as well as their shearing by high-velocity gas to produce entrainment. Similarly, the capability to analytically predict liquid droplets behavior and their entrainment or deposition is rather limited. Also, there are significant uncertainties in some of the empirical correlations developed from test data. This is particularly true for the interfacial friction factor for flooding waves, the entrainment and deposition rate of the liquid droplets, and their impact on suppressing turbulence in the gas core. Here again, as for previous flow patterns, the primary shortcomings arise at the liquid–gas interfaces, due to difficulties in describing their behavior.

2. Alternative empirical correlations are available for many of the parameters significant to annular two-phase flow (e.g., interfacial friction or entrainment rate). Evaluation of the correlations to identify preferred choices would be helpful to

system computer codes. Furthermore, judicious use of the available correlations in system computer codes continues to be an issue. For example, under reflooding conditions, the flooding wave interfacial friction correlation should be employed instead of the disturbance wave correlation.

3. Annular flow with entrained liquid is represented best by a three-field model consisting of a liquid film, a liquid droplet field, and a gas core. Such a description is necessary to deal with such bottom-up phenomena as the occurrence of critical heat flux due to depletion of the liquid film. However, most present system computer codes employ only a two-fluid model. They could be modified to a three-field model in the annular flow region, but it is not clear whether the benefit of this added complexity has been established or can be justified.

4. The representation of annular flow in system computer codes has benefited greatly from separate studies of that flow pattern. Additional improvements would result from additional fundamental studies on flooding waves, liquid droplet size spectrum and distribution, and basic understanding of the mechanisms of entrainment and deposition. The results may turn out to be too unwieldy to be incorporated in system computer codes and may not improve sufficiently the accuracy of the predictions. Unfortunately, definitive conclusions cannot be drawn at this time because such sensitivity assessments have not been performed with available system computer codes.

5. Heat transfer tests in annular two-phase flow show that the interfacial heat transfer rate is reduced by comparison to standard logarithmic velocity and temperature predictions. This is due to overestimating the interfacial momentum transfer and/or to distortion of the temperature profiles due to recirculating flow within the waves. Complex system computer codes to date have failed to recognize this shortcoming.

4.7 MIST/DISPERSED LIQUID DROPLET FLOW

During this flow regime, gas is in contact with the channel walls and liquid is in the form of droplets traveling within the gas core. It is customary to start from the motion of a single droplet in an infinite (i.e., frictionless) medium of gas and to proceed from there to a frictionless multidroplet system, and finally, to a dispersed liquid–gas flow with wall friction. Ishii and Chawla [4.6.36] recognized that the liquid droplets would tend to be distorted by the gas flow and that the drag coefficient would be governed by that distortion and the swerving motion of the liquid droplets. For that reason, the drag coefficient would not depend on the viscosity but would be proportional to the radius of the particle. For a single droplet of diameter D_P, the drag coefficient C_D would be equal to

$$C_D = \frac{2}{3} D_P \sqrt{\frac{g(\rho_L - \rho_G)}{\sigma}} \qquad (4.7.1)$$

256 FLOW PATTERNS

If we set the drag force on the projected area of the single liquid droplet, $\pi D_P^2/4$, equal to the buoyancy force,

$$\frac{1}{2}\left(\frac{\pi D_P^2}{4}\right) C_D \rho_G (\bar{u}_G - u_{D_P})^2 = \pi \frac{D_P^3}{6}(\rho_L - \rho_G)g \tag{4.7.2}$$

$$\frac{\overline{U}_{GJ}}{1-\overline{\alpha}} = (\bar{u}_G - u_{D_P}) = \sqrt{2}\left[\frac{g\sigma(\rho_L - \rho_G)}{\rho_G^2}\right]^{1/4}$$

where $\overline{U}_{GJ}$ is the drift velocity. According to Eq. (4.6.70), the drag coefficient increases with droplet diameter until the droplet becomes unstable and disintegrates into smaller drops. This limit for the maximum droplet size $D_{P,\max}$ is associated with a critical Weber number We_c of about 12, or

$$We_c = \frac{\rho_G D_{P,\max}(\bar{u}_G - u_{D_{P,\max}})^2}{\sigma} = 12 \tag{4.7.3}$$

which yields

$$D_{P,\max} = 6\sqrt{\frac{\sigma}{g(\rho_L - \rho_G)}} \tag{4.7.4}$$

$$C_{D,\max} = 4$$

In the case of a cloud of droplets, the closure equation (4.3.19) with no friction in a vertical pipe becomes for an average drop $\overline{D}_P$, and its corresponding average velocity $\bar{u}_{\overline{D}_P}$,

$$\frac{P_i \tau_i}{A} = \frac{6(1-\overline{\alpha})}{\overline{D}_P}\frac{1}{8} C_{\overline{D}_P} \rho_G (\bar{u}_G - \bar{u}_{\overline{D}_P})^2 = g(\rho_L - \rho_G)(1-\overline{\alpha})\overline{\alpha} \tag{4.7.5}$$

which simplifies to

$$C_{\overline{D}_P} = \overline{\alpha} C_{\overline{D}_{P,\text{single drop}}} = 4\overline{\alpha}\frac{\overline{D}_P}{D_{P,\max}} \tag{4.7.6}$$

In other words, the drag coefficient for a droplet cloud with an average diameter $\overline{D}_P$ is equal to the drag coefficient of a single droplet of the same diameter $\overline{D}_P$ multiplied by the gas volume fraction $\overline{\alpha}$. It is also equal to $4\overline{\alpha}$ multiplied by the ratio of the Sauter mean droplet-diameter droplet to that of the maximum droplet. Ishii and Chawla obtained comparable answers for mist flow, except for including the impact of Eq. (4.6.54) on the gas viscosity. It was excluded in this derivation because as noted in Reference 4.6.36, the form of the drag coefficient (4.7.1) presumed no viscosity role. It is also interesting to note that if one used $D_P D_{P,\max} = 6.67$, as might be inferred from an upper-limit lognormal droplet distribution and $\bar{\alpha} \approx 0.75$, $C_{\bar{D}_P}$ would be equal to 0.45 or equal to the drag coefficient suggested for a solid particle according to Eq. (4.6.53); this tends to support the simplified methodology of Moeck and Stachiewicz.

Let us next consider mist flow with wall friction in a vertical channel. According to Eq. (4.3.28), there results

$$\frac{P_i C_{\bar{D}_P}}{A} \frac{1}{8} \rho_G (\bar{u}_G - \bar{u}_{\bar{D}_P})^2 = g(\rho_L - \rho_G)\bar{\alpha}(1 - \bar{\alpha}) - \left(-\frac{dp}{dz}\right)_{FG} \frac{1 - \bar{\alpha}}{\bar{\alpha}^2} \quad (4.7.7)$$

where the gas wall shear was written in terms of the single-phase gas pressure loss, $(-dp/dz)_{FG}$, augmented by the square of the actual gas velocity, or $1/\bar{\alpha}^2$. Equation (4.6.76) can be rewritten as

$$C_{\bar{D}_P} = \bar{\alpha} C_{\bar{D}_{P,\text{single drop}}} \left[1 - \frac{(dp/dz)_{FG}}{\bar{\alpha}^3 g(\rho_L - \rho_G)} \right] \quad (4.7.8)$$

Equation (4.6.77) shows that the droplet drag coefficient decreases when wall friction is taken into account as long as the relative velocity of Eq. (4.6.71) and the presumed gaseous wall friction relation are applicable. The decrease is relatively small at reasonable gas velocities and becomes significant only at very high gas velocities. The primary purpose of Eqs. (4.6.76) and (4.6.77) was again to show that there is a closure expression between the gaseous wall friction, the drag force on the liquid droplets, and the gravitational force, and that they cannot all be specified independently. With the gas volumetric fraction $\bar{\alpha}$ specified, there is an interaction between the gaseous wall friction and the drag force, and one or the other must decrease slightly to satisfy the closure equation (4.3.28).

It should be noted that the cross flow of droplets to the channel wall did not enter the momentum solution in mist flow. However, if there is heat transfer at the wall surfaces, the arrival of liquid droplets enhances the wall transfer rate and they need to be taken into account by the methodology outlined under annular flow with liquid entrainment. Finally, if the gas velocity is high, a homogeneous model with the gas and liquid having the same velocity is often used for simplification purposes.

Mist or dispersed liquid droplet flow is the least complicated of all the prevailing flow patterns. The key issue is the distribution and shape of the droplets and whether they have a distorted or an undistorted configuration. Another issue is the role of liquid droplets on the gas viscosity. As for all other two-phase flow patterns, there is a closure equation involving interfacial forces, wall friction losses, and gravitational forces. The closure equation accounts for and establishes the interaction between those forces. Even though this interaction is small in mist or dispersed liquid droplet flow, it still readily illustrates that these forces cannot be specified independently.

4.8 SUMMARY

1. Considerable progress has been made and continues to be made in detailed observations, measurements, and phenomenological models of two-phase flow patterns. However, as stated by Hewitt [4.6.45], generalized constitutive relationships, particularly at the gas–liquid interfaces, "have proved to be an elusive goal."

2. Empirical models and correlations are still the dominant choice in two-phase-flow industrial applications, and it is most likely that they will maintain their dominance for many years to come in complex system computer codes.

3. It is hoped that this integrated, and in some cases different, treatment of all two-phase-flow patterns will help in making progress toward the development of improved phenomenological models which will find their way into complex system computer codes.

REFERENCES

4.3.1   Zuber, N., and Hench, J., *Steady State and Transient Void Fraction of Bubbly Systems and Their Operating Limits, 1: Steady State Operation*, GE Report 62GL100, July 1962.

4.3.2   Haberman, W. L., and Morton, R. K., *An Experimental Investigation of the Drag and Shape of Air Bubbles Rising in Various Liquids*, David W. Taylor Model Basin Report 802, 1953.

4.3.3   Peebles, F. N., and Garber, H. J., "Studies of the Motion of Gas Bubbles in Liquids," *Chem. Eng. Prog.*, **49**, 88–97, 1953.

4.3.4   Harmathy, T. Z., "Velocity of Large Drops and Bubbles in Media of Infinite and Restricted Extent," *A.I.Ch.E. J.*, **6**, 281, 1969.

4.3.5   Davidson, J. F., and Harrison, D., *Fluidized Particles*, Cambridge University Press, New York, 1963.

4.3.6   Kuo, J. T., Wallis, G. B., and Richter, H. J., *Interphase Momentum Transfer in the Flow of Bubbles through Nozzles*, EPRI Report NP-980, 1979.

4.3.7   Wallis, G. B., *One-Dimensional Two-Phase Flow*, McGraw-Hill, New York, 1969.

4.3.8   Zuber, N., "On the Dispersed Two-Phase Flow in the Laminar Flow Regimes," *Chem. Eng. Sci.*, **19**, 1964.

4.3.9   Rouhani, S. T., *Modified Correlation for Void and Pressure Drop*, AB Atomenergi, Report AE-RTV 841, Stockholm, Sweden, March 1969.

4.3.10 Zuber, N. and Findlay, J. A., "Average Volumetric Concentration in Two-Phase Flow Systems," J. Heat Transfer, **87**, 453–468, 1965.

4.3.11 Andersen, J. G. M., and Chu, K. H., *BWR Refill–Reflood Program, Task 4.7: Constitutive Correlations for Shear and Heat Transfer for the BWR Version of TRAC*, NUREG/CR-2134, EPRI NP-1582, GEAP-24940, December 1981.

4.3.12 Wallis, G. B., "Some Hydrodynamic Aspects of Two-Phase Flow and Boiling," *Proceedings of the International Heat Transfer Conference*, Boulder, Colo., Paper 38, 1961.

4.3.13 Rose, S. C., Jr., and Griffith, P., *Some Hydrodynamic Characteristics of Bubbly Mixtures Flowing Vertically Upwards in Tubes*, Report 5003-30, MIT, 1964.

REFERENCES

4.3.14 Meyer, P. E., and Wallis, G. B., *Bubbly Flow in Straight Pipes*, Report NYO-3114-12 (EURA EG 1530), Dartmouth College, 1965.

4.3.15 Delhaye, J. M., and Brichard, P., "Interfacial Area in Bubble Flow: Experimental Data and Correlations," *Nucl. Eng. Des.*, **151**, 65–77, 1994.

4.3.16 Bensler, H. P., "Détermination de l'aire interfaciale, due taux de vide et du diamètre moyen de Sauter dans un écoulement a bulles a partir de l'atténuation d'un faisceau d'ultrasons," Ph.D thesis, Institut National Polytechnique de Grenoble, France, 1990.

4.3.17 Fritz, W., "Maximum Volume of Vapor Bubbles," *Phys. Z.*, **36**, 379–384, 1935.

4.3.18 Park, J. W., "Void Wave Propagation in Two-Phase Flow," Ph.D. thesis, Rensselaer Polytechnic Institute, 1992.

4.3.19 Kalrach-Navarro, S. et al., "Analysis of the Bubbly/Slug Flow Regime Transition," *Nucl. Eng. Des.*, **151**, 15–39, 1994.

4.3.20 Serizawa, A. and Katoake, I., "Phase Distribution in Two-Phase Flow," *Proceedings of the ICHMT Conference on Transport Phenomena in Multiphase Flow*, Dubrovnik, Yugoslavia, 1987.

4.3.21 Wang, S. K., et al., "Turbulence Structure and Phase Distribution Measurements in Bubbly Two-Phase Flows," *Int. J. Multiphase Flow*, **3**, 327–343, 1987.

4.3.22 Liu, T. J., "The Effect of Bubble Size on Void Fraction Distribution in a Vertical Channel," *Proceedings of the Conference on Multiphase Flow*, Tsukuba, Japan, Vol. I, pp. 453–457, 1991.

4.3.23 Lahey, R. T., Jr., "The Analysis of Phase Separation and Phase Distribution Phenomena Using Two-Fluid Models," *Nucl. Eng. Des.*, **122**, 17–40, 1990.

4.3.24 Serizawa, A., et al., "Dispersed Flow," paper presented at the 3rd International Workshop on Two-Phase Flow Fundamentals, London, June 1992.

4.3.25 Zun, I., "Mechanism of Bubble Non-Homogeneous Distribution in Two-Phase Shear Flow," *Nucl. Eng. Des.*, **118**, 155–162, 1990

4.3.26 Kocamustafaogullari, G., and Huang, W. D., "Internal Structure and Interfacial Velocity Development for Bubbly Two-Phase Flow," *Nucl. Eng. Des.*, **151**, 79–101, 1994.

4.3.27 Furukawa, T. et al., "Phase Distribution for Air–Water Two-Phase Flow in Annuli," *Proceedings of the 17th National Heat Transfer Symposium*, pp. 349–351, 1980.

4.3.28 Sadatomu, K., et al., "Two-Phase Flow in Vertical Non-circular Channels," *Int. J. Multiphase Flow*, **8**(6), 641–655, 1982.

4.3.29 Sim, S. K., and Lahey, R. T., Jr., "Measurement of Phase Distribution in a Triangular Conduit," *Int. J. Multiphase Flow*, **12**(3), 405–425, 1986.

4.3.30 Jones, D. C., and Zuber, N., "Use of a Cylindrical Hot-Film Anemometer for Measurement of Two-Phase Void and Volume Flux Profiles in a Narrow Rectangular Channel," *A.I.Ch.E. Symposium Series*, 1978.

4.3.31 Radovcich, N. A., and Moissis, R., *The Transition from Two-Phase Bubble Flow to Slug Flow*, Report 7-7673-22, MIT, 1962.

4.3.32 Moissis, R., and Griffith, P., "Entrance Effects in a Two-Phase Slug Flow," *Trans. ASME*, Ser. C, **28**, 1962.

4.3.33 Schwartzbeck, R. K., and Kocamustafaogullari, G., "Two-Phase Flow Pattern Transition Scaling Studies," *ANS Proceedings of the 1988 National Heat Transfer Conference*, Houston, Texas, pp. 387–398, July 24–27, 1988.

4.3.34 Taitel, Y., et al., "Modeling Flow Pattern Transitions for Steady Upwards Gas–Liquid Flow in Vertical Tubes," *A.I.Ch.E. J.*, **26**, 345–354, 1980.

4.3.35 Mishima, K., and Ishii, M., "Flow Regime Transition Criteria for Upward Two-Phase Flow in Vertical Tubes," *Int. J. Heat Mass Transfer*, **27**, 723–737, 1984.

4.3.36 Lu, Z., and Zhang, X, "Identification of Flow Patterns of Two-Phase Flow by Mathematical Modeling," *Nucl. Eng. Des.*, **149**, 111–116, 1994.

4.3.37 Lahey, R. T., Jr., et al., "The Effect of Virtual Mass on the Numerical Stability of Accelerating Two-Phase Flow," *Int. J. Multiphase Flow*, **6**, 281–294, 1980.

4.3.38 Kelly, J. E., and Kazimi, M. S., "Interfacial Exchange Relations for Two-Fluid Vapor–Liquid Flow: A Simplified Regime-Map Approach," *Nucl. Sci.*, **81**, 305–318, 1982.

4.3.39 Pauchon, C., and Banerjee, S., "Interphase Momentum Effects in the Averaged Multifield Model, I: Void Propagation in Bubble Flows," *Int. J. Multiphase Flow*, **12**, 559–573, 1986.

4.4.1 Dukler, A. E, and Hubbard, M. G., "A Model for Gas–Liquid Slug Flow in Horizontal and near Horizontal Tubes," *Ind. Eng. Chem. Fundam.*, **14**, 337–347, 1976.

4.4.2 Dumitrescu, D. T., "Strömung an einer Luftblase im seukrechteu Rohr," *Z. Angew. Math. Mech.*, **23**(3), 134, 1943.

4.4.3 Davis, R. M., and Taylor, G. I., "The Mechanics of Large Bubbles Rising through Extended Liquids and through Liquids in Tubes," *Proc. R. Soc. (London)*, **200**, Ser. A, 375–390, 1950.

4.4.4 White, E. T., and Beardmore, R. H., "The Velocity of Rise of Single Cylindrical Air Bubbles through Liquids Contained in Vertical Tubes," *Chem. Eng. Sci.*, **17**, 351–361, 1962.

4.4.5 Wallis, G. B., *General Correlations for the Rise Velocity of Cylindrical Bubbles in Vertical Tubes*, GE Report 626L130, 1962.

4.4.6 Dukler, A. E., and Fabre, J., "Gas–Liquid Slug Flow: Knots and Loose Ends," paper presented at the 3rd International Workshop on Two-Phase Flow Fundamentals, Imperial College, London, 1992.

4.4.7 Runge, D. E., and Wallis, G. B., *The Rise Velocity of Cylindrical Bubbles in Inclined Tubes*, Report No. NYD-3114-8 (EURAEC-1416), Dartmouth College, 1965.

4.4.8 Griffith, P., and Wallis, G. B., "Two-Phase Slug Flow," *J. Heat Transfer*, Ser. C, **83**(3), 307–320, 1961.

4.4.9 Nicklin, D. J., Wilkes, J. O., and Davidson, J. F., "Two-Phase Flow in Vertical Tubes," *Trans. Inst. Chem. Eng.*, **40**, 61–68, 1962.

4.4.10 Griffith, P., *The Prediction of Low-Quality Boiling Voids*, ASME Paper 63-HT-20, National Heat Transfer Conference, Boston, 1963. Also MIT Report 7-7673-23, 1963.

4.4.11 Frechon, D., "Étude de l'Éconlement ascendante a trois fluide en conduite vertical," These, Institut Polytechnique, Toulouse, 1986.

4.4.12 Ellis, J. E, and Jones, E. L., "Vertical Gas–Liquid Flow Problems," *Proceedings of the Two-Phase Flow Symposium*, University of Exeter, Vol. 2, 1965.

4.4.13 Ishii, M., and Mishima, K., "Study of Two-Fluid Model and Interfacial Area," ANL 80-111, NUREG/CR-1873, December 1980.

4.4.14 Vermeulen, L. R., and Ryan, J. T., "Two-Phase Slug Flow in Horizontal and Inclined Tubes," *Can. J. Chem. Eng.*, **49**, 195–201, 1976.

4.4.15 Dukler, A. E., and Fabre, J., "Gas–Liquid Slug Flow, Knots and Loose Ends," paper presented at the 3rd International Workshop on Two-Phase Flow Fundamentals, Imperial College, London, 1992.

4.4.16 Benjamin, T. B., "Gravity Currents and Related Phenomena," *J. Fluid Mech.*, **31**, 209–248, 1968.

4.4.17 Webber, M. E., "Drift in Intermittent Two-Phase Flow in Horizontal Pipes," *Can. J. Chem. Eng.*, **59**, 398–99, 1981.

4.4.18 Gregory, G. A., and Scott, D. S., "Correlation of Liquid Slug Velocity and Frequency in Horizontal Cocurrent Gas–Liquid Slug Flow," *A.I.Ch.E. J.*, **15**, 933–935, 1969.

4.4.19 Tronconi, E., "Prediction of Slug Frequency in Horizontal Two-Phase Slug Flow," *A.I.Ch.E. J.*, **36**, 701–709, 1990.

4.4.20 Gregory, G. A., et al., "Correlation of the Liquid Volume Fraction in the Slug for Horizontal Gas–Liquid Slug Flow," *Int. J. Multiphase Flow*, **4**, 33–39, 1978.

4.4.21 Barnea, D., and Brauner, N., "Holdup of the Liquid Slug in Two-Phase Intermittent Flow," *Int. J. Multiphase Flow*, **11**, 43–50, 1985.

4.4.22 Azzopardi, B. J., et al., "Two-Phase Slug Flow in Horizontal Pipes," paper presented at the Pipelines Symposium, Utrecht, Holland, 1985.

4.4.23 Wallis, G. B., *The Transition from Flooding to Upwards Cocurrent Annular Flow in a Vertical Pipe*, AEEW-R142, UKAEA, 1962.

4.4.24 Wallis, G. B., et al., *Two-Phase Flow and Boiling Heat Transfer*, Dartmouth College, Report NYO 3114-14, February 1966.

4.4.25 Griffith, P., and Haberstroh, R. D., *The Transition from the Annular to the Slug Flow Regime in Two-Phase Flow*, Report 5003-28, Mechanical Engineering Department, MIT, 1964.

4.4.26 Griffith, P., *The Slug-Annular Flow Regime Transition at Elevated Pressure*, ANL-6796, 1963.

4.4.27 Taitel, Y., and Dukler, A. E., "A Model for Predicting Flow Regime Transition in Horizontal and Near Horizontal Gas–Liquid Flow," *A.I.Ch.E. J.*, **22**(1), 1976.

4.4.28 Taitel, Y., "Stability of Severe Slugging," *Int. J. Multiphase Flow*, **12**(2), 203–217, 1986.

4.4.29 Fernandez, R. C., et al., "Hydrodynamic Model for Gas–Liquid Slug Flow in Vertical Tubes, *A.I.Ch.E. J.*, **29**, 981–989, 1983.

4.5.1 Andritsos, N., and Hanratty, T. J., "Interfacial Instabilities for Horizontal Gas–Liquid Flows in Pipelines," *Int. J. Multiphase Flow*, **13**, 583–603, 1987.

4.5.2 Shi, J., and Kocamustafaogullari, G., "Interfacial Measurements in Horizontal Stratified Flow Patterns," *Nucl. Eng. Des.*, **149**, 81–96, 1994

4.6.1 Levy, S., "Theory of Pressure Drop and Heat Transfer for Annular Steady-State Two-Phase Two-Component Flow in Pipes," *Proceedings of the 2nd Midwestern Conference on Fluid Mechanics*, p. 337, 1952.

4.6.2 Calvert, S., and Williams, B., "Upwards Co-current Annular Flow of Air and Water in Short Tubes," *A.I.Ch.E. J.*, **1**, 78, 1955.

4.6.3 Anderson, G. H., and Mantzouranis, G. B., "Two-Phase (Gas–Liquid) Flow Phenomena, I: Pressure Drop and Hold-up for Two-Phase Flow in Vertical Tubes," *Chem. Eng. Sci.*, **12**, 109, 1960.

4.6.4 Dukler, A. E., *Hydrodynamics of Liquids Film in Single and Two-Phase Flow*, Ph.D. Thesis. Univ. of Delaware, 1951.

4.6.5 Dukler, A. E., "Fluid Mechanics and Heat Transfer in Falling Film Systems," *Chem. Eng. Prog. Symp. Ser.*, **56**(30), 1–10, 1960.

4.6.6 Hewitt, G. F., *Analysis of Annular Two-Phase Flow; Application of the Dukler Analysis to Vertical Upward Flow in a Tube*, AERE-R3680, 1961.

4.6.7 Levy, S., and Healzer, J. M., "Application of Mixing Length Theory to Wavy Turbulent Liquid–Gas Interface," *J. Heat Transfer*, **103**, 492–500, 1981.

4.6.8 Gill, L. E., and Hewitt, G. F., *Further Data on the Upwards Annular Flow of Air–Water Mixtures*, AERE-R3935, 1962.

4.6.9 Hewitt, G. F., and Hall-Taylor, N. S., *Annular Two-Phase Flow*, Pergamon Press, Oxford, 1970.

4.6.10 Schlichting, H., *Lecture Series, Boundary Layer Theory, II: Turbulent Flows*, NACA TM 1218, p. 27, 1949.

4.6.11 Wallis, G. B., "Annular Two-Phase. 1: A Simple Theory; 2L: Additional Effects," *J. Basic Eng.*, **60**, 59–82, 1970.

4.6.12 Chien, S., and Ibele, W., "Pressure Drop and Liquid Film Thickness of Two-Phase Annular and Annular-Mist Flows," *J. Heat Transfer*, Ser. C, **86**, 80–96, 1964.

4.6.13 Ueda, T., and Tanaka, T., "Studies of Liquid Film Flow in Two-Phase Annular and Annular-Mist Flow Regions; 1: Downflow in a Vertical Tube," *Bull. Jpn. Soc. Mech. Eng.*, **17**, 603–613, 1974.

4.6.14 Ueda, T., and Tanaka, T., Studies of Liquid Film Flow in Two-Phase Annular and Annular-Mist Flow Region, 2: Upflow in a Vertical Tube," *Bull. Jpn. Soc. Mech. Eng.*, **17**, 614–624, 1974

4.6.15 Moeck, E. D., and Stachiewicz, J. W., "A Droplet Interchange Model for Annular-Dispersed, Two-Phase Flow," *Int. J. Heat Mass Transfer*, **15**, 637–653, 1972.

4.6.16 Dobran, F., "Hydrodynamic and Heat Transfer Analysis of Two-Phase Annular Flow with a New Liquid Film Model of Turbulence," *Int. J. Heat Mass Transfer*, **26**(8), 1159–1171, 1983.

4.6.17 Cousins, L. B., et al., *Liquid Mass Transfer in Annular Two-Phase Flow*, AERE R-4926, 1965.

4.6.18 Asali, J. C., and Hanratty, T. J., "Interfacial Drag and Film Height for Vertical Annular Flow," *A.I.Ch.E. J.*, **31**(6), 895–902, 1985.

4.6.19 Nedderman, R. M., and Shearer, C. J., "The Motion and Frequency of Large Disturbance Waves in Annular Two-Phase Flow of Air–Water Mixtures," *Chem. Eng. Sci.*, **18**, 661–670, 1963.

4.6.20 Hall-Taylor, N. S., et al., "The Motion and Frequency of Large Disturbance Waves in Annular Two-Phase Flow of Air–Water Mixtures," *Chem. Eng. Sci.*, **18**, 537, 1963.

4.6.21 Hewitt, G. F., "In Search of Two-Phase Flow," *J. Heat Transfer*, **118**, 518–527, 1996.

4.6.22 Srivastoa, R. P. S., "Liquid Film Thickness in Annular Flow," *Chem Eng. Sci.*, **28**, 819–824, 1973.

4.6.23 Chung, H. S., and Murgatroyd, W., "Studies of the Mechanism of Roll Wave Formation on Thin Liquid Film," *Proceedings of the Symposium on Two Phase Flow*, University of Exeter, Vol. 2, 1965.

4.6.24 Levy, S., "Prediction of Two-Phase Annular Flow with Liquid Entrainment," *Int. J. Heat Mass Transfer*, **9**, 171–188, 1966.

4.6.25 Nigmatulin, B., et al., "Experimental Investigation of the Hydrodynamics of Equilibrium Dispersed-Annular Steam-Water Flow," *Teplofiz. Vys. Temp.*, **16**, 1258, 1978.

4.6.26 Subbotin, V. I., et al., "Integrated Investigation into Hydrodynamic Characteristics of Annular Dispersed Steam–Liquid Flows," *Int. Heat Transfer Conf.*, **1**, 327, 1978.

4.6.27 Whalley, P. B., and Hewitt, G. F., *The Correlation of Liquid Entrainment Fraction and Entrainment Rate in Annular Two-Phase Flow*, AERE-R9187, 1978.

4.6.28 Bharathan, D., and Wallis, G. B., Air–Water Countercurrent Flow, EPRI-NP-1165, 1979.

4.6.29 Abe, Y., "Estimation of Shear Stress in Countercurrent Annular Flow," *J. Nucl. Sci. Technol.*, **28**(3), 208–217, 1991.

4.6.30 Govan, A., et al., "Flooding and Churn Flow in Vertical Pipes," *Int. J. Multiphase Flow*, **17**, 27–44, 1991.

4.6.31 Kelly, J. M., and Freitas, R. L., "Interfacial Friction in Low Flowrate Vertical Annular Flow," *Proceedings of the 6th International Meeting on Nuclear Reactor Thermal Hydraulics*, Vol. 1, pp. 154–160, 1993.

4.6.32 Wicks, M., "Liquid Film Structure and Drop Size Distribution in Two-Phase Flow," Ph.D. thesis, Chemical Engineering Department, University of Houston, 1967.

4.6.33 James, P. W., et al., *Droplet Motion in Two-Phase Flow*, NUREG/CP-0014, Vol. 2, pp. 1484–1503, 1980.

4.6.34 Azzopardi, B. J., et al., *Drop Sizes and Deposition in Annular Two-Phase Flow*, AERE-R9634, 1979.

4.6.35 Nukiyama, S., and Tanasawa, I., *Trans. Soc. Mech. Eng. Jpn.*, **4**(14), 86, 1938.

4.6.36 Ishii, M., and Chawla, T. C., *Local Drag Laws in Dispersed Two-Phase Flow*, ANL-79-105, 1979.

4.6.37 Wallis, G., *The Onset of Droplet Entrainment in Annular Gas–Liquid Flow*, GE Report 626L 127, 1962.

4.6.38 Ishii, M., and Mishima, K., "Liquid Transfer and Entrainment Correlations for Droplet-Annular Flow," *Proceedings of the 7th International Heat Transfer Conference*, Munich, Vol. 5, pp. 307–312, 1982.

4.6.39 Hewitt, G. F., and Govan, A. H., "Phenomenological Modeling of Non-equilibrium Flows with Phase Change," *Int. J. Heat Mass Transfer*, **33**, 229–242, 1990.

4.6.40 McCoy, D. D., and Hanratty, T. J., "Rate of Deposition of Droplets in Annular Two-Phase Flow," *Int. J. Multiphase Flow*, **3**, 319–331, 1977.

4.6.41 Owen, D. G., "An Experimental and Theoretical Analysis of Equilibrium Annular Flow," Ph.D. thesis, University of Birmingham, England, 1986.

4.6.42 Petukhov, B. S., "Heat Transfer and Friction in Turbulent Pipe Flow with Variable Physical Properties," in *Advances in Heat Transfer*, Academic Press, San Diego, Calif., 1970.

4.6.43 Jayanti, S., and Hewitt, G. F., "Hydrodynamics and Heat Transfer in Wavy Annular Gas–Liquid Flow: A CFD Study," *Int. J. Heat Mass Transfer* (to be published).

4.6.44 Pushkina, O. L., and Sorokin, Y. L., "Breakdown of Liquid Film Motion in Vertical Tubes," *Heat Transfer Sov. Res.*, **1**, 56, 1969.

4.6.45 Hewitt, G. F., "In Search of Two-Phase Flow," *J. Heat Transfer*, **118**, 518–527, 1996.

CHAPTER 5

LIMITING MECHANISM OF COUNTERCURRENT FLOODING

5.1 INTRODUCTION

In many practical applications, the downward flow of liquid may be limited by the upward flow of gas. The phenomenon, generally referred to as *countercurrent flooding*, has been studied for over 60 years. Initially, the mechanism received attention in the chemical process industries because the most common countercurrent flow condition is found in packed towers, where it can lead to loading and flooding problems. Loading occurs as the gas flow rate is increased and the pressure drop across the packed bed rises sharply. Flooding corresponds to when undesirable liquid carryover or liquid flow reversal takes place. Similarly, in some reflux condensers, the vapor entering at the bottom of the condenser can be harmful if it is at a value high enough to limit or stop the condensed liquid from falling up countercurrently to the rising vapor.

This early interest in the mechanism of flooding increased significantly with the advent of light-water-cooled nuclear power plants because it could limit the ingress of water into the nuclear reactor core during a hypothetical loss-of-coolant accident (LOCA). During such an accident, emergency core cooling systems (ECCSs) are provided to reflood the reactor core with cold water through either top spraying or bottom reflooding, but the flow of this cooling water can be reduced and even diverted temporarily by flashing water and uprising steam generated within the reactor core or from other hot surfaces. This possibility led to intensive and substantial analytical and experimental development programs to study countercurrent flooding. However, the mechanisms for flooding are still not fully understood today, and testing and empirical correlations are the primary means for predicting their occurrence and behavior. This is particularly true in such complex two-phase-flow systems as nuclear power plants, where the situation is much more complicated due

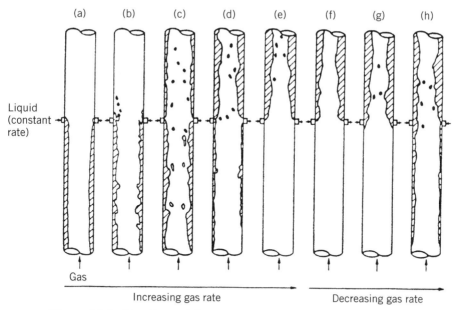

Figure 5.1.1 Flooding and flow reversal in a tube. (From Ref. 5.1.1.)

to unsteady-state flow conditions, condensation of the steam due to quenching by cold water, flashing of input water from its contact with hot surfaces, merging of flow streams, and the presence of pressure pulses.

Figure 5.1.1 shows countercurrent annular flow in a small-diameter tube with a falling liquid film along the wall and an upward flow of gas in the center [5.1.1]. At no-gas or low-gas flow, there is a falling liquid film with very few ripples on its surface. As the gas flow rate is increased, the liquid film becomes more disturbed and liquid begins to be carried above the injection point. Generally known as the *flooding point*, this is shown in Figure 5.1.1b. With further increases in gas flow, the direction of more of the liquid flow is reversed, and some of the liquid is flowing co-currently with the gas above the injection point while countercurrent flow still occurs below the injection (Figure 5.1.1c and d). At a high enough gas flow rate, all the flow is upward co-current, as shown in Figure 5.1.1e. If the gas velocity is now reduced, a point is reached where the liquid starts to flow downward. This condition, referred to as *flow reversal*, is illustrated in Figure 5.1.1g.

Conversely, if the gas flow rate is kept constant and the liquid flow is increased, all the liquid will flow downward at low liquid rates. By increasing the liquid flow rate, a maximum value for the downflow film is reached. Beyond that liquid-film flow, the excess of liquid moves upward. Therefore, there exists a maximum value for the liquid flow rate that can move downward countercurrent to the afore-prescribed gas flow rate. This downward rate of liquid flow cannot be increased without decreasing the gas flow rate. In that sense, the flooding phenomenon is a process that limits the possible countercurrent liquid flow rate.

Extensive reviews of the flooding phenomena have been published. They include the analytical and test survey by Tien and Liu [5.1.2], the mechanisms discussion by Dukler et al. [5.1.3] and by Moalem and Dukler [5.1.4], the literature reviews by Bankoff and Lee [5.1.5, 5.1.6], and the most recent papers by Hewitt [5.1.1, 5.1.7]. In this chapter we first discuss countercurrent flooding in simplified tubular and annular geometries and describe the available test results and the corresponding applicable models; next, nonthermal equilibrium conditions and more complicated configurations are dealt with; and finally, multidimensional and transient tests are covered to provide a perspective of our knowledge of flooding in complex two-phase-flow systems and of the approximations employed to represent it in system computer codes.

5.2 COUNTERCURRENT FLOODING IN SIMPLIFIED GEOMETRIES

5.2.1 Experimental Setups

Two basic test arrangements [5.1.2] have been employed and they are illustrated in the upper portion of Figure 5.2.1. The principal difference is whether the liquid is introduced at the top of the tube (Figure 5.2.1a) or through porous injection close to the middle of the tube (Figure 5.2.1b). The porous injection produces the minimum disturbance in the liquid film while top flood entry (Figure 5.2.1c) tends to have entry effects that vary with the entry configuration and generally accelerate the occurrence of flooding. There are also two different modes of providing gas into the lower plenum and they are shown in Figures 5.2.1d. In one case gas is introduced into the same bottom plenum reached by the water, while in the other, gas is required to flow through a nozzle placed at the center of the end of the tube undergoing flooding. Again, the gas entry and the tube end geometry affect the gas velocity at which flooding occurs. Generally speaking, increased gas turbulence produced by poor entry gas or liquid modes will promote flooding. An important advantage of Figure 5.2.1b, worth noting, is that it provides a good opportunity to observe and measure flow reversal of the liquid film.

Another contributor to poor agreement among test results is the fact that there is no agreement about the definition for the inception of flooding. Some researchers have associated flooding with the first liquid being carried up beyond the injector, others have adopted the criterion of significant liquid flow reversal or no liquid penetration at all, while still others have identified the start of liquid entrainment for the onset of flooding. Two other means have also been employed to define the occurrence of flooding: One relies on visual observations and determines when the liquid flow becomes chaotic with large interfacial wave formation; the other method uses pressure drop measurement and employs a sharp rise in the pressure gradient to identify flooding. Both of these techniques are related to the phenomenon governing transition between slug and churn flows and were discussed in Section 4.4.7.

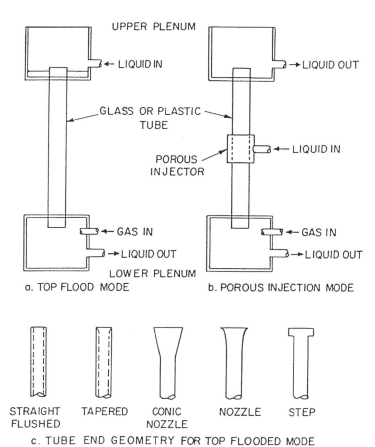
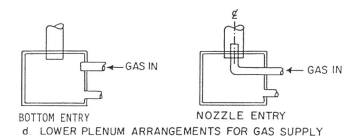

Figure 5.2.1 Typical apparatus arrangement for vertical channel flooding experiments. (From Ref. 5.1.2.)

5.2.2 Principal Empirical Correlations in Tube Geometries

Empirical correlations are derived by plotting the test data against appropriate nondimensional groupings. Two different sets of applicable appropriate nondimensional groupings have been proposed. Wallis [5.2.1] employed a balance between inertial forces in the gas and hydrostatic forces to develop the following gas and liquid nondimensional average volumetric fluxes:

$$\bar{j}_G^* = \bar{j}_G \sqrt{\frac{\rho_G}{gD(\rho_L - \rho_G)}} \qquad \bar{j}_L^* = \bar{j}_L \sqrt{\frac{\rho_L}{(\rho_L - \rho_G)gD}} \qquad (5.2.1)$$

where $\bar{j}_G$ and $\bar{j}_L$ are the gas and liquid volumetric fluxes, ρ_L and ρ_G the liquid and gas densities, D the tube diameter, and g the gravitational constant.

The alternative set of flooding nondimensional groupings employs the Kutateladze numbers, or

$$\mathrm{Ku}_G = \bar{j}_G \left[\frac{\rho_G^2}{\sigma g(\rho_L - \rho_G)}\right]^{1/4} \qquad \mathrm{Ku}_L = \bar{j}_L \left[\frac{\rho_L^2}{\sigma g(\rho_L - \rho_G)}\right]^{1/4} \qquad (5.2.2)$$

where σ is the fluid surface tension. The Kutateladze number has been derived from considerations of the stability of the liquid film or from the gas flow needed to suspend the largest stable liquid drop. This approach was proposed first by Pushkina and Sorokin [5.2.2] to correlate their data.

The Wallis and Kutateladze groupings are similar: Both do not involve the viscosity of the fluids and they differ only because of the presence of the channel diameter in the Wallis grouping and of the surface tension in the Kutateladze number. There is a relation between the two groupings:

$$\mathrm{Ku}_k = D^{*1/2} \bar{j}_k^* \qquad k = L, G \qquad (5.2.3)$$

where D^* is the square root of the Bond number:

$$\mathrm{Bo}^{1/2} = D^* = D \left[\frac{(\rho_L - \rho_G)g}{\sigma}\right]^{1/2} \qquad (5.2.4)$$

Wallis empirical flooding correlations have the general form

$$\bar{j}_G^{*1/2} + m \bar{j}_L^{*1/2} = C \qquad (5.2.5)$$

where m and C are constants that depend on the test setup and geometry. The value of m is usually between 0.8 and 1.0, while C lies between 0.7 and 1.0.

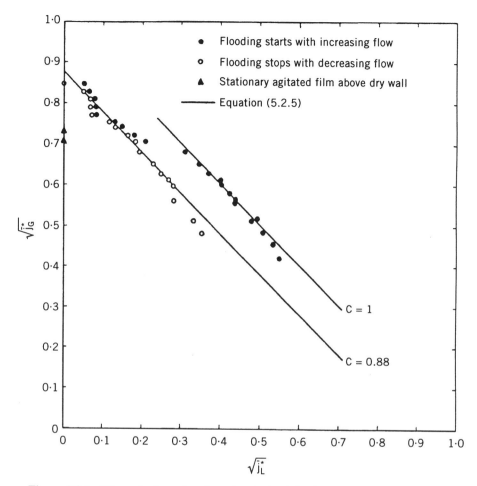

Figure 5.2.2 Dimensionless plot of results obtained for flooding. (From Ref. 5.2.3.)

The data obtained by Hewitt and Wallis [5.2.3] with a porous liquid injector in a 1.25-in. tube are reproduced in Figure 5.2.2. The value of m was set equal to 1.0, and so was the value of C, to fit the data at high liquid rates and increasing gas flows. At low liquid flow rates ($\bar{j}_L^* < 0.3$), the constant C decreases to 0.88 to match the data. The line with $C = 0.88$ also fits the data at higher liquid flows, when the gas flow is reduced to produce flow reversal. In other words, there is some degree of hysteresis in that region since the gas velocity at flooding is greater than at flow reversal. Figure 5.2.2 also shows that the agitated liquid film hangs above the drywall at the liquid inlet point when $\sqrt{\bar{j}_G^*}$ is between 0.7 and 0.72.

Equation (5.2.5) has been employed by many experimenters to plot their flooding data. For example, the data obtained by Lobo et al. [5.2.4] and Sherwood et al. [5.2.5] in packed towers is correlated satisfactorily in Figure 5.2.3 with $m = 1.0$ and $C = 0.775$. Similarly, Dukler and Smith [5.2.6] employed a smooth nozzle

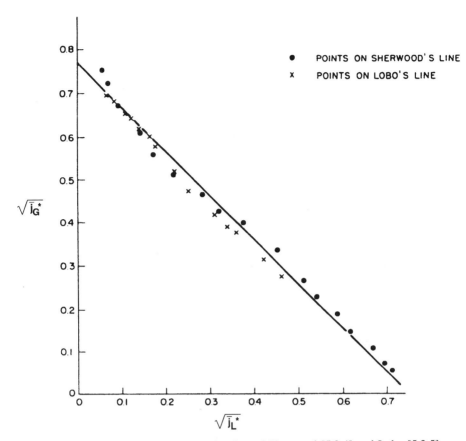

Figure 5.2.3 Flooding correlations for data of Sherwood [5.2.4] and Lobo [5.2.5].

entry on both the top and bottom ends of their test section and obtained good agreement with Wallis correlation, with $C = 0.88$ for air–water data in a 2-in. tube. However, considerable variations in the constants m and C have been reported by other investigators and they have been traced to a great extent to different inlet and exit conditions. This is illustrated in Figure 5.2.4, where flooding data are shown for a square-edged and a bell-mouth (tapered) exit and are compared to a porous-wall system [5.2.7]. Hewitt [5.1.1] concluded that end effects can be substantial and significant.

Another area of disagreement about the Wallis correlation is its prediction of increased flooding gas velocity at constant liquid velocity when the tube diameter increases. Several investigators have reported no dependence upon channel size, especially when the channel diameter exceeds 5 cm [5.2.2, 5.2.8]. The Kutateladze groupings have been employed under those circumstances to predict flooding from

$$\mathrm{Ku}_G^{1/2} + \mathrm{Ku}_L^{1/2} = C \quad (5.2.6)$$

where C is usually ascribed the value of 1.79.

5.2 COUNTERCURRENT FLOODING IN SIMPLIFIED GEOMETRIES 271

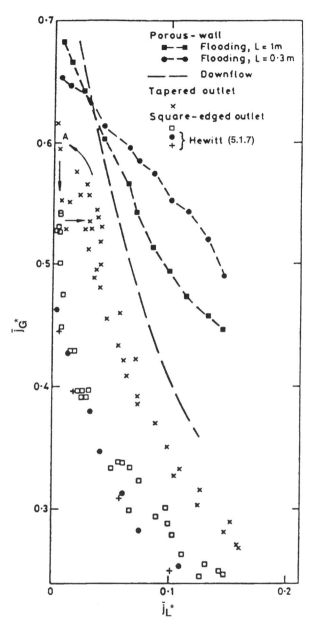

Figure 5.2.4 Comparison of bell-mouth and porous wall and sharp-edge inlet for flooding. (Reprinted from Ref. 5.2.7. with permission from Elsevier Science.)

Interpolations between the Wallis and Kutateladze channel diameter treatment have been suggested to resolve the role of tube diameter. For instance, Wallis and Makkencherry [5.2.8] have proposed the following variation of the constant C in Eq. (5.2.6) with the Bond number:

D^*	$\leqslant 2$	4	10	14	20	28	$\geqslant 50$
C	0	1.0	1.45	1.58	1.67	1.73	1.79

This compromise has been used successfully in the RELAP-UK code [5.2.9]. Another correlation has been proposed by McQuillan and Whalley [5.2.10] to deal with tube diameter and fluid viscosity. Its form is

$$\mathrm{Ku}_G = 0.286 \mathrm{Bo}^{0.26} \mathrm{Fr}^{-0.22} \left(1 + \frac{\mu_L}{\mu_W}\right)^{-0.18} \tag{5.2.7}$$

where Bo is the Bond number given by Eq. (5.2.4) and Fr is the Froude number, obtained from

$$\mathrm{Fr} = \frac{\bar{u}_{LS} \pi}{4} \left[\frac{g(\rho_L - \rho_G)}{\sigma}\right]^{1/4} \tag{5.2.8}$$

In Eq. (5.2.8), $\bar{u}_{LS}$ is the liquid superficial velocity, and in Eq. (5.2.7), μ_L corresponds to the liquid viscosity, while μ_W corresponds to the water viscosity at room temperature.

Most recently, Hewitt [5.1.1] has offered a reasonable explanation of why the Wallis and Kutateladze correlations may be valid over different channel diameters. In a small channel the waves can grow much more relative to the channel span and even bridge the tube to produce flooding. In a large channel, the waves can remain small by comparison to the channel diameter, and they become standing waves near the bottom of the tube, where they break up due to gas flow to produce droplet entrainment and cause flooding beyond the injector.

In summary, there is an abundant set of steady-state flooding data in simple tubular geometries. The entrance and exit conditions have a significant impact on the test results as well as the lack of a good definition for the initiation of flooding. This leads to considerable data scatter and variations in the applicable empirical flooding correlations. It also means that prototype tests are needed to duplicate the entrance and exit conditions. There is a growing agreement that the role of tube diameter decreases as it grows in size and that flooding becomes independent of channel size when the diameter exceeds about 5 cm or the Bond number approaches 30. As noted already in the case of RELAP-UK, the system computer codes tend to favor simplified flooding empirical correlations of the type of Eq. (5.2.6).

5.2.3 Flooding Analysis in Tube Geometries

Several possible mechanisms have been suggested to explain the occurrence of flooding in tubes. They include:

- The use of density (kinematics) or continuity waves by Wallis [5.2.11] to obtain a simplified explanation of the limiting condition for countercurrent flow
- The instability of the liquid-film waves or the formation of a standing wave to justify liquid bridging across the tube cross section
- The occurrence of upward film flow due to increased shear stress at the wavy gas–liquid interface
- The start of liquid entrainment from the wave crests and the transport of the liquid drops above the point of injection

The analytical treatment of each of these potential mechanisms is covered next. However, it should be noted that under all the preceding mechanisms, the presence of waves and preferably large waves at the gas–liquid interface is essential to cause flooding. The primary difference among the various available models is how the liquid in the interfacial waves is lifted above the injection point.

5.2.3.1 Mechanism of Kinematic Waves

Wallis [5.2.11] was the first to realize that the large kinematic waves formed at the gas–liquid interface can account for the mechanism of flooding. In two-phase flow, flow-limiting phenomena are associated with the propagation of two different types of waves: dynamic waves produced by pressure perturbations, which propagate at the sonic velocity of the phases or mixture; and kinematic waves generated by perturbations in the two-phase density (i.e., its gas volume fraction), which are created by perturbations in other parameters of the flowing stream, such as temperature or enthalpy. The kinematic waves are often referred to as *density* or *continuity waves* because they depend only on the continuity equation and do not consider the momentum equation and therefore acceleration or frictional pressure losses.

Wallis employed the drift flux model discussed in Section 3.5.4 to demonstrate the occurrence of kinematic waves and countercurrent flood limiting conditions. According to Eq. (3.5.36), the drift flux relation can be written as

$$\bar{j}_G = (C_o \bar{j} + \overline{U}_{Gj})\bar{\alpha} \tag{5.2.9}$$

where $\bar{\alpha}$ is the average gas volume fraction, C_o is the flux concentration parameter, usually set at 1.2, and $\overline{U}_{Gj}$ is the drift velocity, which can be specified from Eqs. (3.5.37) for different flow patterns. As long as the drift velocity $\overline{U}_{Gj}$ is only dependent on fluid properties and the average gas volume fraction $\bar{\alpha}$, there exists a definite relation between the specified flow of the two phases and the average gas volume fraction $\bar{\alpha}$, as discussed in Section 3.5. It is interesting to note that if we introduce Eq. (3.5.37) for slug flow into Eq. (5.2.9), one would recreate the following nondimensional group $\bar{j}_G^*, \bar{j}_L^*$, and $(\rho_L/\rho_G)^{1/2}$. Similarly, if we put Eq. (3.5.37) for bubble flow into Eq. (5.2.9), the nondimensional groupings Ku_G, Ku_L, and $(\rho_L/\rho_G)^{1/2}$ would emerge.

274 LIMITING MECHANISM OF COUNTERCURRENT FLOODING

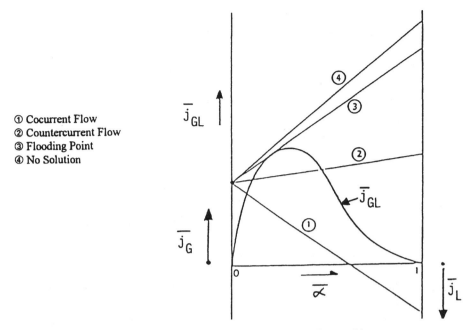

① Cocurrent Flow
② Countercurrent Flow
③ Flooding Point
④ No Solution

Figure 5.2.5 Graphical solution of the flooding problem.

The drift flux velocity, $\bar{j}_{GL}$ corresponds to the volumetric flux of the gas with respect to the total volumetric flux velocity $\bar{j}$, or

$$\bar{j}_{GL} = \bar{\alpha}(\bar{u}_G - \bar{j}) = (1 - \bar{\alpha})\bar{j}_G - \bar{\alpha}\bar{j}_L \tag{5.2.10}$$

If we utilize Eq. (5.2.9) and set $C_o = 1.0$ for simplification purposes, there results

$$\bar{j}_{GL} = \overline{U_{Gj}}\bar{\alpha} \tag{5.2.11}$$

Equation (5.2.11) can be represented by a plot of $\bar{j}_{GL}$ versus $\bar{\alpha}$ if we employ Eqs. (3.5.37) and if we know the transition value of $\bar{\alpha}$ to go from one flow pattern to another. Also, $\bar{j}_{GL}$ equals zero when $\bar{\alpha} = 0$ and the same is true when $\bar{\alpha} = 1$ if we employ the expression for $\bar{j}_{GL}$ for annular flow. This is illustrated in Figure 5.2.5. On the same figure, Eq. (5.2.10) is plotted as a straight line, labeled line 1, starting at $\bar{j}_G$ at $\bar{\alpha} = 0$ and ending at $-\bar{j}_L$ at $\bar{\alpha} = 1.0$ in the case of cocurrent flow. For countercurrent flow, $\bar{j}_L$ is negative and the corresponding line is labeled as line 2. For line 4 there is no intersection between it and the representation of Eq. (5.2.11). Line 3 in Figure 5.2.5, in fact, corresponds to the flooding condition because increasing the liquid flow rate beyond that corresponding to line 3 would not allow the additional liquid to flow down; thus, the limit of countercurrent liquid flow rate is reached as discussed in Section 5.1. Kinematic or density waves are, therefore, capable of providing a good qualitative physical picture of the mechanism of countercurrent flooding. However, if accurate quantitative predictions are desired, they can be obtained only if the curve of $\bar{j}_{GL}$ versus $\bar{\alpha}$ shown in Figure 5.2.5 is determined experimentally.

5.2.3.2 Mechanism of Sudden Change in Wave Motion

Once a finite wave is formed, the wave will grow on the falling liquid film as gas flow is increased. Two types of wave instability have been reported: Large waves grow first at the bottom of the test section and move upward to produce flooding, or waves become unstable locally and grow to bridge the tube and cause flooding. In both cases, liquid is transported upward by the gas flow. This concept of wave growth instability was first suggested by Schutt [5.2.12] and was pursued by many subsequent investigators, including Cetinbudakhar and Jameson [5.2.13], Imura et al. [5.2.14], and Chung [5.2.15]. Most of these models produce the appropriate nondimensional groupings but do not always match the test data without some empirical adjustments. Also, while such visualization studies as those performed by McQuillan et al. [5.2.16] support the wave growth theory as flooding is approached, other tests, such as those carried out by Dukler et al. [5.1.3], show that flooding can occur without the formation of large, rapidly growing waves.

The wave interface instability models are covered in Reference 5.1.2 and are not repeated here. Instead, the focus is placed on the formation of a standing wave as suggested by Shearer and Davidson [5.2.17] and the simplified model proposed by Whalley [5.2.18] to describe this mechanism. Figure 5.2.6 shows a control volume in a pipe of cross-sectional area A, diameter D, and an ideal circular wave of height

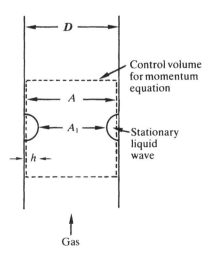

Figure 5.2.6 Control volume for momentum equation calculation on a stationary wave. (Reprinted from Ref. 5.2.18 by permission of Oxford University Press.)

h. The liquid-film thickness is small and the gas behaves as if it flows through a contraction–expansion created by the wave. The pressure drop Δp across the wave is

$$\Delta p = \frac{1}{2}\rho_G \bar{u}_G^2 \left[1 - \frac{A}{(\pi/4)(D-2h)^2}\right]^2 \tag{5.2.12}$$

and if $h/D \ll 1$, Eq. (5.2.12) can be simplified with $A = (\pi/4)D^2$ to

$$\Delta p = \frac{1}{2}\rho_G \bar{u}_G^2 \frac{16h^2}{D^2} \tag{5.2.13}$$

The momentum equation can be used across the control volume shown in Figure 5.2.6 to get

$$\Delta p\, A = (\rho_L - \rho_G)V_w g \tag{5.2.14}$$

where V_W is the wave volume, or

$$V_W = \pi h^2 \left(\frac{\pi D}{2} - \frac{4h}{3}\right) \approx \frac{\pi^2 h^2 D}{2} \tag{5.2.15}$$

as long as $h/D \ll 1$. If we substitute Eqs. (5.2.13) and (5.2.15) into Eq. (5.2.14), there results

$$\frac{\bar{u}_G^2 \rho_G}{D(\rho_L - \rho_G)g} = \frac{\pi}{4} \qquad j_G^* = \sqrt{\frac{\pi}{4}} = 0.89 \tag{5.2.16}$$

Equation (5.2.16) is a simplified form of Wallis flooding empirical equation (5.2.5). Its constant, 0.89, is very close to the 0.88 slope fitting most of the data in Figure 5.2.2. Also, it is about 20% above the 0.72 constant listed in the same figure for a hanging liquid film. One could explain this latter difference by introducing into the pressure drop relations of Eqs. (5.2.12) and (5.2.16) a parameter to take into account the shape of the standup wave and its circumferential variation; however, such a modification could be confirmed accurately only by tests.

5.2.3.3 Mechanism of Upward Liquid-Film Flow

Under this mechanism, suggested by Dukler et al. [5.1.3], there may be upward flow in the liquid film brought about by an increase in the interfacial gas shear. As the gas rate is increased,

5.2 COUNTERCURRENT FLOODING IN SIMPLIFIED GEOMETRIES

more pronounced wave action takes place. At or near the condition for flooding, the gas pressure gradient jumps up and there is a sudden and sufficient increase in the interfacial shear stress to cause flow reversal in the liquid film. For this mechanism to work, the gas pressure gradient or shear stress after the jump must be used [5.1.3]. A model describing those circumstances was developed in Reference 5.1.4. It presumes a thin liquid film of thickness δ in a pipe of diameter D with laminar flow in the film. At a distance y from the wall and for $y \ll D/2$, a force balance on the liquid film gives

$$\tau = \mu_L \frac{du}{dy} = \tau_i - \left(\frac{dp}{dz} + \rho_L g\right)(\delta - y) \tag{5.2.17}$$

where τ is the shear stress within the film, u the local velocity, μ_L the liquid viscosity, τ_i the interface shear stress, and dp/dz the pressure gradient.

Equation (5.2.17) can be solved for the liquid velocity u, which can be integrated to obtain the average liquid superficial velocity:

$$\bar{u}_{LS} = \frac{4}{\mu_L D}\left[\frac{1}{2}\tau_i \delta^2 - \frac{1}{3}\left(\frac{dp}{dz} + \rho_L g\right)\delta^3\right] \tag{5.2.18}$$

A force balance on the gas core yields

$$\frac{dp}{dz} = -\frac{4\tau_i}{D - 2\delta} - \rho_G g \quad \text{or} \quad \tau_i = \frac{\bar{u}_{LS}\mu_L D}{2\delta^2} + \frac{2}{3}(\rho_L - \rho_G)g\delta \tag{5.2.19}$$

The interface shear, τ_i, is represented in terms of a friction factor, f_i:

$$\tau_i = \frac{1}{8}\rho_G f_i \frac{\bar{u}_{GS}^2}{(1 - 2\delta/D)^4}$$

where $\bar{u}_{GS}$ is the gas superficial velocity and f_i is obtained from the empirical correlation of Bharathan (3.6.5) described in Eqs. (3.6.37) and (3.6.38). It is also discussed in Section 4.5.4.4. Bharathan's correlation is of the form

$$f_i = 0.020 + 4B\left(\frac{\delta}{D}\right)^n \tag{5.2.21}$$

where B and n vary with the size of the pipe. For example, for a pipe diameter of 2.5 cm, $B = 280$ and $n = 2.13$. In dimensionless form, there results

$$\delta' = \frac{\delta}{D}$$

$$\bar{u}'_{LS} = \frac{\mu_L}{D^2(\rho_L - \rho_G)g}\bar{u}_{LS}$$

$$\bar{u}'_{GS} = \left[\frac{\rho_G}{(\rho_L - \rho_G)gD}\right]^{1/2} \bar{u}_{GS} \qquad (5.2.22)$$

$$\tau'_i = \frac{\tau_i}{(\rho_L - \rho_G)gD}$$

$$\bar{u}'_{GS} = \left[\frac{2(1 - 2\delta')^4}{0.020 + 4B\delta'^n}\left(\frac{u'_{LS}}{2\delta'^2} + \frac{2}{3}\delta'\right)\right]^{1/2}$$

With the superficial gas and liquid velocity specified, one can calculate $\bar{u}'_{LS}$ and $\bar{u}'_{GS}$ and solve for δ' from the last of Eqs. (5.2.22). The results are shown in Figure 5.2.7, where it should be noted that $\bar{u}'_{LS}$ is negative for falling film flow and $\bar{u}'_{LS}$ is positive for net upflow. For a specified liquid flow rate, there are three possible liquid-film thicknesses for the same gas flow rate. Two of the liquid-film thicknesses are for downflow and one is for upflow.

Figure 5.2.7 provides a good explanation for the behavior of a countercurrent flow system:

- At zero gas flow, the liquid film flows downward and its thickness is given by the intersections of the constant liquid flow curves at $\bar{u}'_{GS} = 0$.
- As the gas flow rate is increased, the film thickness grows slowly until it reaches a maximum value. This corresponds to the initial flooding condition.
- If the gas flow rate increases further, the liquid flow downward decreases and the system moves along the locus of the maxima or the dotted curve in Figure 5.2.7.
- Eventually, there is no liquid downflow as the gas flow rate reaches the maximum, corresponding to $\bar{u}'_{LS} = 0$ and $\bar{u}'_{GS} = 0.87$.
- If the gas flow continues to be raised, all the liquid flow continues to be upward and its film thickness decreases.

It should be noted that portions of the curves in Figure 5.2.7 are drawn solid, indicating that the flow is stable, while other segments of the curves are drawn dashed, indicating flow instability. Stability occurs when the film thickness δ' returns to its original value after it is perturbed. Figure 5.2.8 reproduces the plots of the nondimensional shear stress τ'_i versus the nondimensional film thickness δ' at constant values of $\bar{u}'_{LS}$ and $\bar{u}'_{GS}$ for a pipe with a 2.5-cm diameter. At low gas flow rates (e.g., $\bar{u}'_{GS} = 0.5$), a small increase in the film thickness at point A would result in a gas shear stress below that needed to maintain this new film thickness, and it would return to its original value, and the solution at point A is stable. On the other hand, at points B and C, a small increase in film thickness would produce a gas shear stress in excess of that required to maintain the perturbed

5.2 COUNTERCURRENT FLOODING IN SIMPLIFIED GEOMETRIES

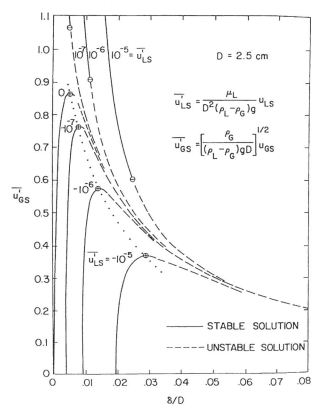

Figure 5.2.7 Film thickness for falling and climbing films in a 2.5-cm-diameter pipe. (From Ref. 5.1.3.)

film thickness and the solutions at points *B* and *C* are unstable because the system will continue to move to a new state. Similarly, at increased gas flow rates (e.g., $\bar{u}'_{GS} = 0.9$) point *D* is stable while point *F* is unstable and point *E* is at neutral stability. The rule is that the flow is stable when an increase in δ' occurs at locations where the curve of τ'_i versus δ' at a specified liquid rate is above the curve of τ'_i versus δ' at the gas flow rate of interest. The flow is unstable when the reverse is true for the two curves of τ'_i versus δ'. Using the information in Figure 5.2.8, one can establish the start of instability as shown in Figure 5.2.7. When the flow becomes unstable, film flow can no longer exist and large disturbance waves are formed. Also, for small liquid flow rates as the gas rate is reduced, the film becomes unstable in Figure 5.2.7 above the critical gas velocity of $\bar{u}'_{GS} = 0.87$, for which there is no downflow solution. The flow reversal path is then identical to the flooding case with increasing gas flow. At high liquid flow rates, the solution instability occurs below the critical gas velocity of $\bar{u}'_{GS} = 0.87$. This means that the liquid film can exist at a reduced value of $\bar{u}'_{GS}$ when the gas flow rate is being decreased and that flooding will occur only at the instability point. This explains the hysteresis in the flow reversal process which was illustrated in Figure 5.2.2.

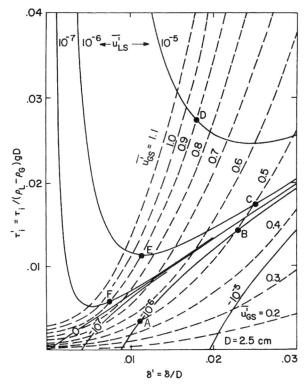

Figure 5.2.8 Stability analysis for falling and climbing films. (From Ref. 5.1.3.)

Equations (5.2.22) can be employed to predict the flow rate at which flooding and flow reversal begins and ends. The calculated results are shown in Figure 5.2.9 for a 2.5-cm-diameter pipe, and they are compared to Wallis correlations with values of 1.0, 0.88, and 0.82 for C as well as data obtained by Hewitt et al. [5.2.19]. Predicted flow reversal conditions are also shown in Figure 5.2.9, together with reversal data from the same Hewitt reference.

In summary, the mechanism of upward liquid flow provides an excellent description of the behavior of flooding. In many ways, its performance is similar to Wallis's mechanism of kinetic waves, except that wall friction and interfacial shear are now considered. The mechanism of upward liquid flow also gives acceptable agreement with experimental flooding and flow reversal data. Finally, the predicted maximum critical gas velocity of $\bar{u}'_{GS} = 0.87$ is nearly identical to the standing-wave value of 0.89 proposed by Whalley. This type of overall agreement is not overly surprising since the Dukler et al. derivations employed the empirical correlation for interface friction factor by Bharathan [3.6.5], which, in turn, was obtained from flooding tests. Another shortcoming of the proposed model is the use of a laminar liquid film, but its impact is not expected to be significant, as demonstrated in several other downward liquid flow analyses. A final concern is that the proposed mechanism of upward liquid flow may not be easy to extrapolate to different inlet and exit test setups.

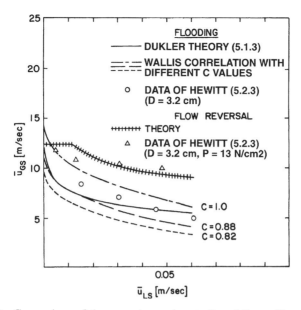

Figure 5.2.9 Comparison of theory and experiment: $D = 2.5$ cm. (From Ref. 5.1.3.)

5.2.3.4 Mechanism of Droplet Entrainment

Under this mechanism, flooding occurs when the gas flow rate is sufficient to lift the largest stable liquid drops upward. In this case, the waves serve as the source of entrainment and it is presumed that the gas velocity required to form the large interfacial waves and to tear off liquid droplets from the top of the waves is less than the velocity to lift the liquid drops out of the tube. This mechanism was first suggested by Dukler and Smith [5.2.6], and it is controlled more by droplet rather than wave mechanics. It is important to note that this mechanism was covered in Section 4.5.6.3, where we discuss the case of annular flow with a gas core containing entrained liquid drops. In this section a simple balance between the gravity force and the drag produced by gas flow are employed to develop the applicable entrainment parameter. The drag force F_d on a liquid drop of diameter d_d is given by

$$F_d = \frac{\pi d_d^3}{6}(\rho_L - \rho_G)g \quad (5.2.23)$$

That same force can be expressed in terms of a drag coefficient C_d and the average gas velocity $\bar{u}_G$, or

$$F_d = \tfrac{1}{8}\rho_G C_d \pi d_d^2 \bar{u}_G^2 \quad (5.2.24)$$

The liquid drop diameter d_d is usually obtained by employing the associated Weber number, or

$$We = \frac{\rho_g \bar{u}_G^2 d_d}{\sigma} \quad (5.2.25)$$

Combining Eqs. (5.2.23), (5.2.24), and (5.2.25) yields

$$\bar{u}_G = \left(\frac{4\text{We}}{3C_d}\right)^{1/4} \left[\frac{\sigma(\rho_L - \rho_G)g}{\rho_G^2}\right]^{1/4} \tag{5.2.26}$$

For large drops, $C_d = 0.44$. Also, Hinze [5.2.20] recommended We = 12 for a drop suddenly accelerated. Finally, the superficial gas velocity or flux $\bar{u}_{GS} = \overline{\alpha u_G}$, so that

$$\bar{u}_{GS} = 2.46\bar{\alpha}\left[\frac{\sigma(\rho_L - \rho_G)g}{\rho_G^2}\right]^{1/4} \quad \text{or} \quad \text{Ku}_{GS} = 2.46\bar{\alpha} \tag{5.2.27}$$

As discussed in Chapter 4, disturbance waves and entrainment will tend to occur at $\bar{\alpha} \approx 0.7$, which gives $\text{Ku}_{GS} = 1.72$ or a value very close to the 1.79 recommended by Pushkina and Sorokin [5.2.2] and the Kutateladze equation for flooding. It is anticipated that the appropriate value of the average gas volume fractions to start entrainment will depend on pipe size, gas and liquid velocities, and their fluid properties, which explains the variability measured in tests for the constant of Eq. (5.2.27). It is also worth noting that the Hinze value for the Weber number of 12 to 13 is valid for the entrainment of the largest stable liquid drop diameter. Entrainment of smaller droplets may, in fact, take place sooner and that case is discussed in more detail in Section 4.5.6.3. The analysis in Chapter 4 is more complete, but it is also much more complicated and requires several empirical relations (e.g., droplet distribution, interfacial shear, etc.). For that reason, the simplified treatment of Eq. (5.2.27) has considerable merit while giving credible answers. It also confirms that there might be variability in the Kutateladze constant.

5.2.3.5 Discussion of Flooding Mechanisms

1. Several potential mechanisms for flooding have been proposed and they provide a good physical picture of the flooding phenomenon.

2. The Wallis continuity waves solution is valid when the wall and interfacial shear forces can be neglected with respect to gravity. The upward liquid-film flow model of Duckler et al. is an extension of the Wallis solution, which includes wall and interfacial shear pressure losses. The standing-wave solution of Whalley corresponds to a single point among the upward liquid-film flow results, while the entrainment model of Kutateladze deals with flooding when the disturbance at the gas–liquid interface are waves of limited size. The various available models, in fact, complement each other, and ideally, their combination into a single model would provide the best prediction of flooding.

3. All the proposed models require empirical constants or relations to fit the data. Tests are needed particularly when the entrance and exit conditions involve extra gas or liquid turbulence.

4. Simplified models coupled with prototypical tests continue to be favored because of our inadequate knowledge and understanding of countercurrent gas–liquid interfaces.

5.2.3.6 Flooding in Annular Geometries
This geometry is of great importance because it is prototypical of a nuclear reactor downcomer in which the emergency cooling water is injected and flows countercurrently to the steam. Air–water and steam–water tests have been performed in simulated pressurized water reactor (PWR) geometries. Most tests employed an annular geometry consisting of two concentric cylinders with various hot and cold pipes attached to the outer cylinder. In a few experiments, the annulus was unwrapped and the test section was composed of two flat plates with appropriate pipe attachments. The early air–water tests were correlated by a Wallis type of equation (5.2.5). For example, Naff and Whitebeck [5.2.21] used radial gaps of 0.9, 1.2, and 1.7 cm and downcomer lengths of 15.2, 63.5, 86.4, and 172.7 cm and obtained a good correlation of their air–water test data with

$$\bar{j}_G^{*1/2} + 0.8 \bar{j}_L^{*1/2} = 0.9 \quad (5.2.28)$$

Cudnik and Morton [5.2.22] employed a $\frac{1}{30}$-scale model PWR downcomer and correlated their air–water data with

$$\bar{j}_G^{*1/2} + \bar{j}_L^{*1/2} = 0.9 \quad (5.2.29)$$

Corresponding data for a $\frac{1}{15}$-scale model were fitted with

$$\bar{j}_G^{*1/2} + 0.77 \bar{j}_L^{*1/2} = 0.8 \quad (5.2.30)$$

Crowley et al. [5.2.23] performed air–water tests in a $\frac{1}{30}$-scale model unwrapped annulus and correlated their data with

$$\bar{j}_G^{*1/2} + \bar{j}_L^{*1/2} = 0.94 \quad (5.2.31)$$

Equation (5.2.31) is in good agreement with Eq. (5.2.29).

Steam–water tests were also performed in $\frac{1}{30}$-, $\frac{1}{15}$-, and $\frac{1}{5}$-scale models, and the results showed that the annulus width is not the appropriate characteristic length to be introduced in the Wallis groupings. Instead, the average annulus circumference had to be used in the nondimensional parameter $\bar{j}^*$. An alternative is to utilize the Kutateladze groupings [5.2.24], which do not require a characteristic length.

As in the case of pipe configurations, the tests described in References 5.2.21 to 5.2.24 find it difficult to satisfy the Wallis type of flooding correlations over a wide range of annular geometries. An explanation for those difficulties is found in the multidimensional air–water tests performed by Dempster and Abouhadra [5.2.25], who studied the two-phase-flow regimes prevailing in a transparent $\frac{1}{10}$-scale PWR downcomer. The flow patterns observed are illustrated in Figures 5.2.10a to c. The flow-patterns are presented by unwrapping the downcomer and they extend from the top to

284 LIMITING MECHANISM OF COUNTERCURRENT FLOODING

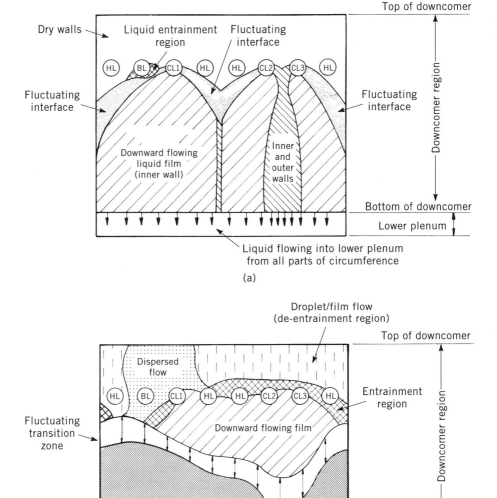

Figure 5.2.10 Flow patterns for air–water tests. (Reprinted from Ref. 5.2.25 with permission from Elsevier Science.)

the bottom of the downcomer. The various pipe attachments are also shown, and they are labeled HL for the hot loop piping, CL for the cold loop, and BL for the broken loop pipes. Figure 5.2.10a deals with the start of bypass (SOB). The majority of the liquid is seen to flow to the lower plenum and to be distributed over the entire downcomer surface shortly below the injection level. A disturbed region exists near the

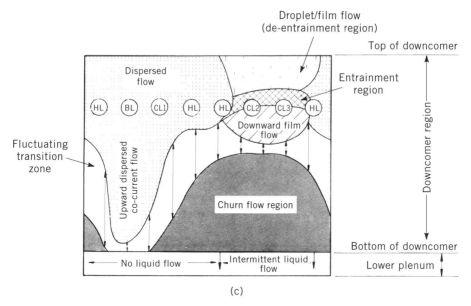

Figure 5.2.10 *(continued)*

outer edges of the liquid film. The start of bypass was associated with liquid being torn from the liquid arc originating from the cold leg and being transported and entrained to the break by the high-velocity airflow. Figure 5.2.10b illustrates the flow patterns when the airflow is increased by 330%. Four flow patterns are seen to exist, with the largest region being associated with churn flow near the bottom plenum. Another large region is associated with downward film flow; it is located below the injection level and its size is the largest in the regions farthest away from the broken cold leg. Above the injection level, there is a region of droplet flow and downward-moving liquid rivulets on the wall. This liquid comes from entrainment close to the injection points by upward-flowing gas. The liquid, however, is deentrained and falls back in the form of rivulets. The last flow pattern occurs closest to the exit point. It consists of dispersed droplet flow and film shear flows similar to an annular mist flow pattern. The flow bypass in Figure 5.2.10b is low and on the order of 15%.

The flow conditions, however, are quite disturbed and the gas flow is too low to transport the liquid over the entire length of the downcomer. This leads to water being allowed to flow upward and subsequently downward and to producing fluctuating transition zones. Figure 5.2.10c corresponds to 50% bypass of the inlet flow. The liquid leaving the downcomer comes mostly from cold leg 1, which is located nearest the break. The flow patterns are similar to those in Figure 5.2.10b except for an additional and extensive region near the break of wispy annular flow, which extends over a large portion of the downcomer.

Two principal flooding processes were observed to occur in Figure 5.2.10: a liquid entrainment mechanism taking place near the liquid injection levels, and a film flow reversal mechanism existing in the low downcomer regions. Dempster

286 LIMITING MECHANISM OF COUNTERCURRENT FLOODING

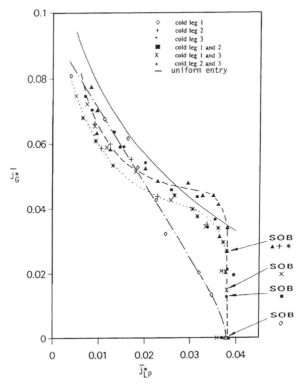

Figure 5.2.11 Flooding curves for different injection modes. (Reprinted from Ref. 5.2.25 with permission from Elsevier Science.)

and Abouhadra developed flooding curves by using Wallis nondimensional groupings and by plotting $\bar{j}_G^*$ versus $\bar{j}_{Lp}^*$, where $\bar{j}_{Lp}$ is the liquid flux reaching the lower plenum. The characteristic dimension employed was the mean downcomer circumference. The resulting flooding curves are reproduced in Figure 5.2.11 for different cold-leg entry conditions. They show a dependence on the water inlet configurations and are different from the case of uniform liquid entry, which was correlated by

$$\bar{j}_G^{*1/2} + 0.96 \bar{j}_L^{*1/2} = 0.375 \tag{5.2.32}$$

Equation (5.2.32) for uniform liquid injection is quite similar to Eqs. (5.2.29) and (5.2.31) and the difference in the constant C on the right-hand side comes primarily from using the average circumference of the annulus instead of the hydraulic diameter based on the annular spacing.

In summary, flooding in an annular geometry is expected to be multidimensional and to exhibit a variety of flow regimes, which vary spatially. More of the liquid is bypassed for nonuniform liquid injection. Analytical predictions of the flow patterns or of the effect of liquid inlet configuration are far from being available.

Approximate flooding curves can be formulated by averaging the flow conditions over the circumference of the annulus and by using that dimension for the characteristic length. Even under those conditions, the flooding correlations are accurate only if they are based on test results.

5.3 COUNTERCURRENT FLOODING FOR NONEQUILIBRIUM THERMAL CONDITIONS AND MORE COMPLEX GEOMETRIES

5.3.1 Nonequilibrium Thermal Conditions

Nonequilibrium thermal conditions can occur in single-component countercurrent flow when the liquid is subcooled and the vapor is at saturation temperature or when there is an additional vapor flux due to evaporation of the liquid on superheated walls. Such conditions can have a significant impact on the preceding treatment of flooding, which was based exclusively upon hydrodynamic considerations. Tien [5.3.1] considered the effect of vapor condensation by calculating the steam flow reduction due to balancing the vapor condensation latent heat to the heat capacity required to raise the temperature of the subcooled liquid to the saturation temperature. Tien defined an effective vapor flow based on the Wallis or Kutateladze number such that

$$\bar{j}^*_{Ge} = \bar{j}^*_G - F\bar{j}^*_{L,\text{in}} \frac{C_{pL}(T_{\text{sat}} - T_{\text{in}})}{H_{fg}} \left(\frac{\rho_L}{\rho_G}\right)^{1/2}$$

$$\text{Ku}_{Ge} = \text{Ku}_G - F\text{Ku}_{L,\text{in}} \frac{C_{pL}(T_{\text{sat}} - T_{\text{in}})}{H_{fg}} \left(\frac{\rho_L}{\rho_G}\right)^{1/2} \quad (5.3.1)$$

where C_{pL} is the liquid heat capacity, T_{in} its inlet temperature, T_{sat} its saturation temperature, H_{fg} the heat of vaporization, $\bar{j}^*_{L,\text{in}}$ and $\text{Ku}_{L,\text{in}}$ correspond to the liquid inlet flow, and F represents the condensation efficiency and is determined empirically. Similarly, the vapor produced on superheated walls can be accounted for by adding the term $\bar{j}^*_{G,HW}$ or $\text{Ku}_{G,HW}$ to the effective groupings of Eqs. (5.3.1), to get

$$\bar{j}^*_{Ge} = \bar{j}^*_G + \bar{j}^*_{G,HW} - F\bar{j}^*_{L,\text{in}} \frac{C_{pL}(T_{\text{sat}} - T_{\text{in}})}{H_{fg}} \left(\frac{\rho_L}{\rho_G}\right)^{1/2}$$

$$\text{Ku}_{Ge} = \text{Ku}_G + \text{Ku}_{G,HW} - F\text{Ku}_{L,\text{in}} \frac{C_{pL}(T_{\text{sat}} - T_{\text{in}})}{H_{fg}} \left(\frac{\rho_L}{\rho_G}\right)^{1/2} \quad (5.3.2)$$

5.3.2 Boiling Water Reactor Upper Plenum Reflooding

Let us consider next countercurrent flow-limiting phenomena in boiling water reactors (BWRs), where low-pressure subcooled emergency cooling water is injected into the upper plenum from core spray spargers located around the periphery of the reactor core. Because BWR fuel bundles are enclosed in shrouds or channels, full-scale tests of a single fuel bundle can be performed to establish the countercurrent flow limitations that occur at its upper tie plate. Tobin [5.3.2] performed experiments with steam and water in a fuel bundle containing 7×7 fuel rods, and his data were correlated by Sun and Fernandez [5.3.3] by the standard Eq. (5.2.6) with $C = 1.79$. Sun [5.3.4] subsequently recommended a value of $C = 2.08$ for the upper tie plate based on his review of fuel bundle tests with 7×7 and 8×8 fuel rods. The system computer code for BWRs, TRAC BD1/MOD [5.3.5], employs a comparable value of $C = 2.05$. It is worth noting that most of the full-scale fuel bundle tests employed a small amount of subcooling, which explains its minimal role in the proposed correlations.

BWR fuel bundles are subject to another countercurrent flow-limiting condition, which occurs at the bottom side orifice through which water must escape to reach the bottom plenum after it enters at the top of the fuel bundle. The flooding correlation for the side-entry orifice was obtained from tests and is

$$\mathrm{Ku}_G^{1/2} + 0.59\, \mathrm{Ku}_L^{1/2} = (2.14 - 0.008\, \sqrt{\mathrm{Bo'}}) \qquad (5.3.3)$$

where Bo$'$ is the Bond number as defined in Eq. (5.2.4), with the characteristic dimension being the wetted perimeter of the bottom orifice. There is no subcooling effect in Eq. (5.3.3) because the injection water comes in contact with fuel rods as it travels downward and should be at saturation temperature when it reaches the orifice.

Figure 5.3.1 shows schematically the very complex distribution of steam and water in an actual BWR after a loss-of-coolant accident (LOCA). Several parallel

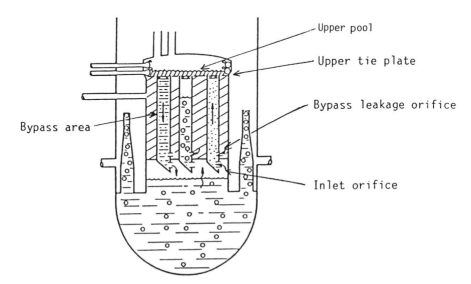

Figure 5.3.1 BWR vessel showing schematically the flow and coolant distributions during refilling of the vessel. (From Ref. 5.3.6.)

flow paths occur, and they are illustrated. The water injected at the top can penetrate a large number of parallel fuel bundles which produce different amounts of power (e.g., bundles at the edge of the reactor core produce less power, i.e., less steam, than fuel in the center of the core). Water can also flow downward in the interstitial spaces between the fuel channels, which are referred to as the bypass area, and that water can enter the fuel bundles at their bottom through a bypass leakage orifice. In Figure 5.3.1, upper tie plates, fuel bundle inlet, bypass leakage orifices, an upper partially subcooled pool of water, and a bottom plenum partially full of saturated water are shown. Prototypical tests have been performed in that multichannel configuration and they have shown that they can be quite beneficial in cooling the fuel. As reported by Dix [5.3.6]:

- The upper plenum pool of water is subcooled over the peripheral fuel bundles because they are closest to the sources of the cold emergency cooling water. That subcooled water condenses the reduced amount of steam exiting from the peripheral fuel bundles and it causes rapid breakdown of the countercurrent flow-limiting condition in that region to allow large flows of water into the lower plenum. Two rows away from the periphery of the reactor core the top pool of water is saturated, the steam flow is higher, and flooding is predictable from Eq. (5.2.6) with $C \approx 2.0$ as obtained from full-scale bundle tests.
- A residual pool of water covers all the fuel bundles and distributes water to them, but the behavior in the upper plenum is cyclic with periods with and without localized flooding.
- Water penetrating through the upper tie plates into the fuel channels drains very slowly at the bottom due to countercurrent flow-limiting conditions at the inlet bottom orifice. Water also enters the fuel through the bypass leakage orifice. Sufficient cooling of the fuel is therefore provided shortly after initiation of the emergency cooling systems.
- Vapor produced in the lower plenum can vent through a few channels, producing co-current steam–water droplet upflow, which is also capable of cooling the fuel in those few fuel bundles.

These experiments helped dispel the fears that the fuel upper tie plates would work in a homogeneous way to prohibit the reflooding of the lower plenum and of the core. The tests also served to demonstrate how difficult it would be to predict such multichannel flooding behavior and why the use of Eq. (5.2.6) for the upper tie plate and Eq. (5.3.3) for the side-entry orifice tend to be employed in simplified models in BWR computer system codes to obtain conservative predictions of fuel temperature during a LOCA.

5.3.3 Upper Plenum Injection in Pressurized Water Reactors

As discussed in detail in Chapter 1, similar types of flow conditions exist in PWRs when subcooled emergency cooling water is injected into the hot legs or the upper plenum of a PWR. Mohr and Jacoby [5.3.7] investigated the flooding phenomena through the end box and upper core support plate for German PWR designs. The results were identical to those reported by Sun [5.3.4] for BWR upper tie plates [i.e., Eq. (5.2.6) applies, with the constant C having values between 2 and 2.12]. Full-scale tests [5.3.8] have also been performed in the full-scale Upper Plenum Test Facility (UPTF), where it was found that:

- The emergency core cooling water delivery to the upper plenum forms a water pool in the upper plenum. Penetration to the core occurs near the injecting hot leg, where a local subcooled pool forms and encourages water breakthrough without delay.
- When the emergency core cooling delivery is intermittent, the water breakthrough is intermittent, but its true average value is not affected by this time-varying behavior.
- Ascending or descending steam–water upflow rates have a different impact on the water breakthrough, and the water downflow is much higher for ascending upflow steam–water flows than for the descending upflow rates.
- The flow is highly heterogeneous and the Kutateladze groupings cannot be applied without recognizing the local variations in pool temperature and height as well as in the local upflow rates.

Here again, tests (preferably full-scale tests) have been employed to get a realistic and accurate treatment of cold water injection into the upper plenum of PWRs [5.3.8].

5.3.4 Behavior of Core Cooling Injection Ports

Pressure and flow oscillations have been reported in pipes where the subcooled emergency cooling water is injected. These oscillations are produced mainly by direct condensation of steam on the injected subcooled water. Several small-scale tests of this injection behavior were performed by Rothe et al. [5.3.9] and by Crowley et al. [5.3.10], and they showed that complete emergency cooling-water flow reversal and plug formation can occur at high steam flow and water injection rates. Full-scale UPTF tests were carried out subsequently [5.3.8] to quantify the conditions leading to pressure and flow oscillations in full-scale piping. The UPTF tests also showed that the mode of emergency core cooling delivery depends on the flow pattern in the pipes around the injection ports. Both stable and unstable plug flow occur in the pipe adjacent to the emergency core cooling injection port. Unstable plug flow is associated with oscillation of large amplitude in the pipe associated with water hammer. The other flow pattern identified in the pipe is stratified flow, where water moves at the bottom and steam at the top of the pipe. Stratified flow leads to continuous coolant delivery to the reactor core. It will be co-current with injection in the cold leg and countercurrent with injection in the hot leg. The region of stratified flow can be specified by comparing the steam condensation potential of the injected coolant or $\rho_L \bar{j}_{L,in} C_{pL}(T_{sat} - T_{in})$ to the steam flow latent heat $\rho_G \bar{j}_G H_{fg}$. When that ratio is below 1, excess steam is present and stratified flow will

5.3 COUNTERCURRENT FLOODING FOR NONEQUILIBRIUM THERMAL CONDITIONS

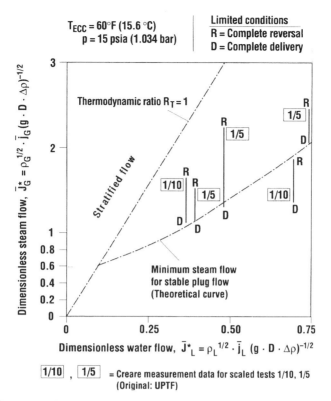

Figure 5.3.2 Comparison of Creare countercurrent flow test results for scaled hot leg with theoretical curve for stable water plug developed from UPTF tests. (Reprinted from Ref. 5.3.8 with permission from Elsevier Science.)

result, to provide a flow path at the top of the pipe for the excess steam. With some temperature stratification in the water flowing in the bottom of the pipe, particularly in the case of injection into the cold leg, stratified flow can exist at values of the ratio $R_T = \rho_L \bar{j}_{L,\text{in}} C_{PL}(T_{\text{sat}} - T_{\text{in}})/\rho_G \bar{j}_G H_{fg}$ slightly in excess of 1. The occurrence of stable plug flow was specified from stability considerations by Emmerling and Weiss [5.3.11], and their prediction for water injection into the hot leg is reproduced in Figure 5.3.2 in terms of the Wallis grouping based on the pipe diameter D. The corresponding $\frac{1}{5}$ and $\frac{1}{10}$ small-scale test results obtained by Crowley et al. [5.3.10] are plotted in Figure 5.3.2 in the form of vertical bars, with the symbols D being used to represent complete delivery and R complete flow reversal. The theoretical stable plug curve tends to coincide with the start of flow reversal in the small-scale tests, which means that the formation of a stable water plug would take much higher steam velocities in full-scale than in small-scale facilities.

5.3.5 Injection in Downcomer of Pressurized Water Reactors

Small-scale tests of cold water injection into downcomers of PWRs have been performed and correlated using the effective nondimensional groupings of Eqs.

(5.3.2). Block and Crowley [5.3.12] recommended the following flooding correlation:

$$(\bar{j}_{Ge}^*)^{1/2} + m(\bar{j}_L^*)^{1/2} = C \qquad (5.3.4)$$

with

$$\bar{j}_{Ge}^* = \bar{j}_G^* + \bar{j}_{G,\mathrm{HW}}^* - F\bar{j}_{L,\mathrm{in}}^* \frac{C_{pL}(T_{\mathrm{sat}} - T_{\mathrm{in}})}{H_{fg}} \left(\frac{\rho_L}{\rho_G}\right)^{1/2}$$

$$F = \left(\frac{p}{p_{\mathrm{atm}}}\right)^{1/4} \frac{1}{1 + 30\bar{j}_{L,\mathrm{in}}^*} \qquad (5.3.5)$$

with p corresponding to the test pressure and p_{atm} representing atmospheric pressure. In Eq. (5.3.4), C was set at 0.32 and m was to be given by $\exp(-5.6\,\bar{j}_{L,\mathrm{in}}^*)^{0.6}$. It is important to note the highly empirical nature of Eqs. (5.3.4) and (5.3.5). Also, it should be noted that tests in downcomers of different size could be correlated only by using the average annulus circumference, not their annular spacing. Many other similar correlations have been developed. They rely on several empirical parameters, which are obtained from a variety of downcomer tests and which are all functions of scale [5.3.13].

Extrapolation of such flooding correlations tends to be conservative, and full-scale tests at UPTF were performed to investigate the degree of conservatism. Figure 5.3.3 shows a plot of temperatures in the unwrapped downcomer of UPTF with water injection through cold legs 1, 2, and 3. This two-dimensional illustration shows strongly heterogeneous conditions not previously observed or reported in small-scale experiments [5.3.8]. The injection of cold water in the adjoining legs 2 and 3 creates a large subcooled zone, which leads to early water delivery into the lower plenum without being seriously affected by the rising steam. The flooding-data for UPTF and the $\frac{1}{5}$-scale test facility [5.3.10] are plotted in Figure 5.3.4. A condensation efficiency of 0.6 to 0.8 derived from UPTF test 6 was utilized in calculating the effective Wallis steam grouping $\bar{j}_{Ge}^*$. Also, the average downcomer annulus circumference was used as the characteristic length. Due to the heterogeneous conditions prevalent in UPTF, the dimensionless water downflows are much larger at UPTF than at Creare for $\bar{j}_{Ge}^* > 0.2$. For $\bar{j}_{Ge}^* < 0.2$, the water delivery to the lower plenum is less than in the reduced-scale facility, but this is only an anomaly produced by the lower-scale water injection rates employed at UPTF. The downward water flow would have been higher if the injection water rates had been raised. The primary conclusions derived from the full-scale tests with downcomer injection were:

- The heterogeneous flow conditions resulting from the loop arrangement are beneficial to water delivery to the lower plenum.

5.3 COUNTERCURRENT FLOODING FOR NONEQUILIBRIUM THERMAL CONDITIONS

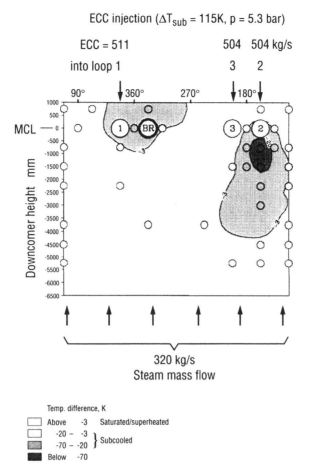

Figure 5.3.3 Countercurrent flow conditions in full-scale downcomer for strongly subcooled cold leg ECC injection: distribution of subcooling (UPTF test 5B). (Reprinted from Ref. 5.3.8 with permission from Elsevier Science.)

- Steam and subcooled water interact more efficiently in the full-scale downcomer, and the increased steam condensation is beneficial to coolant delivery to the reactor core.
- The small-scale flooding correlations are conservative and would predict too little water transport to the lower plenum.

5.3.6 Prediction of Countercurrent Flooding in Complex System Computer Codes

Many system computer codes use simplified flooding correlations. The use of Eq. (5.2.6) in the RELAP-UK code was noted previously. So was the application of Eq. (5.2.6) with $C = 2.05$ for the upper tie plate of BWR fuel bundles in TRAC BD1/

294 LIMITING MECHANISM OF COUNTERCURRENT FLOODING

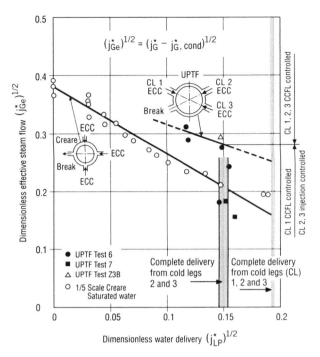

Figure 5.3.4 Effect of scale on water delivery to lower plenum in countercurrent flow. (Reprinted from Ref. 5.3.8 with permission from Elsevier Science.)

MOD1 and the utilization of Eq. (5.3.3) for the side entry orifice. Such a simplified approach [5.3.14] fails to recognize:

- The validity of employing flooding correlations for nearly steady-state to transient conditions. For instance, the mechanisms of flooding in PWR downcomers last about 20 seconds and inertial effects need to be considered.
- Steam condensation instabilities and pressure oscillations as the steam comes into contact with subcooled water.
- Nonuniform multidimensional flow conditions in the downcomer, which are enhanced by the arrangement of water delivery piping.

The difficulties of dealing with such considerations are discussed thoroughly in Reference 5.3.14, which concludes that at least a two-dimensional downcomer model is needed for the system computer code CATHARE to come close to the test data. Still, it was necessary to be selective about the nodalization (i.e., a 16 × 8 scheme for the downcomer was used as shown in Figure 5.3.5). Also, the gas–liquid interfacial shear correlation had to be changed and be based on the Laplace scale instead of the downcomer annulus, and a new discretization was employed for the acceleration or momentum transport terms to get a satisfactory comparison with the full-scale nearly saturated UPTF tests performed with water injection in one cold

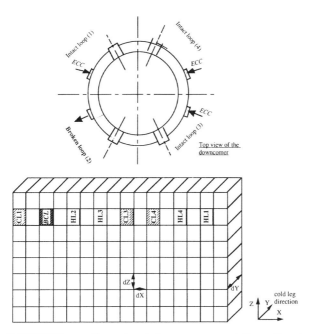

Figure 5.3.5 CATHARE two-dimensional downcomer modeling. (Reprinted from Ref. 5.3.14 with permission from Elsevier Science.)

leg. This comparison is shown in Figure 5.3.6, where the Wallis groupings are employed with the average annular circumference as the characteristic length. One can expect further adjustments to the model as cold water enters the remaining cold legs.

When subcooled water was injected in the UPTF tests, the preceding model yielded reduced water delivery to the lower plenum and the discrepancy increased as the water subcooling rose. An improved downflow prediction was obtained when the CATHARE condensation coefficient was increased by a factor of 20, but the validity of this factor is far from being established fully. It is clear from all the required changes to the system computer code CATHARE that it became relevant only if it was adjusted to match test data, particularly full-scale tests.

5.4 SUMMARY

1. The basic mechanisms and applicable nondimensional groupings are well defined for countercurrent flooding in simplified geometries under nearly steady-state and thermal equilibrium conditions.

2. Entrance and exit configurations have a significant impact on flooding behavior, and their influence can be established only from tests and empirical correlations.

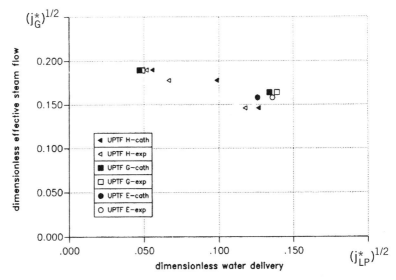

Figure 5.3.6 Flooding curve for UPTF nearly saturated tests: comparison of experimental data with predictions of modified CATHARE code. (Reprinted from Refs. 5.3.14 with permission from Elsevier Science.)

Similarly, as the channel size increases, the characteristic length changes and becomes less important.

3. Geometries found in such complex systems as nuclear power plants create nonuniform multidimensional flow conditions that tend to increase water downward delivery. Such heterogeneous behavior can be specified only from tests, preferably full-scale tests.

4. The impact of nonthermal equilibrium conditions is understood. Subcooling of the liquid increases the liquid flooding velocities, while vapor formed at superheated walls will reduce them. Subcooled water coming into contact with steam facilitates water delivery but can also generate flow oscillations, pressure pulses, and even significant water hammer, which are not readily predictable. Condensation effects are generally underpredicted and the condensation coefficient needs to be empirically adjusted to match the data.

5. System computer codes employ simplified empirical correlations which tend to underestimate the water delivery. For complex geometries, no scale extrapolation of a physical or empirical correlation is reliable. Only full-scale data are relevant and accurate.

6. The primary difficulties in modeling countercurrent flooding can be traced to the same issues noted in previous chapters: the inability to model interfacial shear and heat transfer for the large number of potential and varying flow patterns.

REFERENCES

5.1.1 Hewitt, G. F., "In Search of Two-Phase Flow," *J. Heat Transfer*, **118**, 518–527, 1996.

5.1.2 Tien, C. L., and Liu, C. P., *Survey on Vertical Two-Phase Countercurrent Flooding*, EPRI Report NP-984, 1979.

5.1.3 Dukler, A. E., et al., *Two-Phase Interactions in Countercurrent Flow*, Report, Department of Chemical Engineering, University of Houston, 1980.

5.1.4 Moalem, M. D., and Dukler, A. E., "Flooding and Upwards Film Flow in Vertical Tubes, II: Speculations on Film Flow Mechanisms," *Int. J. Multiphase Flow*, **10**, 599–621, 1984.

5.1.5 Bankoff, S. G., and Lee, S. C., "A Brief Review of Countercurrent Flooding Models Applicable to PWR Geometries," *Nucl. Saf.*, **26**, 139–152, 1985.

5.1.6 Bankoff, S. G., and Lee, S. C., "A Critical Review of the Flooding Literature," in *Multiphase Science and Technology*, G. F. Hewitt, J. M. Delhaye, and N. Zuber (eds.), Vol. 2, pp. 95–180, 1986.

5.1.7 Hewitt, G. F., "Countercurrent Two-Phase Flow," NURETH-4, *Proceedings of the 4th International Topical Meeting on Nuclear Reactor Thermal-Hydraulics*, Karlsruhe, Germany, pp. 1129–1144, October 1984.

5.2.1 Wallis, G. B., "Flooding Velocities for Air and Water in Vertical Tubes," UKAEA Report AEEW-R-123, 1961.

5.2.2 Pushkina, O. L., and Sorokin, Y. L., "Breakdown of Liquid Film Motion in Vertical Tubes," *Heat Transfer Sov. Res.*, **1**, 56, 1969.

5.2.3 Hewitt, G. F., and Wallis, G. B., *ASME Multi-Phase Flow Symposium*, Philadelphia, Nov. 1963.

5.2.4 Lobo, W. E., et al., "Limiting Capacity of Dumped Tower Packings," *Trans. A.I.Ch.E.*, **41**, 693, 1945.

5.2.5 Sherwood, T. K., et al., "Flooding Velocities in Packed Columns," *Ind. Eng. Chem.*, **30**, 765, 1938.

5.2.6 Dukler, A. E., and Smith, L., *Two-Phase Interactions in Counter-Current Flow Studies of the Flooding Mechanism*, NUREG-0214, 1977.

5.2.7 Govan, A. H., et al., "Flooding and Churn Flow in Vertical Pipes," *Int. J. Multiphase Flow*, **10**, 585–597, 1991.

5.2.8 Wallis. G. H., and Makkencherry, S., "The Hanging Film Phenomenon in Vertical Annular Two-Phase Flow," *J. Fluids Eng.*, **96**, 297, 1974.

5.2.9 Holmes, J. A., "Description of the Drift Flux Model in the LOCA Code RELAP-UK," *J. Mech. Eng.*, C 206/77, 1977.

5.2.10 McQuillan, K. W., and Whalley, P. B., "A Comparison Between Flooding Correlations and Experimental Flooding Data," *Chem. Eng. Sci.*, **40**, 1425–1440, 1984.

5.2.11 Wallis, G. B., *One-Dimensional Two-Phase Flow*, McGraw-Hill, New York, 1969.

5.2.12 Schutt, J. B., "A Theoretical Study of the Phenomenon of Bridging in Wetted Wall Volumes," B.S. thesis, University of Rochester, 1959.

5.2.13 Cetinbudakhar, A. G., and Jameson, G. J., "The Mechanism of Flooding in Vertical Counter-Current Two-Phase Flow," *Chem. Eng. Sci.*, **24**, 1669, 1969.

5.2.14 Imura, H., et al., "Flooding Velocity in a Counter-Current Annular Two-Phase Flow," *Chem. Eng. Sci.*, **32**, 78, 1977.

5.2.15 Chung, K. S., "Flooding Phenomena in Counter-Current Two-Phase Flow Systems," Ph.D. thesis, University of California, Berkeley, 1978.

5.2.16 McQuillan, K. W., et al., "Flooding in Vertical Two-Phase Flow," *Int. J. Multiphase Flow*, **11**, 741–760, 1985.

5.2.17 Shearer, C. J., and Davidson, T. F., "The Investigation of a Standing Wave Due to Gas Blowing Upward over a Liquid Film: Its Relation to Flooding in Wetted Wall Columns," *J. Fluid Mech.*, **22**, 321–335, 1965.

5.2.18 Whalley, P. B., *Boiling, Condensation, and Gas-Liquid Flow*, Oxford Science Publications, Oxford, 1987.

5.2.19 Hewitt, G. F., et al., *Transition in Film Flow in a Vertical Tube*, AERE-R 4614, 1965.

5.2.20 Hinze, J. O., "Fundamentals of the Hydrodynamic Mechanism of Slitting in Dispersion Process," *A.I.Ch.E. J.*, **1**, 289, 1955.

5.2.21 Naff, S. A., and Whitebeck, J. F., "Steady State Investigation of Entrainment and Counter-Current Flow in Small Vessels," Topical Meeting on Water Reactor Safety, Conference Paper F30304, 1973.

5.2.22 Cudnick, R. A., and Woton, R. O., *Penetration of Injected ECC Water through Downcomer Annulus in the Presence of Recovered Core Steam Flow*, Batelle Columbus Laboratories Report, 1974.

5.2.23 Crowley, C. J., et al., *Steam/Water Interaction in a Scaled Pressurized Water Reactor Downcomer Annulus*, Report COO-2294-4, Dartmouth College, 1994.

5.2.24 Block, J. A., and Schrock, V. E., "Emergency Cooling Water Delivery to the Core Inlet of PWR's During LOCA," *Proceedings of the ASME Symposium on the Thermal and Hydraulic Aspects of Nuclear Reactor Safety*, Vol. 1, p. 109, 1977.

5.2.25 Dempster, W. M., and Abouhadra, D. S., "Multidimensional Two-Phase Flow Regime Distribution in a PWR Downcomer during an LBLOCA Refill Phase," *Nucl. Eng. Des.*, **149**, 153–166, 1994.

5.3.1 Tien, C. L., "A Simple Analytical for Counter-Current Flow Limiting Phenomena with Condensation," *Lett. Heat Transfer Mass Transfer*, **4**, 231, 1977.

5.3.2 Tobin, R. J., "CCFL Test Results, Phase 1: TLTA 7 × 7 Bundle, GE BWR/ECC Program," *7th Monthly Report*, GE 1977.

5.3.3 Sun, K. H., and Fernandez, R. T., "Counter-Current Flow Limitation Correlation for BWR Bundles during LOCA," *ANS Trans.*, **27**, 625, 1977.

5.3.4 Sun, K. H., "Flooding Correlation for BWR Bundle Upper Tie Plates and Bottom-Side Entry Orifices," *Procedings of the 2nd Multi-phase Flow and Heat Transfer Symposium Workshop*, Miami Beach, Fla., 1979.

5.3.5 Borkowski, J. A., *TRAC-BF1/MOD1 Models and Correlations*, NUREG/CR-4391, EGG-2860, Rev. 4, 1992.

5.3.6 Dix, G. E., "BWR Loss of Coolant Technology Review," *Proceedings of the Conference on Thermal-Hydraulics of Nuclear Reactors, 2nd International Topical Meeting*, Vol. 1, ANS, January 1983.

5.3.7 Mohr, C. M., and Jacoby, J. K., *Quick Look Report on KUU Support Column Air–Water Experiment*, EG&G Report RDW-23-78, 1978.

5.3.8   Weiss, P., et al., "UPTF Experiment Refined PWR LOCA Thermal-Hydraulic Scenarios: Conclusions from a Full-Scale Experimental Program," *Nucl. Eng. Des.*, **149**, 335–347, 1994.

5.3.9   Roth, P. H., et al., "Flow Oscillations," Symposium on Thermal and Hydraulic Defects of Nuclear Reactor Safety, Proceedings prepared by Brookhaven Nat. Lab. Vol. 1, pp. 133–150, 1977.

5.3.10 Crowley, C. J., et al., *Summary Report: Hot Leg ECC Flow Reversal Experiments*, Creare TN-870, 1982.

5.3.11 Emmerling, R., and Weiss, P., "UPTF-Experiment-Analysis of Flow Pattern in Pipes of Large Diameter with Subcooled Water Injection," paper presented at the Two-Phase Flow Meeting, Paris, June 1989.

5.3.12 Block, J. A., and Crowley, C. J., *Effect of Steam Upflow and Superheated Walls on ECC Delivery in a Simulated Multiloop PWR Geometry*, Creare TN-210, 1975.

5.3.13 Richter, H. J., and Zuber, N., *Review of Flooding and ECC Bypass Correlations*, NRC Reactor Systems Branch, May 1986.

5.3.14 Dor, I., "Analysis of UPTF Downcomer with the CATHARE Multidimensional Model," *Nucl. Eng. Des.*, **149**, 129–140, 1994.

CHAPTER 6

CRITICAL FLOW

6.1 INTRODUCTION

Critical or choked flow is obtained at the discharge of a system or a component when a further reduction in downstream pressure does not increase the mass flow rate. For adiabatic, frictionless, single-phase, compressible, steady-state flow, the critical conditions are well understood and predictable analytically. They occur at the minimum cross-sectional area and they are associated with reaching the isentropic sonic velocity at that location. Two-phase critical flow is much more complicated because of rapid formation or changes in the gas–liquid interfaces and in the corresponding mass, momentum, and heat transfer exchanges at those interfaces. Flow pattern changes now take place over very small periods of time or very short distances. Furthermore, if the condition of interest involves single-component flow, as in the case of a loss-of-coolant accident in a light-water nuclear reactor, vaporization of the liquid will occur and nonequilibrium thermal circumstances are bound to arise. It is therefore not surprising that despite extensive experimental and analytical studies over several decades, a complete theory describing two-phase critical flow is not yet available.

There are several good reviews of two-phase critical flow in the literature. Available models have been summarized and assessed, for example, by Saha [6.1.1], Wallis [6.1.2], Abdollahian et al. [6.1.3], and by a panel of experts sponsored by the Nuclear Energy Agency [6.1.4]. The experimental data are listed in References 6.1.4 and 6.1.5. A recent publication by Elias and Lellouche [6.1.6] provides a systematic evaluation of 10 different two-phase critical flow models against an extensive set of data from critical flow tests with water. In this chapter, the goal is to deal with two-phase critical flow in complex systems. As in the case of Chapter 3, we start with homogeneous two-phase flow because it provides a default and often adequate predictive approach for complex system computer codes. Next, separated gas and liquid models are covered without specifically dealing with their interfaces or prevailing flow patterns. This is followed by flow pattern models and by an assessment of nonthermal equilibrium conditions. Finally, a few specific applications, such as two-phase critical flow through pipe cracks and through a small break in a horizontal pipe, are discussed to illustrate the present state of the art and to identify remaining shortcomings for conditions of practical interest.

The material presented herein emphasizes the critical discharge rate of subcooled water and steam–water mixtures for two reasons. First, the dominant share of available studies has been carried out with water, and second, the critical flow phenomena are of great importance during a loss-of-coolant accident in light-water nuclear reactors. The discharge rates through a break not only determine the nuclear reactor pressure and water inventory as a function of time but also influence the structural integrity of the reactor core and the workability of the emergency core cooling systems provided to cope with such emergency situations. Critical flow circumstances, however, are just as important in the chemical process, refrigeration, cryogenics, and other power generation industries. In most instances, safety considerations require knowledge of the critical flow rates for numerous different geometries and circumstances.

6.2 SINGLE-PHASE FLOW

This section provides a recap of critical flow in compressible single-phase flow. It is purposely limited in scope. Further details may be found in classical textbooks on gas dynamics, such as those authored by Shapiro [6.2.1] and Binder [6.2.2].

6.2.1 Propagation of Disturbances in a Compressible Gas

Let us consider the propagation of a disturbance created by the movement of a piston at one end of a pipe of area A, as illustrated in Figure 6.2.1. Let us assume that the motion is one-dimensional with no friction and no heat transfer. Some distance ahead of the piston, the gas is at rest, its pressure is p_1, its density ρ_1, and its velocity $u_1 = 0$. The piston moves with a velocity u_2 and just ahead of it the gas pressure is p_2 and the density is ρ_2. Because of the piston motion, there is a pressure wave traveling to the right at velocity c, and across this pressure wave there is a pressure discontinuity from p_2 to p_1, as shown in Figure 6.2.1a. In Figure 6.2.1b, the fluid is shown to approach the pressure wave with velocity c to the left, and the velocity is $c - u_2$ after the wave. Continuity requires that

$$Ac\rho_1 = A\rho_2(c - u_2) \tag{6.2.1}$$

The corresponding momentum equation gives

$$(p_1 - p_2)A = A\rho_1 c[(c - u_2) - c] \tag{6.2.2}$$

302 CRITICAL FLOW

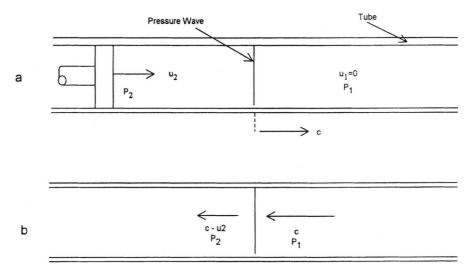

Figure 6.2.1 Propagation of wave.

If the gas satisfies the perfect gas law,

$$p = R\rho T \quad \text{and} \quad p\left(\frac{1}{\rho}\right)^\gamma = \text{constant} \tag{6.2.3}$$

where R is the universal gas law constant and T is the gas temperature. If the flow is adiabatic and if we neglect head losses [6.2.2], application of the energy equation gives

$$\frac{(c - u_2)^2 - c^2}{2} = \frac{\gamma}{\gamma - 1}\left(\frac{p_1}{\rho_1} - \frac{p_2}{\rho_2}\right) \tag{6.2.4}$$

where γ is the ratio of specific heat, or $\gamma = c_p/c_v$. Combining Eqs. (6.2.1), (6.2.2), and (6.2.4) to solve for the velocity c yields

$$c = \sqrt{\frac{\gamma p_1}{\rho_1}}\left[\frac{\gamma - 1}{2} + \left(\frac{p_2}{p_1}\right)\left(\frac{\gamma + 1}{2}\right)\right]^{1/2} \frac{1}{\sqrt{\gamma}} \tag{6.2.5}$$

If the disturbance is infinitely small so that $p_2 \to p_1$, Eq. (6.2.5) gives

$$c = \sqrt{\frac{\gamma p_1}{\rho_1}} = \sqrt{\gamma R T_1} \tag{6.2.6}$$

and its propagation velocity c is the isentropic acoustic velocity for the gas at pressure p_1, temperature T_1, and density ρ_1.

6.2.2 Critical Flow

Let us next consider the release of a compressible fluid from a tank at constant pressure p_0 to an environment at pressure p_e as illustrated in Figure 6.2.2. If the process is adiabatic and frictionless (i.e., isentropic), continuity in the discharge pipe of constant area requires that the mass flow rate per unit area G be constant, or

$$G = \rho u = \text{constant} \qquad (6.2.7)$$

The corresponding momentum equation at an axial position z is

$$\rho u \frac{du}{dz} + \frac{dp}{dz} = 0 \qquad (6.2.8)$$

If the fluid is incompressible as in the case of liquid flow, Eq. (6.2.8) can be integrated to give

$$G_L = \sqrt{2\rho_L(p_0 - p_e)} \qquad (6.2.9)$$

Equation (6.2.9) is often referred to as the *Bernoulli relation*. As discussed later, an empirical coefficient is usually added to it to take into account the vena contracta of the liquid flow field occurring at the pipe exit.

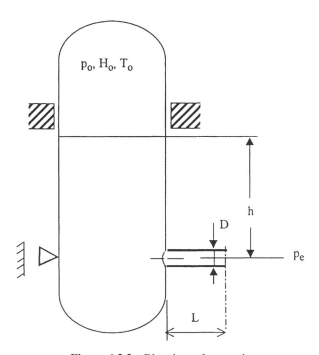

Figure 6.2.2 Blowdown from tank.

304 CRITICAL FLOW

For a compressible fluid, Eq. (6.2.8) can be rewritten for a constant-entropy process:

$$\rho u \frac{du}{dz} = G^2 \frac{d}{dz}\left(\frac{1}{\rho}\right) = -\frac{G^2}{\rho^2}\frac{d\rho}{dz} = -\frac{G^2}{\rho^2}\left(\frac{\partial \rho}{\partial p}\right)_s \frac{dp}{dz} = -\frac{dp}{dz} \quad (6.2.10)$$

where s represents the fluid entropy. Equation (6.2.10) can be expressed as

$$-\left[1 - \frac{G^2}{\rho^2}\left(\frac{\partial \rho}{\partial p}\right)_s\right]\frac{dp}{dz} = 0 \quad (6.2.11)$$

According to Eq. (6.2.11), $-dp/dz$ can have a maximum value different from zero when a critical velocity u_c is reached so that

$$\left(\frac{G_c}{\rho_c}\right)^2 = u_c^2 = \left(\frac{\partial p}{\partial \rho}\right)_s = \gamma RT \quad (6.2.12)$$

As the pressure decreases along the discharge pipe, the density decreases and the fluid velocity increases. This means that the maximum or critical velocity for isentropic flow is reached at the pipe exit and that it is equal to the acoustic velocity at that location. The same answer is obtained when a hydrostatic and/or frictional loss is added to Eq. (6.2.8) except that now $-(dp/dz) \to \infty$.

In the case where the pipe cross section changes along the flow direction (e.g., a converging–diverging nozzle), maximum critical flow at the acoustic velocity will again occur at the throat of the nozzle as long as the pressure p_c at that location is equal to

$$\frac{p_c}{p_0} = \left(\frac{2}{\gamma + 1}\right)^{\gamma/\gamma - 1} \quad (6.2.13)$$

Typical values for the critical pressure ratio are

$$\begin{array}{lll} \text{Air} & \gamma = 1.40; & p_c/p_0 = 0.528 \\ \text{Wet steam} & \gamma = 1.13; & p_c/p_0 = 0.578 \\ \text{Dry steam} & \gamma = 1.33; & p_c/p_0 = 0.540 \end{array} \quad (6.2.14)$$

If the environment pressure p_e is reduced below p_c, the pressure at the throat remains at p_c and the discharge rate remains constant and controlled by the acoustic velocity at that location. This is explained by the fact that the environmental pressure $p_e < p_c$ cannot be telegraphed back into the throat of the converging nozzle because the fluid in the throat is moving with the acoustic velocity of pressure propagation. According to Sallet [6.2.3], the theoretical mass flow rate for choked flow is

$$G_c = \frac{p_0}{(ZRT_0)^{1/2}} \left[\gamma \left(\frac{2}{\gamma + 1} \right)^{(\gamma+1)/(\gamma-1)} \right]^{1/2} \tag{6.2.15}$$

while for subcritical flow, the theoretical mass flow rate becomes

$$G = \frac{p_0}{(ZRT_0)^{1/2}} \left[\frac{2\gamma}{\gamma - 1} \left(\frac{p}{p_0} \right)^{2/\gamma} - \left(\frac{p}{p_0} \right)^{(\gamma+1)/\gamma} \right]^{1/2} \tag{6.2.16}$$

In Eqs. (6.2.15) and (6.2.16), Z is the compressibility factor and is equal to $p/R\rho T$, so that for an ideal gas, Eq. (6.2.15) predicts the isentropic gas sonic velocity at the throat. In Eq. (6.2.16), p is the local pressure along the nozzle.

6.2.3 Nonideal Compressible Gas Flow

Up to this point, ideal flow conditions (e.g., frictionless, adiabatic) were assumed and perfect gas law and simple flow geometries have been employed. A few typical nonideal circumstances are covered next.

6.2.3.1 Critical Flow with Friction In this particular case, the conservation of mass, momentum, and energy equations are

$$\rho u A = \text{constant}$$

$$dp + \rho u \, du = -\frac{P}{A} \tau \, dz \tag{6.2.17}$$

$$dH + u \, du = 0$$

where H is the fluid enthalpy, A the flow area, τ the wall shear stress, and P the fluid wetted perimeter. In addition, the fluid satisfies appropriate equations of state such as

$$\begin{aligned} p &= p(\rho, T) \\ H &= H(\rho, T) \end{aligned} \tag{6.2.18}$$

For a perfect gas, $p = \rho RT$ and $H = c_p T$. The conservation equations (6.2.17) have often been written in matrix form:

$$[B] \frac{d\mathbf{X}}{dz} = \mathbf{C}$$

with

$$B \equiv \begin{pmatrix} 1/p & 1/u & -1/T \\ 1 & \rho u & 0 \\ 0 & u & c_p \end{pmatrix} \quad C \equiv \begin{pmatrix} -A'/A \\ -\tau^* \\ 0 \end{pmatrix} \quad X \equiv \begin{pmatrix} p \\ u \\ T \end{pmatrix} \quad (6.2.19)$$

where $A' \equiv dA/dz$ and $\tau^* = P\tau/A$. Equations (6.2.19) yield a solution as long as the determinant of $[B]$ is different from zero. If the determinant becomes zero, the critical flow condition is reached so that for a perfect gas

$$\det[B] = 0 = \frac{c_p}{u}\left(\frac{u^2}{\gamma RT} - 1\right) \quad \text{or} \quad u = \sqrt{\gamma RT} \quad (6.2.20)$$

and the velocity u becomes the sonic velocity at that location. The pressure drop (dp/dz) is given by

$$\frac{dp}{dz} = \frac{-(1/A)(dA/dz)\rho c_p u + (P\tau/A)(c_p/u + u/T)}{(c_p/u)\,[(u^2/\gamma RT) - 1]} \quad (6.2.21)$$

These equations have been solved for Fanno flow, which corresponds to adiabatic flow conditions with friction in a constant area pipe of length L and diameter D. As shown in Figure 6.2.3, the flow will proceed to a mass limiting value of $G_c\sqrt{RT_0}/p_0$ for each value of the term fL/D, where f is the friction factor and the subscript 0 defines stagnation conditions [6.2.1]. Figure 6.2.3 shows that friction reduces the critical mass flow rate and the critical pressure ratio p_c/p_0.

Similar results could have been derived for diabatic flow if a specified heat addition term had been added to the energy equation.

6.2.3.2 Nonideal Area and Gas Conditions
Departures from theoretical predictions are known to occur for complex flow configurations or deviations from the perfect gas flow. For instance, Ducoffe et al. [6.2.4] plotted their data for air through various flow devices by plotting $G_c(\sqrt{T_0}/p_0)$ versus p_c/p_0, and their results are shown in Figure 6.2.4. Their data indicate that the critical mass flow decreases as the geometry departs from the ideal nozzle and even for the ideal nozzle the critical pressure ratio is somewhat lower than the value of 0.53 predicted for ideal gases. The flow rates are even less predictable when dealing with such complex components as valves that require prototypical tests to establish the valve coefficient,

$$C_v = \frac{G(\text{measured})}{G(\text{theoretical})} \quad (6.2.22)$$

Figure 6.2.5 shows the test results obtained by Sallet [6.2.3] for choked airflow tests in a three-dimensional full-scale valve model of a safety valve. The valve

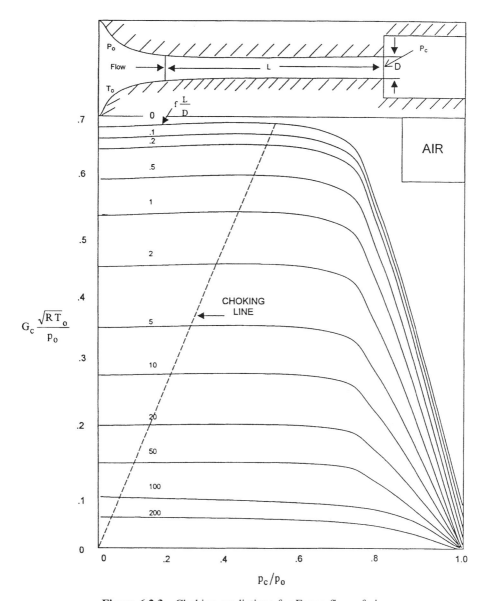

Figure 6.2.3 Choking predictions for Fanno flow of air.

coefficient C_v is seen to depend on the valve disk position, and it is not readily predictable because it does not vary linearly with the degree of disk opening. Figure 6.2.5 also shows a dependence of the valve coefficient on the position of the valve ring adjustment. The solid lines correspond to tests performed with a low position for the adjusting ring and the dashed lines refer to tests with the valve ring in the high position. Such valve adjustments are made routinely at the valve fabrication

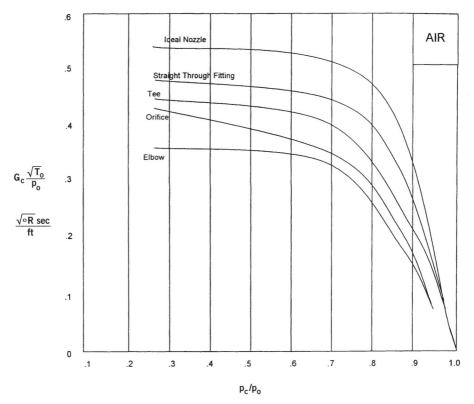

Figure 6.2.4 Flow rate studies of air through standard fittings compared to ideal. (From Ref. 6.2.4.)

and test sites, and they clearly influence the valve performance. Finally, contrary to the ideal gas theory, there is a variation in Figure 6.2.5 of the valve coefficient with the pressure ratio. This comes about because the choking flow area of this particular valve is not the minimum area deduced from the valve geometric design, but instead, it is the minimum flow area established by the flow field, which, in turn, changes with the pressure ratio.

This influence of the pressure ratio is even more striking in Figure 6.2.6, which provides experimental results of several investigators [6.2.5 to 6.2.8] for compressible flow through sharp-edged orifices with a stagnation pressure p_0 and a reservoir pressure p_r. The flow coefficient is seen to increase from $C_v = C_{v\ell} = 0.61$ at p_r/p_0 close to 1 to a value of $C_{v\ell} = C_{vv} = 0.84$ for p/p_0 approaching zero. The symbols C_{vv} and $C_{v\ell}$ were introduced by Sallet to denote the flow coefficient for liquid flow ($C_{v\ell}$) and the limiting value for choked vapor flow (C_{vv}). Sallet correlated the data of Figure 6.2.6 by

$$C_v = C_{vv} - C_{v\ell}\left(\frac{p_r}{p_0}\right)^2 + (2C_{v\ell} - C_{vv})\left(\frac{p_r}{p_0}\right)^2 \qquad (6.2.23)$$

6.2 SINGLE-PHASE FLOW 309

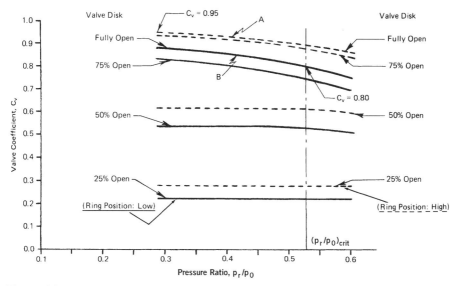

Figure 6.2.5 Valve coefficient dependence on pressure ratio p_r/p_0, valve disk position, and adjusting ring position. (High ring position is indicated by the dashed line, while the solid line indicates low ring position.) (From Ref. 6.2.3.)

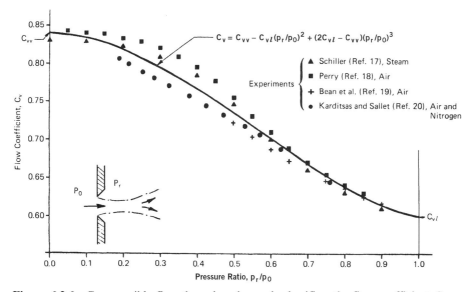

Figure 6.2.6 Compressible flow through a sharp-edged orifice: the flow coefficient C_v as a function of the pressure ratio p_r/p_0. (From Ref. 6.2.3.)

Because a sharp-edged orifice can be considered to be an idealized representation of many valves, Sallet recommended the use of Eq. (6.2.23) to predict the valve coefficient C_v as a function of p_r/p_0, where C_{vv} and $C_{v\ell}$ are known or measured experimentally. Sallet also reported an underprediction of the choking steam flow rate through valves when the absolute pressure is above 400 MPa or 1600 psia. This difference was traced to steam not being able to behave as a perfect gas over a wide pressure range, and it is correctable by employing Eq. (6.2.15) with the compressibility factor Z being calculated from actual steam property data. Also, this is an appropriate location for a reminder that a coefficient of the type $C_{v\ell}$ needs to be introduced into the Bernoulli relation [Eq. (6.2.4)] in order to deal with vena contracta or other real rather than ideal liquid flow conditions.

6.2.4 Summary: Single-Phase Critical Flow

1. Single-phase critical flow is reasonably well understood, and good theoretical models are available to predict mass flow rates for ideal compressible gas flow conditions.

2. Nonideal gas behavior and complex geometries require adjustments to the theoretical prediction and they are best obtained from prototypical tests and measurements. They are especially needed when the flow fields can influence the location and size of the minimum choking flow area or when the gas behavior is nonideal.

3. Variations in component geometries can arise due to operation, manufacturing, or test adjustments, and they need to be considered.

6.3 TWO-PHASE CRITICAL FLOW MODELS

The available number of two-phase critical flow models is very large, and this can be traced primarily to our inadequate understanding of the phenomena involved. The variations are excessive and include the use of isothermal, isenthalpic, isentropic, polytropic, or frozen flow processes. In those cases where the liquid–gas interfaces are not considered, both equal and unequal gas and liquid velocities with and without thermal equilibrium conditions have been assumed. For two-fluid models with exchanges at the gas–liquid interfaces, the number of conservation equations and the interfacial closure equations are seldom the same. Another contributor to this unsatisfactory situation is the fact that critical flow tests have tended to emphasize critical flow measurements at the expense of information about the flow structure and the local behavior of the gas and liquid streams. Such local measurements are not available because they are quite difficult to make, due to the high velocities involved and the rapid changes in the geometry and transfer rates at the gas–liquid interfaces.

It is not our purpose in this section to cover all available two-phase critical flow models but instead, to emphasize the principal models used in complex system computer codes. We begin with homogeneous flow, proceed to gas slip with respect to the liquid, then to thermal nonequilibrium conditions, and finally, to two-fluid models.

6.3.1 Homogeneous Equilibrium Critical Flow Models

This model assumes a velocity ratio of unity and thermal equilibrium between the phases. If one assumes no friction and heat addition (i.e., isentropic conditions), the critical mass flux and the acoustic velocity are obtained from straightforward extensions of one-dimensional single-phase results developed in Section 6.2. The properties of the homogeneous mixtures are

$$\frac{1}{\rho_H} = \frac{1-\bar{x}}{\rho_L} + \frac{\bar{x}}{\rho_G} = \frac{1}{\rho_L} + \bar{x}\left(\frac{1}{\rho_G} - \frac{1}{\rho_L}\right)$$

$$H = H_L(1-\bar{x}) + H_G\bar{x} = H_L + \bar{x}(H_G - H_L) \tag{6.3.1}$$

$$s = s_L(1-\bar{x}) + s_G\bar{x} = s_L + \bar{x}(s_G - s_L)$$

where $\bar{x}$ represents the equilibrium quality, ρ the density, H the enthalpy, and s the entropy at saturation temperature, and subscripts L, G, and H are used for the liquid, gas, and homogeneous flow, respectively.

According to the energy equation, for an upstream stagnant condition defined by the subscript 0,

$$\frac{1}{2} u_H^2 = H_0 - H = \frac{1}{2}\left(\frac{G}{\rho_H}\right)^2 \quad \text{or} \quad G = [2(H_0 - H)]^{1/2}\rho_H \tag{6.3.2}$$

If the process is isentropic,

$$s_0 = s \quad \text{or} \quad \bar{x} = \frac{s_0 - s_L}{s_G - s_L} \tag{6.3.3}$$

If the stagnation conditions s_0, H_0, and p_0 are specified for one-component flow, such as water–steam, one can assume a pressure p at the exit or minimum flow cross-sectional area and calculate $\bar{x}$ from Eq. (6.3.3) and H, ρ_H, and G from Eqs. (6.3.1) and (6.3.2), and the calculation is repeated until the flow reaches a maximum or critical value G_c. These critical mass flow rates have been calculated by Moody [6.3.1] for steam–water mixtures and are shown in Figure 6.3.1 as a function of the stagnation enthalpy H_0 and stagnation pressure p_0. Figure 6.3.1 shows no dependence on pipe diameter and pipe length because ideal flow conditions were assumed to exist as for the case of perfect gas flow.

We can also write a momentum equation for the isentropic homogeneous equilibrium model and find that Eqs. (6.2.11) apply except for replacing ρ by the homoge-

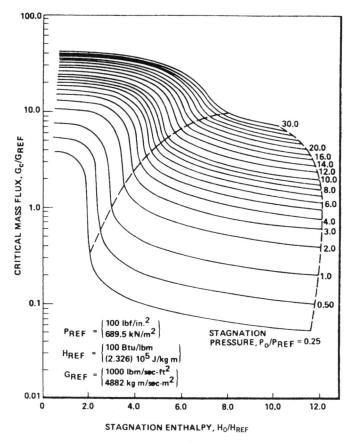

Figure 6.3.1 Steam–water critical mass flux according to the homogeneous model of Moody. (From Ref. 6.3.1.)

neous density ρ_H so that the maximum or equivalent sonic velocity can be obtained from Eq (6.2.12), or

$$\left(\frac{G}{\rho_H}\right)^2_c = \left(\frac{\partial p}{\partial \rho_H}\right)_s \quad \text{or} \quad \left[-\frac{\partial(1/\rho_H)}{\partial p}\right]_s = \frac{1}{G_c^2} \qquad (6.3.4)$$

so that

$$G_c^2 = \left[(1-\bar{x}_c)\left[\frac{\partial(1/\rho_L)}{\partial p}\right]_s + \bar{x}_c\left[\frac{\partial(1/\rho_G)}{\partial p}\right]_s + \left(\frac{1}{\rho_G}-\frac{1}{\rho_L}\right)\left(\frac{\partial \bar{x}_c}{\partial p}\right)_s\right]^{-1} \qquad (6.3.5)$$

The derivative $(\partial \bar{x}_c/\partial p)_s$ can be obtained from Eq. (6.3.3), and the calculated critical flow rates for the homogeneous equilibrium model for steam–water mixtures

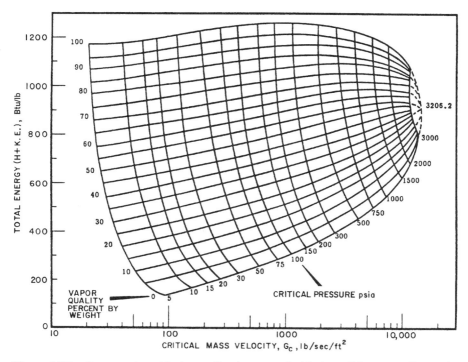

Figure 6.3.2 Steam–water critical mass flux in terms of critical conditions according to the homogeneous model. (From Ref. 6.3.2.)

are shown in Figure 6.3.2 in terms of the conditions at the critical location [6.3.2].

As pointed out by Wallis [6.1.2], "the homogeneous equilibrium model is not a bad way of predicting the critical mass flux, G_c, in long pipes where there is sufficient time for equilibrium to be achieved and when the flow pattern is conducive to interphase forces that are adequate to repress relative motion." This conclusion tends to be supported by the homogeneous model comparison to available steam–water data by Elias and Lellouche [6.1.6] shown in Figure 6.3.3. A close look at the model-test deviations would show that the homogeneous equilibrium model tends to be inaccurate and low for short pipes, in which there is insufficient time for the vapor and liquid to reach equilibrium and for annular flow regimes that produce unequal gas and liquid velocities. The differences are greatest at low steam quality and they decrease as the steam quality increases. Still, because of its simplicity, the homogeneous equilibrium model was and remains a possible option for calculating critical flow in several complex system computer codes.

Another important advantage of the homogeneous equilibrium model is its ability to utilize single-phase compressible results such as for frictional or orifice or valve

314 CRITICAL FLOW

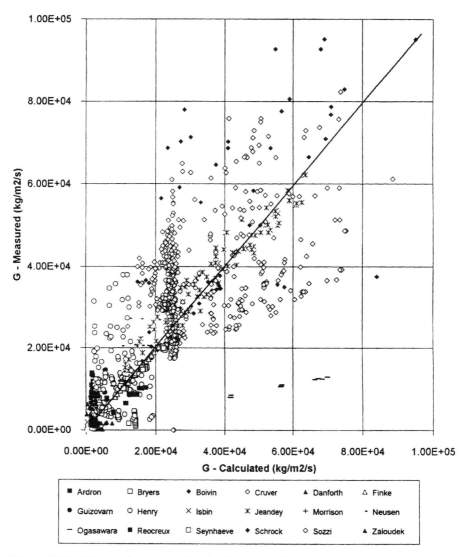

Figure 6.3.3 Measured versus predicted critical mass flux using the homogeneous equilibrium model. (Reprinted from Ref. 6.1.6 with permission from Elsevier Science.)

flow. For example, consider the water–steam flow system depicted in Figure 6.3.4. The calculated mass flow rates depend on whether the liquid temperature is high enough to produce flashing at or ahead of position 3. A typical pressure profile in the duct of hydraulic diameter D_H is depicted in Figure 6.3.4. The liquid pressure starts at p_1 and decreases due to entrance and friction losses until it reaches saturation pressure at position F. Beyond F, flashing occurs and two-phase flow exists. If the liquid temperature always remains below saturation, the rate of liquid flow per unit

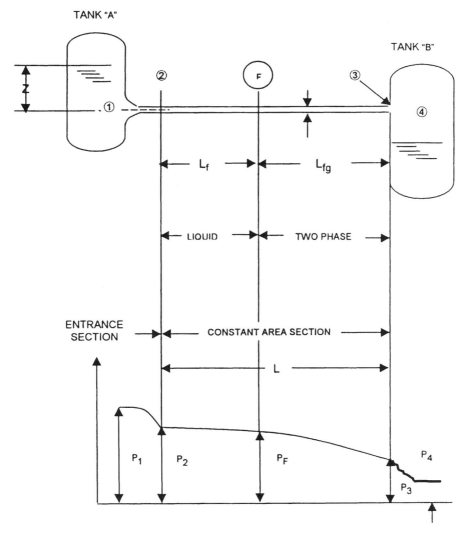

Figure 6.3.4 Pressure profile for liquid–vapor escape.

area, G, is obtained by equating the available driving head to the losses in the flow circuit,

$$G = \sqrt{\frac{2(p_1 - p_3)\rho_L}{1 + fL/D_H}} \qquad (6.3.6)$$

where f is the friction factor for liquid flow and D_H the hydraulic diameter of the duct.

Another possibility is to reach critical flow and saturation conditions simultaneously at position 3. For homogeneous equilibrium flow, the maximum coolant escape

CRITICAL FLOW

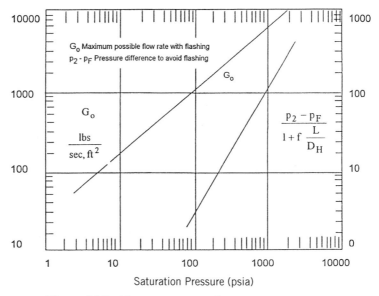

Figure 6.3.5 Flow rate to avoid flashing in water escape.

rate per unit area G_0 is obtained from Figure 6.3.5, where G_0 is plotted versus $p_f = p_3$ for water. Figure 6.3.5 also gives the value of the pressure difference $(p_2 - p_f)/(1 + fL/D_H)$ required to suppress flashing in the duct.

Finally, in the case where saturated conditions are reached before the pipe exit, the flow per unit area G between position 1 and F is obtained from an equation of type (6.3.6), or

$$G = \sqrt{\frac{2g_c(p_1 - p_F)\rho_L}{1 + f\, L_f/D_H}} \qquad (6.3.7)$$

where g_c is the gravitational constant to deal with the British units used in Figure 6.3.5. Beyond the saturation point F, two-phase flow exists and the pressure drop is obtained by integrating the momentum equation, including friction for homogeneous flow (see Chapter 3). The corresponding energy equation reduces to Eq. (6.3.2). This relation and the momentum equation give two expressions that can be integrated for the variables ρ_H and p as a function of position z. Typical results obtained by Sajben [6.3.3] for homogeneous flow of steam–water mixtures at 1000 psia are reproduced in Figure 6.3.6, a plot of the flow ratio G/G_0 in terms of the pressure parameter $(p_F - p_3)/p_F$ and the homogeneous nondimensional grouping $f_H/(L_{fg}/D_H)$. The friction factor f_H is calculated for homogeneous flow as discussed in Chapter 3. For given values of p_1, p_F, p_3, and L, the value of G_0 is first obtained from Figure 6.3.5. A value of G is next assumed and L_f is computed from Eq. (6.3.7). The distance L_{fg} is also obtained from Figure 6.3.6, and the assumed value of G is correct only if the sum of the distances L_f and L_{fg} adds up to L. Figure 6.3.6 shows that the presence of frictional losses reduces the critical flow rates.

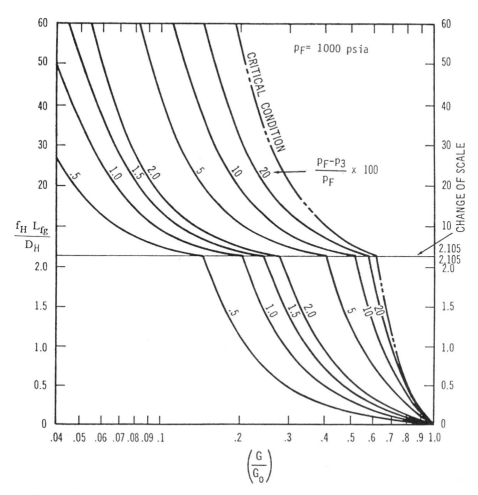

Figure 6.3.6 Escape rate of steam–water mixtures at 1000 psia. (From Ref. 6.3.3.)

Let us next consider homogeneous critical steam–water flow across an orifice as illustrated in Figure 6.2.6. If the stagnation conditions p_0, H_0, and T_0 are specified, Figure 6.3.1 allows the theoretical critical flow rate to be determined, and the critical pressure p_c is obtained from Figure 6.3.2. Equation (6.2.23) is employed next to calculate the coefficient C_v for the orifice based upon the pressure ratio p_c/p_0 obtained from the homogeneous equilibrium model. The same approach can be used to derive the loss coefficients C_v for such components as valves if the values of C_v are known for pure steam (C_{vv}) and for water flow ($C_{v\ell}$) and if Eq. (6.2.23) is utilized. It is important that the *contraction coefficients* C_{vv} and $C_{v\ell}$ be based on physical grounds, preferably upon measurements with single-phase gas and liquid flows. This has not always been the case in many model comparisons to test data where the contraction coefficients have been used as adjustable constants to match test data. For example, a loss coefficient of 1.0 has been employed for the homogeneous equilibrium model for a pipe break during a loss-of-coolant accident in light-water nuclear reactors, while a loss coefficient of 0.6 has been applied to the same geometry in the Moody model, which assumes gas slip with respect to the liquid.

6.3.2 Moody Critical Flow Model

Moody is one of many models where the gas and liquid are presumed to have different velocities but to be in thermal equilibrium. The energy Eq. (6.3.2) now becomes

$$H_0 = H_L(1-\bar{x}) + H_G\bar{x} + \frac{G^2}{2}\left[\frac{\bar{x}^3}{\rho_G^2\bar{\alpha}^2} + \frac{(1-\bar{x})^3}{\rho_L^2(1-\bar{\alpha})^2}\right] \quad (6.3.8)$$

where $\bar{\alpha}$ is the volume fraction occupied by the gas. Moody [6.3.4] employed Eqs, (6.3.3) and (6.3.8) to get

$$G^2 = \frac{2\left[H_0 - H_L - \dfrac{H_G - H_L}{s_G - s_L}(s_0 - s_L)\right]}{\left(\dfrac{\bar{u}_G}{\bar{u}_L}\dfrac{s_G - s_0}{s_G - s_L}\dfrac{1}{\rho_L} + \dfrac{s_0 - s_L}{s_G - s_L}\dfrac{1}{\rho_L}\right)^2\left[\dfrac{s_0 - s_L}{s_G - s_L} + \dfrac{s_G - s_0}{s_G - s_L}\left(\dfrac{\bar{u}_L}{\bar{u}_G}\right)^2\right]} \quad (6.3.9)$$

where $\bar{u}_G$ and $\bar{u}_L$ represent the average gas and liquid velocity, respectively. If $\bar{u}_G/\bar{u}_L$ and p are assumed to be independent, G has a maximum value G_c if $dG/dp = 0$, or

$$\left[\frac{\partial G}{\partial(\bar{u}_G/\bar{u}_L)}\right]_p = \left[\frac{\partial G}{\partial p}\right]_{\bar{u}_G/\bar{u}_L} = 0 \quad (6.3.10)$$

The first part of Eq. (6.3.10) leads to

$$\frac{\bar{u}_G}{\bar{u}_L} = \sqrt[3]{\frac{\rho_L}{\rho_G}} \quad \text{and} \quad \bar{\alpha} = \left[1 + \frac{1-\bar{x}}{\bar{x}}\left(\frac{\rho_G}{\rho_L}\right)^{2/3}\right]^{-1} \quad (6.3.11)$$

If the stagnant conditions are known, the critical flow rates can be calculated as for the homogeneous model. A critical pressure is assumed, G^2 is calculated from Eq. (6.3.9) with $\bar{u}_G/\bar{u}_L$ specified from Eq. (6.3.11), and this process is repeated until the flow rate is maximized or $G = G_c$.

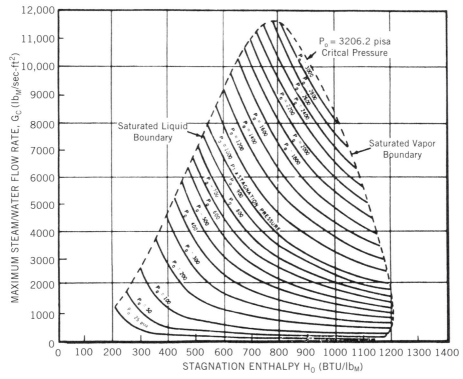

Figure 6.3.7 Maximum steam–water flow rate and local stagnation properties. (From Ref. 6.3.4.)

The maximum steam–water flow rates obtained from the Moody model are shown in Figure 6.3.7 in terms of the stagnation enthalpy and pressure. An algebraic fit to the Moody model has been developed by McFadden et al. [6.3.5] to facilitate system computer code calculations of critical flow. The form of this fit is an expansion in terms of stagnation pressure and specific enthalpy:

$$G_c(p,H) = \begin{cases} \exp\left(\sum_{j=0}^{5}\sum_{i=0}^{5} M_{1i,j} p^j H^i\right) & 15 < p < 200 \text{ psia} \\ \exp\left(\sum_{j=0}^{5}\sum_{i=0}^{5} M_{2i,j} p^j H^i\right) & 200 < p < 3000 \text{ psia} \end{cases} \quad (6.3.12)$$

The constant coefficients $M_{1i,j}$ and $M_{2i,j}$ are listed in Table 6.3.1 and are valid only

Table 6.3.1 Constant Coefficients in Expressions for Moody Critical Flow as a Function of Stagnation Pressure and Specific Enthalpy

i	0	1	2	3	4	5
			$M_{1_{i,j}}$			
0	0.75398883×10^{1}	0.48635447×10^{0}	$-0.14054847 \times 10^{-1}$	$0.18252651 \times 10^{-3}$	$-0.10492510 \times 10^{-5}$	$0.21537617 \times 10^{-8}$
1	$-0.27349762 \times 10^{-1}$	$-0.25172525 \times 10^{-2}$	$-0.90215743 \times 10^{-4}$	$-0.12541056 \times 10^{-5}$	$0.74318133 \times 10^{-8}$	$-0.15500316 \times 10^{-10}$
2	$0.77033614 \times 10^{-4}$	$0.66198222 \times 10^{-5}$	$-0.25893147 \times 10^{-6}$	$0.36860723 \times 10^{-8}$	$-0.22032243 \times 10^{-10}$	$0.46118179 \times 10^{-13}$
3	$-0.11597165 \times 10^{-6}$	$-0.84665377 \times 10^{-8}$	$0.35752133 \times 10^{-9}$	$-0.51915992 \times 10^{-11}$	$0.31249946 \times 10^{-13}$	$-0.65501247 \times 10^{-16}$
4	$0.85314613 \times 10^{-10}$	$0.52492322 \times 10^{-11}$	$-0.23790918 \times 10^{-12}$	$0.35151649 \times 10^{-14}$	$-0.21287482 \times 10^{-16}$	$0.44796907 \times 10^{-19}$
5	$-0.24171538 \times 10^{-13}$	$-0.12636194 \times 10^{-14}$	$0.61256651 \times 10^{-16}$	$-0.91925726 \times 10^{-18}$	$0.55968113 \times 10^{-20}$	$-0.11802990 \times 10^{-22}$
			$M_{2_{i,j}}$			
0	0.64582892×10^{4}	0.15724915×10^{3}	$-0.16101131 \times 10^{0}$	$-0.14741137 \times 10^{-4}$	$0.61947560 \times 10^{-7}$	$-0.19206668 \times 10^{-10}$
1	$-0.59379818 \times 10^{2}$	$-0.76566784 \times 10^{0}$	$0.11162190 \times 10^{-2}$	$-0.18726780 \times 10^{-6}$	$-0.22806979 \times 10^{-9}$	$0.88455810 \times 10^{-13}$
2	0.19040194×10^{0}	$0.14863253 \times 10^{-2}$	$-0.27575375 \times 10^{-5}$	$0.85169692 \times 10^{-9}$	$0.30284416 \times 10^{-12}$	$-0.16287505 \times 10^{-15}$
3	$-0.27991119 \times 10^{-3}$	$-0.13833255 \times 10^{-5}$	$0.31653794 \times 10^{-8}$	$-0.12626170 \times 10^{-11}$	$-0.16803993 \times 10^{-15}$	$0.15109240 \times 10^{-18}$
4	$0.19327093 \times 10^{-6}$	$0.60309542 \times 10^{-9}$	$-0.17199274 \times 10^{-11}$	$0.79444415 \times 10^{-15}$	$0.29596473 \times 10^{-19}$	$-0.71294533 \times 10^{-22}$
5	$-0.50859277 \times 10^{-10}$	$-0.95031339 \times 10^{-13}$	$0.35699646 \times 10^{-15}$	$-0.18158912 \times 10^{-18}$	$0.17431814 \times 10^{-23}$	$0.13825335 \times 10^{-25}$

Source: Reprinted from Ref. 6.1.6 with permission from Elsevier Science.

6.3 TWO-PHASE CRITICAL FLOW MODELS

for saturated or two-phase inlet conditions and for British units (G in lb/sec-ft^2, p in psia, and H in Btu/lb).

It is also possible to use the second of Eq. (6.3.10) to establish the critical flow rate in terms of the steam–water properties at the critical location. There results

$$G_C^2 = 2 \frac{c}{a(ad + 2be)} \quad (6.3.13)$$

where

$$a = \frac{\bar{u}_G}{\bar{u}_L} \frac{1}{\rho_L} + \bar{x}\left(\frac{1}{\rho_G} - \frac{\bar{u}_G}{\bar{u}_L} \frac{1}{\rho_L}\right)$$

$$b = \left(\frac{\bar{u}_L}{\bar{u}_G}\right)^2 + \bar{x}\left[1 - \left(\frac{\bar{u}_L}{\bar{u}_G}\right)^2\right]$$

$$c = -\frac{1}{\rho_L} - \bar{x}\left(\frac{1}{\rho_G} - \frac{1}{\rho_L}\right) \quad (6.3.14)$$

Also,

$$d = \frac{ds_G/dp}{(\bar{u}_G/\bar{u}_L)^2 (s_G - s_L)} - \frac{ds_L/dp}{s_G - s_L} - \frac{d[(s_G - s_L)(\bar{u}_G/\bar{u}_L)^2]/dp}{(\bar{u}_G/\bar{u}_L)^4 (s_G - s_L)}(1 - \bar{x}) + \frac{\bar{x}[d(s_G - s_L)/dp]}{s_G - s_L}$$

$$e = (s_G - s_L)(1 - \bar{x}) \frac{d}{dp}\left(\frac{\bar{u}_G}{\bar{u}_L} \frac{s_G - s_L}{\rho_L}\right) + \frac{\bar{u}_G}{\bar{u}_L} \frac{1}{\rho_L} \frac{ds_G/dp}{s_G - s_L} - \frac{1}{\rho_L} \frac{ds_L/dp}{s_G - s_L}$$

$$+ \bar{x}(s_G - s_L) \frac{d}{dp}\left[\frac{1}{\rho_G (s_G - s_L)}\right]$$

Total derivatives are used in Eq. (6.3.14) since all the properties are taken at saturation conditions and they are only a function of the pressure p. Figure 6.3.8 gives the value of the steam–water critical flow rates in terms of the critical pressure p_c and the local steam quality $\bar{x}_c$. The critical flow rates calculated by Moody's model are higher than those from the homogeneous equilibrium model because of the slip velocity of the gas with respect to the liquid. A comparison of the available steam–water data versus the predictions from the Moody model was generated by Elias and Lellouche [6.1.6] and is reproduced in Figure 6.3.9. The comparison has improved over the results in Figure 6.3.3 for homogeneous equilibrium flow, but the data scatter is still quite large. A close look at the large deviations in Figure 6.3.9 would show that they continue to occur predominantly for short pipes and subcooled or close to saturated-water inlet conditions.

Here again, frictional pressure losses can be accounted for by referring back to Figure 6.3.4 and substituting the Moody model from the saturation location F to

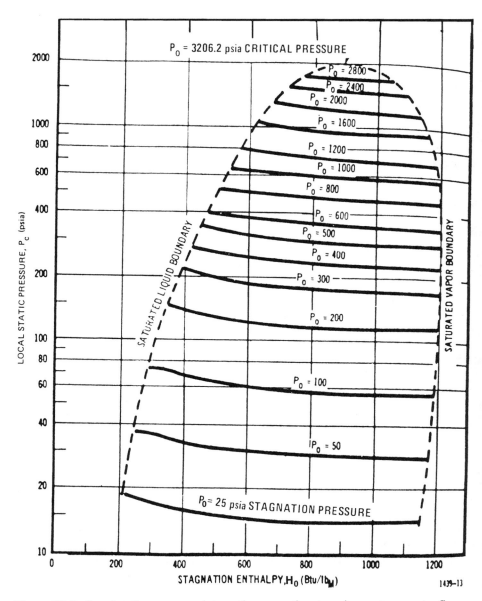

Figure 6.3.8 Local static pressure and stagnation properties at maximum steam–water flow rate. (From Ref. 6.3.4.)

the duct exit at point 3. Moody chose to define the stagnation enthalpy Ho as a function of the pressure p and the steam quality $\bar{x}$. By assuming Ho to be a constant, there results

$$\left.\frac{\partial \text{Ho}}{\partial \bar{x}}\right|_p d\bar{x} + \left.\frac{\partial \text{Ho}}{\partial p}\right|_{\bar{x}} dp = 0 \quad \text{or} \quad \frac{d\bar{x}}{dp} = -\frac{(\partial \text{Ho}/\partial p)_{\bar{x}}}{(\partial \text{Ho}/\partial \bar{x})_p} \quad (6.3.15)$$

Equation (6.3.15) yields a relation between $\bar{x}$ and p, while Eq. (6.3.11) expresses

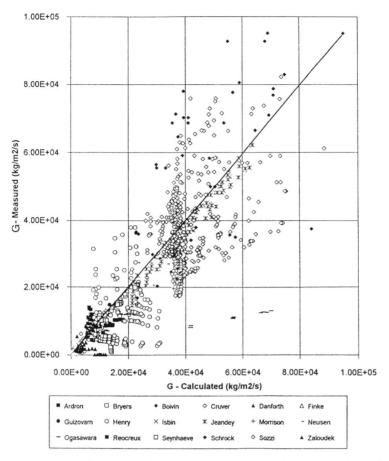

Figure 6.3.9 Measured versus predicted critical mass flux using the Moody model with slip. (Reprinted from Ref. 6.1.6 with permission from Elsevier Science.)

the volume fraction in terms $\bar{x}$ of and p. The corresponding momentum equation used by Moody was

$$-dp = G^2 \left\{ \left[\frac{\partial(1/\bar{\rho}_M)}{\partial \bar{x}}\right]_p \frac{d\bar{x}}{dp} dp + \left[\frac{\partial(1/\bar{\rho}_M)}{\partial p}\right]_{\bar{x}} dp \right\} + \frac{G^2}{2\rho_L} \phi^2_{TPL} \frac{f_L P}{4A} dz \quad (6.3.16)$$

In Eq. (6.3.16) Moody chose to neglect the head losses and to define the two-phase wall shear stress in terms of the multiplier $\phi^2_{TPL} = (1 - \bar{x})^2/(1 - \bar{\alpha})^2$. Also, $\bar{\rho}_M$ is the momentum density:

$$\frac{1}{\bar{\rho}_M} = \left(\frac{\bar{x}}{\rho_G} \frac{\bar{u}_L}{\bar{u}_G} + \frac{1-\bar{x}}{\rho_L}\right)\left[1 + \bar{x}\left(\frac{\bar{u}_G}{\bar{u}_L} - 1\right)\right] = \frac{\bar{x}^2}{\bar{\alpha} \rho_G} + \frac{(1-\bar{x})^2}{(1-\bar{\alpha}) \rho_L} \quad (6.3.17)$$

and f_L is the friction factor, as if the liquid was flowing alone in the duct of cross-sectional area A and wetted perimeter P. Moody integrated Eq. (6.3.16) to calculate the critical flow rates for steam–water mixtures in terms of the stagnation properties at point F and $\overline{f}_L L_{fg}/D_H$, where $\overline{f}_L$ is an average liquid friction factor over the flashing length L_{fg} and D_H is the channel hydraulic diameter or $D_H = 4A/P$. His results are illustrated in Figure 6.3.10 from Reference P-1 for the case of $\overline{f}_L L_{fg}/D_H = 0$ and $\overline{f}_L L_{fg}/D_H = 1$. They again show that friction reduces the critical mass flow rate. Similarly, one could deal with other configurations, such as orifices and valves, by following the methodology of the homogeneous equilibrium model except for replacing its critical flow rate and critical pressure ratio with the values predicted from the Moody predictions of Figures 6.3.7 and 6.3.8.

6.3.3 Other Separated Critical Flow Models

Other slip flow models, which assume thermal equilibrium and do not consider gas–liquid interfacial exchanges, have been proposed. Fauske [6.3.6] defined the volume fraction $\overline{\alpha}$ by maximizing the momentum density $\overline{\rho}_M$ to get

$$\frac{\overline{u}_G}{\overline{u}_L} = \sqrt{\frac{\rho_L}{\rho_G}} \qquad \overline{\alpha} = \left(\frac{1-\overline{x}}{\overline{x}}\sqrt{\frac{\rho_G}{\rho_L}} + 1\right)^{-1} \qquad (6.3.18)$$

and calculated the critical flow G_c from

$$G_c^2 = \left[\left(\frac{\partial p}{\partial(1/\overline{\rho}_M)}\right)\right]_s$$

Levy [6.3.7] defined the volume fraction from a momentum exchange model which neglects frictional and head losses. Levy's results are similar to Fauske's predictions except at low pressures where the slip ratio falls below Eq. (6.3.18).

Cruver [6.3.8] used the same velocity ratio as Moody but calculated the critical flow from

$$G_c^2 = -2\frac{1}{\overline{\rho}_H}\left[\frac{\partial p}{\partial(1/\overline{\rho}_{KE}^2)}\right]_s \qquad (6.3.19)$$

with the kinetic energy density $\overline{\rho}_{KE}$ being equal to

$$\frac{1}{\overline{\rho}_{KE}} = \left(\frac{\overline{x}}{\rho_G}\frac{\overline{u}_L}{\overline{u}_G} + \frac{1-\overline{x}}{\rho_L}\right)\left\{1 + \overline{x}\left[\left(\frac{u_G}{u_L}\right)^2 - 1\right]\right\}^{1/2} \qquad (6.3.20)$$

There are several other separated critical flow models that do not consider the

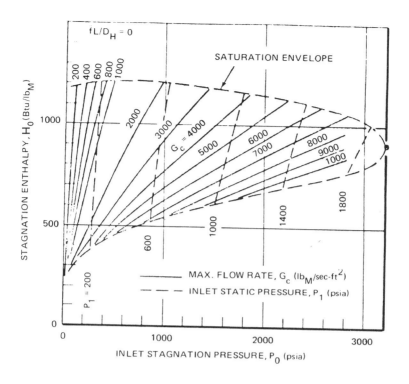

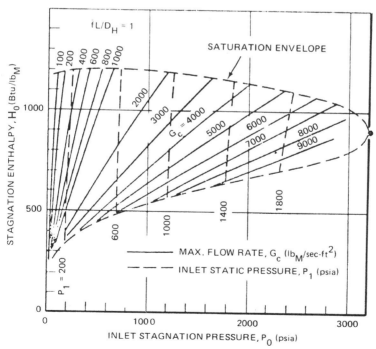

Figure 6.3.10 Pipe maximum steam–water discharge rate. (Reprinted from Ref. P-1 with permission from the American Nuclear Society.) Copyright 1977 by the American Nuclear Society.

gas–liquid interfaces. One approach, for example, has been to use the drift flux methodology to establish the relative motion of the gas with respect to the liquid [6.3.9]. The advantage of all such models is their simplicity and their ability to generate closed-form expressions for the critical flow rate. However, as pointed out by Wallis [6.1.2], "the use of a velocity ratio to maximize the flow rate offers no explanation of how (or if) this condition is actually achieved. In particular, it does not seem possible to relate the predicted velocity ratios to the dynamics of each phase and the interface boundary conditions in a way that is compatible with the other assumptions, such as absence of entropy generation." Similarly, Wallis questioned the use of the drift flux methodology in terms of its applicability to highly accelerating flows. Many other concerns have been raised about the various simplified models. For instance, according to Saha [6.1.1], both Moody [6.3.4] and Levy [6.3.7] employed a homogeneous expression for their local entropy. Cruver [6.3.8] also reported that Fauske's critical slip ratio did not necessarily maximize the pressure gradient at the critical location. Finally, none of the models provide an adequate representation of the detailed behavior of the gas and liquid flows or their interfaces. Still, the predictions of these simplified models tend to match a lot of experimental data as illustrated in Figures 6.3.3 and 6.3.9, and they can be very useful in dealing with friction, heat addition, and more complex geometries.

A comparison of a few of the simplified models is shown in Figure 6.3.11 for water expanding from saturation conditions. Figure 6.3.11 includes the homogeneous model as well as the Fauske and Moody models. A stream tube model is also plotted. It relies on isentropic expansion of individual stream tubes that originate from the vapor–liquid interface as flashing progresses [6.3.10]. The most significant conclusion of Figure 6.3.9 is that slip increases the critical flow rate but its impact is rather small except at low saturation pressures. Also, there appears to be an anomaly in Figure 6.3.11 which shows that the Fauske model with the highest slip falls below the Moody model (which may support the conclusion of Cruver about Fauske's model).

Another important implication derived from Figure 6.3.9 is that all the thermal equilibrium models, irrespective of the degree of slip, are inadequate in predicting critical flow rates as the pipe length is shortened to zero or as the upstream fluid condition is near saturation or subcooled. Nonthermal equilibrium considerations therefore are important under such circumstances and must be taken into account.

6.4 NONTHERMAL EQUILIBRIUM CRITICAL FLOW MODELS

Many models again have been proposed to deal with nonthermal equilibrium phenomena. All the models have empirical features because metastable conditions are very difficult to predict when the liquid is subcooled or near saturation temperature.

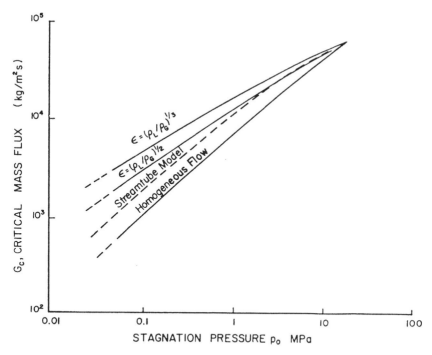

Figure 6.3.11 Critical mass flow versus stagnation pressure for saturated water at inlet, comparison between several "limiting case" models. (Reprinted from Ref. 6.1.2 with permission from Elsevier Science.)

The critical flow rates are also much more dependent on the geometric characteristics. The frozen homogeneous model is the simplest of the proposed nonthermal equilibrium approaches and is covered first.

6.4.1 Frozen Homogeneous Flow

It is assumed that the quality does not change during the flow. The velocities of both phases are presumed to be equal and the vapor to expand isentropically along the flow path. Because the liquid is not allowed to vaporize, the liquid and vapor are not in thermal equilibrium. The frozen model assumptions are extreme since, for example, if the inlet liquid is subcooled or at saturation temperature, only single-phase liquid flow would result. This means that it is best to view the frozen model as a bounding approach to nonthermal equilibrium conditions. If we neglect any change in the liquid enthalpy, Eq. (6.3.2) becomes

$$G = [2(H_0 - H)]^{1/2} \bar{\rho}_H = [2\bar{x}(H_{G0} - H_G)]^{1/2} \bar{\rho}_H \qquad (6.4.1)$$

328 CRITICAL FLOW

where the subscript 0 is used to represent stagnation conditions. If we employ a vapor satisfying the perfect gas law and expanding isentropically, we obtain

$$H_{G0} - H_G = C_P(T_{G0} - T_G) = \frac{\gamma R T_{G0}}{\gamma - 1}\left(1 - \frac{T_G}{T_{G0}}\right)$$

$$= \frac{\gamma p_0}{\rho_{G0}(\gamma - 1)}\left(1 - \frac{T_G}{T_{G0}}\right) \quad (6.4.2)$$

At the critical location denoted by the subscript c, the vapor satisfies the single-phase critical pressure ratio of Eq. (6.2.13) and is presumed to reach sonic velocity. There results

$$\frac{T_{Gc}}{T_{G0}} = \frac{2}{1 + \gamma}$$

$$G_c = \rho_{Hc}\left(\frac{2\gamma p_0}{\rho_{G0}}\frac{\bar{x}}{1 + \gamma}\right)^{1/2} \quad (6.4.3)$$

$$\frac{1}{\rho_{Hc}} = (1 - \bar{x})\frac{1}{\rho_{L0}} + \frac{\bar{x}}{\rho_{G0}}\left(\frac{2}{1 + \gamma}\right)^{1/(1-\gamma)}$$

Equations (6.4.3) have been attributed to Starkman et al. [6.4.1] and Henry and Fauske [6.4.2]. They are valid when the liquid phase does not get a chance to flash. They predict higher critical flow rates than the homogeneous equilibrium model except at very low qualities where $G_c \to 0$ as $\bar{x} \to 0$. This anomaly is caused by the neglect in the derivation of the liquid kinetic energy.

It is interesting to note that if one had started instead from Eq. (6.3.4) and employed the same assumptions of no change in the liquid density ρ_{L0} and in the vapor expanding to the sonic velocity isentropically, the critical flow rate would be

$$G_c^2 = \left\{-\bar{x}\left[\frac{\partial(1/\rho_G)}{\partial p}\right]_s\right\}^{-1} = \frac{p_0 \gamma \rho_{G0}}{\bar{x}}\left(\frac{2}{1 + \gamma}\right)^{(\gamma+1)/(\gamma-1)} \quad (6.4.4)$$

In this case, the critical flow rate goes to infinity ($G_c \to \infty$) as $\bar{x} \to 0$. In fact, the flow rate as $\bar{x} \to 0$ would fall between the prediction of Eqs. (6.4.4) and (6.4.3) and would be obtained from Bournelli's equation (6.2.9) with an appropriate geometry-dependent entrance loss coefficient. The availability of these two critical flow frozen homogeneous models in the literature and the fact that neither is absolutely correct illustrate the importance of employing all three conservation equations and of recognizing all pressure losses from the stagnation point to the critical location, including entrance, acceleration, friction, exit, and other losses.

The homogeneous frozen model increases the critical flow rate and therefore provides improved and bounding answers to nonthermal equilibrium conditions; however, it fails to deal with the impact of geometry and with the actual degree of nonthermal equilibrium. For those reasons it has been supplemented by empirical models that are adjusted to fit the data.

6.4.2 Henry–Fauske Nonequilibrium Model

Henry [6.4.3] and Henry and Fauske [6.4.4] recognized that the actual vapor quality $\bar{x}$ is below the equilibrium quality $\bar{x}_E$ due to thermal nonequilibrium, or

$$\bar{x} = N\bar{x}_E \tag{6.4.5}$$

where N is an empirical parameter that varies with flow geometry. For example, for a long tube having a sharp-edged entrance with $L/D > 12$, it was proposed that

$$\bar{x} = N\bar{x}_E \left\{ 1 - \exp\left[-B\left(\frac{L}{D} - 12\right) \right] \right\}$$

$$N = \begin{cases} 20\, \bar{x}_E & \text{for } \bar{x}_E < 0.05 \\ 1 & \text{for } \bar{x}_E \geq 0.05 \end{cases} \tag{6.4.6}$$

$$B = 0.523$$

For $L/D < 12$ it was suggested that vapor generation from the liquid could be neglected (i.e., the frozen model would be applicable). For smooth inlet configuration, it was recommended that

$$\bar{x} = N\bar{x}_E \left\{ 1 - \exp\left[-B\left(\frac{L}{D}\right) \right] \right\} \tag{6.4.7}$$

with N and B defined from Eq. (6.4.6).

For nozzles, it was best to use Eq. (6.4.5) with

$$N = \begin{cases} \dfrac{\bar{x}_E}{0.14} & \text{for } \bar{x}_E < 0.14 \\ 1 & \text{for } \bar{x}_E \geq 0.14 \end{cases} \tag{6.4.8}$$

Henry and Fauske developed various expressions for the corresponding critical flow rate and the critical pressure ratio. All their derivations had the following common characteristics:

- The liquid and vapor have the same velocity.
- The liquid is assumed to be incompressible.
- The gas expands isothermally, isentropically, or polytropically to the critical location, where it reaches the sonic velocity.

For example, in the case of Eqs. (6.4.6), the critical flow rate G_c was expressed: as a result of an isentropic expansion:

$$G_c^2 = \left[\frac{\bar{x}_c}{\gamma p_c \rho_{Gc}} - \left(\frac{1}{\rho_{Gc}} - \frac{1}{\rho_{L0}}\right)\left(N\frac{d\bar{x}_E}{dp}\right)_c\right]^{-1} \tag{6.4.9}$$

where

$$s_0 = s_{Lc} + \bar{x}_E(s_{GC} - s_{Lc}) \tag{6.4.10}$$

can be used to derive $d\bar{x}_E/dp$. In addition, the pressure drop to the critical location is made of different losses:

$$p_0 - p_c = \Delta p_e + \Delta p_a + \Delta p_f + \Delta p_{aa} \tag{6.4.11}$$

where Δp_e is the entrance loss into the tube, Δp_a corresponds to the acceleration pressure loss, Δp_f represents the frictional pressure loss along the tube, and Δp_{aa} covers pressure losses due to changes in tube configuration (e.g., bends, contraction, or expansion). With appropriate expressions for all these pressure losses, Eqs. (6.4.5) to (6.4.9) can be solved to obtain the critical values of flow G_c and pressure p_c for specified stagnation and nonthermal equilibrium conditions. It is worth noting that for ideal gas flow, no pressure drops, and $\bar{x}_E$ constant, Eq. (6.4.7) reduces to the frozen model, Eq. (6.4.4).

The Henry–Fauske model predicts higher critical flow rates by comparison to the homogeneous equilibrium model. Because of its empirical fitting to test data, it yields improved agreement versus the frozen model. However, Elias and Lellouche [6.1.6] claim that the Henry–Fauske model does not give a good fit to all subcooled and two-phase inlet available data. However, it is not clear to what degree the parameter N was varied in Reference 6.1.6 to match the Henry–Fauske fittings of the data. There is no doubt that the Henry–Fauske parameter N needs adjustment to predict new test data. For example, in the case of the large-scale Marviken tests [6.4.5], the value of N had to be varied from 7 to 100 in the course of a specific blowdown test [6.1.3]. Similarly, adjustments to the parameter N or the effective flow path loss coefficient were necessary to match critical flow data in artificial slits and actual corrosion cracks [6.4.6]. Also, Amos and Schrock [6.4.7] reported that Henry's model for long ducts overpredicted their critical flow results through slits by an order of magnitude.

6.4.3 Water Decompression or Pressure Undershoot Nonequilibrium Model

Critical flow tests with subcoooled liquid show that significant depressurization below the saturation pressure occurs before liquid flashing takes place [6.4.7]. This behavior is shown in Figure 6.4.1, which is taken directly from the steam–water tests of Reocreux [6.4.8] and is reproduced from Levy and Abdollahian [6.4.9]. The local pressure is observed to decrease below the saturation temperature p_{s,T_i}, corresponding to the inlet water temperature T_i for the first bubble to form at the pressure

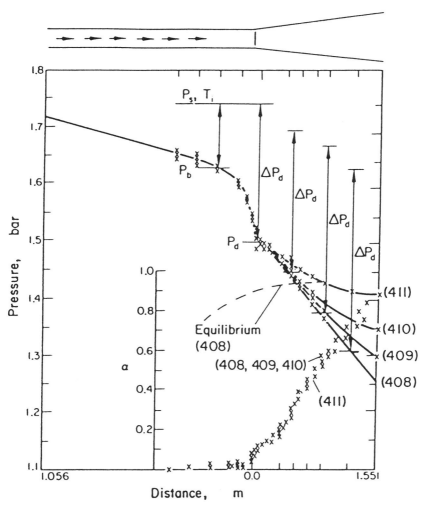

Figure 6.4.1 Characteristics of critical flow with subcooled inlet conditions: run 408 (flow 10,291 kg/m²·s, inlet temp. 115.9°C); run 409 (flow 10,309 kg/m²·s, inlet temp. 115.9°C); run 410 (flow 10,311 kg/m²·s, inlet temp. 115,9°C); run 411 (flow 10,324 kg/m²·s, inlet temp. 116.1°C). (Reprinted from Ref. 6.4.9 with permission from Elsevier Science.)

p_b. The local pressure continues to decrease until it reaches a value p_d, where significant steam is formed. The steam volume fraction, starts to build up beyond that location and it is observed to stay below the corresponding equilibrium volume fraction, plotted as a dashed curve in Figure 6.4.1. At the pressure location, p_d, a decompression pressure difference or pressure undershoot Δp_d exists so that

$$\Delta p_d = p_{s,Ti} - p_d \qquad (6.4.12)$$

Alamgir and Lienhard [6.4.10] analyzed all the available data for rapid depressurization of subcooled water and developed the following correlation for Δp_d:

$$\Delta p_d = \frac{0.258 \sigma^{1.5} T_r^{13.76} (1 + 13.25 \Sigma^{0.8})^{0.5}}{(k_s T_c)^{0.5}(1 - \rho_G/\rho_L)} \qquad (6.4.13)$$

where σ is the surface tension, T_r the reduced temperature or the ratio of initial water temperature to critical temperature, Σ the rate of depressurization in M·atm/s, k_s is Boltzmann's constant, T_c the critical temperature, and ρ_G and ρ_L are the steam and water density. Modified versions of Eq. (6.4.13) have been used by various investigators trying to predict nonthermal equilibrium steam–water critical flow. For example, Levy and Abdollahian replaced $(1 + 13.25\Sigma^{0.8})^{0.5}$ by $(0.49 + 13.25\Sigma^{0.8})^{0.5}$, while Amos and Schrock further modified Levy and Abdollahian's expression by multiplying it by S, where

$$S = a - b\bar{u} \qquad (6.4.14)$$

where $\bar{u}$ is the flow velocity. Values of a and b were chosen by Amos and Schrock to get good agreement between their homogeneous model predictions and their measured critical flow rates through slits. For the largest tested slits, it was found that $a = 0.533$ and $b = 3.4 \times 10^{-3}$ s/m were appropriate. Several explanations have been offered to justify the modifications to the Alamgir and Lienhard equation (6.4.13). They include the possibilities that Alamgir and Lienhard correlated data at high depressurization rates (400 MPa/s to 182 GPa/s) and for relatively stagnant flow conditions. In fact, there are likely to be empirical adjustments to Eq. (6.4.13) to improve the comparison of predictions to test data to account for velocity and different flow geometries.

Once the decompression pressure is specified, the critical flow rate can be calculated from different models. For example, Levy and Abdollahian assumed homogeneous isentropic flow and that as the local pressure decreases below the initial flashing condition enough water will be converted to steam to have its superheat correspond to the pressure decompression Δp_d at that location. Amos and Schrock instead employed a relaxation form of the pressure undershoot at the initial flashing $\Delta p_{d,i}$, or

$$\Delta p_d = \Delta p_{d,i} \exp\left(1 - \frac{t_r}{\tau'}\right) \qquad (6.4.15)$$

where t_r is the residence time of the two-phase mixture in the duct and τ' is the time at which the two-phase mixture relaxes to equilibrium. A value of $\tau' = 0.0011$ s was recommended. Amos and Schrock also employed the drift flow formulation of Kroeger [6.3.9] for critical flow except they assumed no slip of the vapor with respect to the liquid.

The preceding pressure decompression or undershoot models have the advantage of recognizing the need of having the liquid superheated before vapor is formed. They also attempt to take into account the influence of depressurization rate, flow velocity, and configuration upon the degree of required superheat. However, they continue to rely on empirical correlations and it is suspected that their empiricism will need to be adjusted as they are applied to a larger variety of flow geometries.

6.4.4 Other Nonequilibrium Critical Flow Models

Relaxation models for phase change have been used by several authors, as, for example, by Kroeger [6.3.9]. The basic concept is to assume that the rate of change of quality is proportional to the difference between the actual and the equilibrium quality. Empirical proportionality coefficients are utilized to get good agreement with the test data. Several other mechanistic models have been proposed [6.4.11, 6.4.12]. They prescribe the initial characteristics of bubble nucleation (i.e., a density of bubble nuclei of a certain initial radius) coupled with a bubble growth process (see Section 3.6.3.3.2). As pointed out by Wallis [6.1.2], the numbers and processes are selected to correlate the critical flow behavior and they may not be representative of the large pressure gradients and accelerations found in critical flow conditions. His overall impression of such theories is that "they represent no improvement over the grosser empirical methods. There are so many uncertainties about the physics that one merely ends up trying to fit critical flow data by varying adjustable coefficients that are made respectable by being given names implying that they describe physical phenomena." Wallis goes on to state that the "two-fluid" or "separated flow" models offer the best opportunity for dealing with nonequilibrium phenomena and their dependence on geometry. Several such models have been proposed and the emphasis here is put on the model of Richter because of its completeness in terms of interfacial momentum and heat transfer as well as wall friction. This is followed by a discussion of critical flow models employed in system computer codes.

6.5 RICHTER CRITICAL FLOW MODEL

Richter [6.5.1] developed a two-fluid model for unequal liquid and vapor velocities and for nonthermal equilibrium conditions. It employs two mass conservation equations, two momentum conservation equations, and one mixture energy conservation equation. It presumes steady-state one-dimensional flow, and mass transfer is limited to heat transfer between the phases. The applicable equations are

$$\frac{1}{W_L}\frac{dW_L}{dz} = \frac{1}{\rho_L}\frac{d\rho_L}{dz} + \frac{1}{\bar{u}_L}\frac{d\bar{u}_L}{dz} - \frac{1}{1-\bar{\alpha}}\frac{d\bar{\alpha}}{dz} + \frac{1}{A}\frac{dA}{dz} \quad (6.5.1)$$

$$\frac{1}{W_G}\frac{dW_G}{dz} = \frac{1}{\rho_G}\frac{d\rho_G}{dz} + \frac{1}{\bar{u}_G}\frac{d\bar{u}_G}{dz} + \frac{1}{\bar{\alpha}}\frac{d\bar{\alpha}}{dz} + \frac{1}{A}\frac{dA}{dz} \quad (6.5.2)$$

$$\rho_L \bar{u}_L (1-\bar{\alpha})A\frac{d\bar{u}_L}{dz} = -\frac{dp}{dz}(1-\bar{\alpha})A + \tau_{GL}A - \tau_{WL}A - \frac{1}{2}(\bar{u}_G - \bar{u}_L)W\frac{d\bar{x}}{dz}$$
$$- \rho_L g(1-\bar{\alpha})A \cos\theta \quad (6.5.3)$$

$$\rho_G \bar{u}_G \bar{\alpha} A\frac{d\bar{u}_G}{dz} = -\frac{dp}{dz}\bar{\alpha}A - \tau_{GL}A - \tau_{WG}A - \frac{1}{2}(\bar{u}_G - \bar{u}_L)W\frac{d\bar{x}}{dz} \quad (6.5.4)$$
$$- \rho_G g\bar{\alpha}A \cos\theta$$

or

$$W\frac{d\bar{x}}{dz}\left[(H_G - H_L) + \frac{1}{2}(\bar{u}^2{}_G - \bar{u}_L^2)\right] + W_G\left(\frac{dH_G}{dz} + \bar{u}_G\frac{d\bar{u}_G}{dz}\right)$$
$$+ W_L\left(\frac{dH_L}{dz} + \bar{u}_L\frac{d\bar{u}_L}{dz}\right) + Wg\cos\theta = 0 \quad (6.5.5)$$

where W is the mass flow rate, $\bar{\rho}$ the density, $\bar{u}$ the velocity, $\bar{\alpha}$ the gas volume fraction, A the total cross-sectional area, and z the coordinate in the flow direction. The subscripts G and L correspond to gas and liquid. Also, p is the pressure, τ the shear stress per unit volume, H the enthalpy, and θ the inclination angle of the flow duct to the vertical. The subscripts GL correspond to gas–liquid interface, WL to liquid–wall interface, and WG to gas–wall interface.

In addition, the interfacial heat transfer from liquid to the gas must be specified. In the case of bubbly flow, Richter proposed that

$$\frac{6h}{d}(T_L - T_G)\bar{\alpha}A = W\frac{d\bar{x}}{dz}H_{LG} + W_G\frac{dH_G}{dz} \quad (6.5.6)$$

where h is the heat transfer coefficient and d is the bubble diameter. The first term on the right-hand side specifies the energy required to evaporate the mass fraction $d\bar{x}$ over the length dz, while the second term corresponds to the change in bubble enthalpy due to temperature and pressure changes. Because Richter presumed the vapor to be at saturation temperature, one can substitute

$$\frac{dH_G}{dz} = \left(\frac{\partial H_G}{\partial p}\right)_{sat}\frac{dp}{dz} \quad (6.5.7)$$

The heat transfer coefficient h was obtained from the correlation for flow around a single solid sphere, or

$$\frac{hd}{k_L} = 2 + 0.6 \left[\frac{(\bar{u}_g - \bar{u}_L)\rho_L d}{\mu_L}\right]^{0.5} \cdot \Pr_L^{1/3} \qquad (6.5.8)$$

where μ_L is the liquid viscosity, k_L its thermal conductivity, and $\Pr_L$ its Prandtl number.

The gaseous wall shear is set at zero, or $\tau_{WG} = 0$, while the liquid wall shear was taken from Martinelli and Nelson [3.5.2]. The interface shear for bubble flow was written

$$\tau_{GL} = \frac{3}{4}\frac{C_D}{d}\bar{\alpha}(1-\bar{\alpha})^3\rho_L(\bar{u}_G - \bar{u}_L)|\bar{u}_G - \bar{u}_L| + 0.5\rho_L\bar{u}_G\bar{\alpha}\frac{d(\bar{u}_G - \bar{u}_L)}{dz} \qquad (6.5.9)$$

$$C_D = C_{D,SB}(1-\bar{\alpha})^{-4.7}$$

where C_D is the drag coefficient for an array of bubbles and $C_{D,SB}$ is the drag coefficient for a single bubble that was taken as the drag coefficient for a solid sphere of diameter d [see Eq. (4.6.53)]. The second term on the right-hand side in the first of Eqs. (6.5.9) represents the apparent mass force due to the kinetic energy induced by the bubble motion in the liquid.

Richter assumed that bubble flow prevailed to between $\bar{\alpha} = 0.2$ and 0.3, where it transitioned to churn-turbulent flow. Annular flow was assumed to occur beyond $\bar{\alpha} = 0.8$; its interface friction factor was obtained from the Wallis correlation and its interface heat transfer coefficient was calculated from a Colburn correlation:

$$\frac{h}{\rho_L C_{pL}(\bar{u}_G - \bar{u}_L)} \Pr_L^{2/3} = \frac{f_i}{2} = \frac{1}{2}0.005[1 + 75(1 - \bar{\alpha})] \qquad (6.5.10)$$

Between the transition point to churn-turbulent flow and annular flow, Richter calculated the interfacial area, friction factor, and heat transfer by a linear interpolation from bubble to annular flow properties.

Equations (6.5.1) to (6.5.9) are in many ways similar to formulations derived in Chapters 3 and 4, and they suffer from the numerous limitations discussed in those chapters. To initiate bubble nucleation, Richter assumed that the liquid superheat ΔT_{sup} needed for nucleation was given by

$$\Delta T_{\text{sup}} = \frac{(1/\rho_G - 1/\rho_L)4\sigma}{(s_G - s_L)d_i} \qquad (6.5.11)$$

where σ corresponds to surface tension, s to entropy, and d_i the initial bubble diameter, which was taken at $d_i = 2.5 \times 10^{-5}$ m. Equation (6.5.11) corresponds to a pressure difference Δp across the bubble such that $\Delta p = 4\sigma/d_i$.

The final assumption made by Richter was to assume an initial number of nucleation sites N_i and that no further nucleation sites are activated during the flow

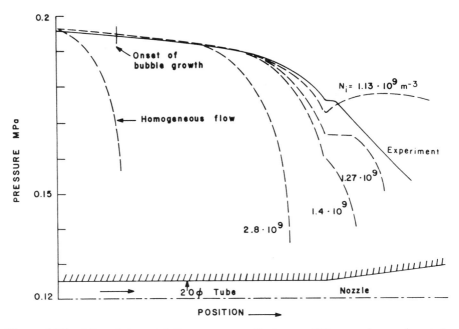

Figure 6.5.1 Attempts to match the pressure profile in one of Reocreux's experiments by varying the nucleation site density (N_i) (P_{sat} = 0.21 MPa, d_i = 18μm). (Reprinted from Ref. 6.5.1 with permission from Elsevier Science.)

process. This last parameter was found to vary with test configuration. In the case of one of the Reocreux's tests [6.4.8], the value of N_i was varied as shown in Figure 6.5.1, and a value of $N_i = 10^9$ m^{-3} was judged to be appropriate. In the case of the nozzle tests of Sozzi and Sutherland [6.5.2], a value of $N_i = 10^{11}$ m^{-3} was selected as shown in Figure 6.5.2. Elias and Lellouche [6.1.6] obtained a similar good comparison of Richter's model with all the Sozzi and Sutherland data. They found that the Richter model was somewhat better for two-phase than for subcooled inlet conditions. This is not too surprising since the Richter model does not consider metastable thermal conditions. Also, Elias and Lellouche were not able to reproduce the good agreement reported by Abdollahian et al. [6.1.3] for the Marviken tests with the Richter model. They attributed their difficulties to the Richter computer code producing three solutions at the critical location and an inability to select one of the roots.

As Wallis [6.1.2] points out, the two-fluid models require numerous assumptions about prevailing flow patterns and the corresponding closure equations to describe the interfacial areas and momentum and heat transfer exchanges. It is not clear that the steady-state correlations for interface friction factor and heat transfer are applicable to highly accelerating flows and that under critical flow conditions the interfacial geometric characteristics are not distorted beyond the idealized spherical bubble geometry and annular geometry employed in the two-fluid models. Furthermore, when all is said and done, the two-fluid model of Richter required an empirical

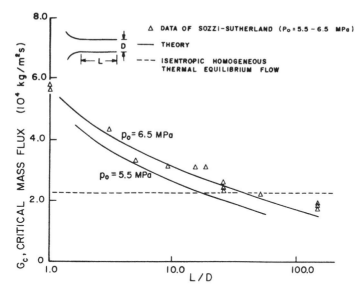

Figure 6.5.2 Effect of nozzle length on critical mass flux predicted by the separated flow model ($N_i = 10^{11}$ m^{-3}, $d_i = 25$ μm.) (Reprinted from Ref. 6.6.2 with permission from Elsevier Science.)

adjustment in the number of nucleation sites to fit the test results. In that sense, the two-fluid models may not be overly superior to simplified critical flow predictions discussed previously. This conclusion is supported further by looking at the critical flow results obtained from system computer codes.

6.6 CRITICAL FLOW MODELS IN SYSTEM COMPUTER CODES

As noted in Chapter 3, most system computer codes now employ two-fluid models. For single- or two-phase flow, their system of conservation equations can be represented by a system of quasi-linear first-order partial differential equations of the form

$$AX'_t + BX'_z = C \tag{6.6.1}$$

where X is a vector of the primary dependent variables and C is a function of X. The prime above X indicates a partial derivative of X, and the subscripts t and z indicate that the partial derivative is with respect to time t and flow direction z. For any flow described by a system of Eqs. (6.6.1) it has been shown under steady-state conditions that it is necessary to have the relationship det$[B] = 0$ for the flow to be critical. This was illustrated for the case of single-phase flow by Eqs. (6.2.17), (6.2.19), and (6.2.20) in Section 6.2.

The same theory of characteristics can be employed to determine the criterion for choked two-phase flow. The propagation velocities of disturbances in the space-time plane are obtained from the real part of the roots, λ_j, of the characteristic equation

$$\det[A\lambda - B] = 0 \tag{6.6.2}$$

with $j \leq n$, the number of differential equations comprising the set of Eq. (6.6.1). In a well-posed hyperbolic system of Eq. (6.6.1), all the roots of Eq. (6.6.2) are real. Critical flow will occur if the solutions of the characteristic equation are zero for one root λ_i and positive for all the others. Under this condition, the signal associated with the root λ_i will propagate with the largest velocity relative to the fluid and will be stationary. Thus no information can propagate backward against the flow upstream from the critical section where an equivalent two-phase "sonic" velocity is reached.

All two-fluid computer system codes are therefore able to calculate a critical flow rate for all scenarios of interest. This approach, however, has met with limited success, for the following reasons:

1. As the critical flow location is approached, it is characterized by a sudden rapid negative increase in the pressure spatial derivative (i.e., $dp/dz \rightarrow -\infty$). Other spatial derivatives change markedly at this point of singularity, and the numerical solutions of Eqs. (6.6.1) become quite difficult. The discrete mesh grid employed in the solution of most computer system codes cannot cope with such a spatial singularity and requires considerably reduced nodalization to get an accurate answer. This is not always economical computer-wise, particularly if the calculation is not fully implicit and the flow geometry involves sudden contractions and expansions.

2. The critical flow rate is dependent on the number of conservation equations employed as well as their numerical treatment and the forms of matrices A and B and the algebraic contents of term C in Eqs. (6.6.1). A complete flow model will use six differential equations, two for each phase for the conservation of mass, momentum, and energy. This is the case for the CATHARE computer system code [6.6.1], which is capable of shifting to a mesh-centered discretization and an implicit numerical treatment. Similarly, RELAP5/MOD3 [3.5.14] usually employs a system of six conservation equations, but it was reduced to five equations for the calculation of critical flow for a two-phase inlet condition. This was accomplished through the substitution of a single energy equation based on the mixture entropy and by assuming the two phases to be in thermal equilibrium. Other attempts have been made to reduce the full two-fluid model to a smaller number of equations to ease the computations. TRAC-BF1/MOD1 [3.5.15] includes a drift flux relation between the relative gas and liquid motion, and it requires only one momentum equation. As discussed in Chapter 3, RELAP and TRAC computer system codes have a discretization with offset meshes, the first set of meshes being used to calculate the scalar

values at the center of the volumes and the second set being offset by half a mesh for calculating vector values. Additional differences in the critical flow models of system computer codes arise in matrices A and B and the term C. In Chapter 3 it was noted that the interface equations and flow pattern maps are seldom the same. It was also pointed out that there is no general agreement about some of the interface transfer laws, such as the virtual mass. When such virtual terms contain derivatives of the principal variables, they modify the matrices A and B and thus enhance the critical flow discrepancies among system computer codes. *All the foregoing differences mean that the predictions of critical flow rates vary from one computer system code to another.* This inconsistency cannot help but affect the credibility of system computer codes.

3. Comparisons of critical flow rates from system computer codes to test data have not been overly satisfactory. For instance, during an exercise of predicting the LOBI-PREX test [6.1.4], the best results were obtained with RELAP4 [6.6.2], which relied upon simplified model options or their fits for the HEM, Moody, and Henry–Fauske formulations. Another excellent example is the case of RELAP5/MOD3, where, after deriving the onset of choking from five conservation equations, it was discarded because of the significant effect the virtual mass coefficient had on the choked-flow dynamics. A simplified choking criterion was selected in the case of a two-phase inlet condition instead:

$$\frac{\overline{\alpha}\rho_L \overline{u}_G + (1-\overline{\alpha})\rho_G \overline{u}_L}{\overline{\alpha}\rho_L + (1-\overline{\alpha})\rho_G} = \pm c_{HE} \qquad (6.6.3)$$

where $\overline{u}_G$ and $\overline{u}_L$ are the gas and liquid velocities, $\overline{\alpha}$ the gas volume fraction, and c_{HE} the HEM speed of sound, discussed in Section 6.3.1. Equation (6.6.3) can be rewritten as

$$\overline{u}_G \frac{1 + (1-\overline{\alpha})\rho_G (\overline{u}_L/\overline{u}_G)/\overline{\alpha}\rho_L}{1 + (1-\overline{\alpha})\rho_G/\overline{\alpha}\rho_L} = \pm c_{HE} \qquad (6.6.4)$$

Since $\overline{u}_L/\overline{u}_G \leq 1$, $\rho_G/\rho_L \ll 1$, $(1-\overline{\alpha})/\overline{\alpha} \leq 1$ for $\overline{\alpha} > 0.5$, Eq. (6.6.4) yields $\overline{u}_G = c_{HE}$ for most practical conditions. Furthermore, because of the staggered grid structure method used in RELAP5/MOD3, which was illustrated in Section 3.7.1, the calculation of Eq. (6.6.3) is not trivial and it might be advantageous simply to utilize the HEM model.

4. The system computer codes have even greater difficulty in dealing with nonthermal equilibrium conditions. The simplified models have had to use empirical parameters that are adjusted with the test geometry to match metastable thermal data, and it is suspected that the system computer codes will have to rely on similar empiricism. For example, for subcooled inlet conditions, RELAP5/MOD3 assumes that the two phases flow at the same velocity and that the difference from the stagnation pressure to the throat pressure is given by a modified form of the Alamgir–

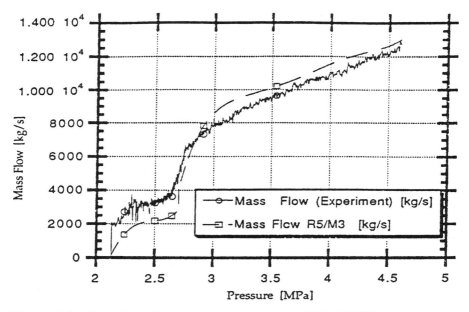

Figure 6.6.1 Comparison of experimental mass flow and RELAP5/MOD3 calculation in the Marviken 22 experiment. (From Ref. 6.6.4.)

Lienhard correlation by Jones [6.6.3]. A comparison of RELAP5/MOD3 predictions of two Marviken tests is reproduced in Figures 6.6.1 and 6.6.2. The comparison is good in the subcooled and transition regions, but it is poor in the thermal equilibrium region, which has been predicted adequately by the HEM model. This discrepancy is probably due to the failure to relax the decompression pressure as a function of time or distance.

In summary, system computer codes have the inherent capability to predict critical flow. Those results have not always been satisfactory because of uncertainties in the formulation of the conservation equations and the interface transfer terms. Also, they are expected to require empirical input to deal with nonthermal equilibrium conditions, and their numerical solutions can become quite tedious near the critical location. Still, they offer the best future opportunity of specifying choked two-phase-flow conditions if their formulation can be improved from increased fundamental understanding of various components of the model and from detailed measurements of critical flow behavior.

6.7 APPLICATIONS OF CRITICAL FLOW MODELS

New practical applications of two-phase flow continue to happen and to involve different geometries or flow circumstances from those considered previously. When such opportunities arise, they offer an excellent chance to test the available models.

6.7 APPLICATIONS OF CRITICAL FLOW MODELS

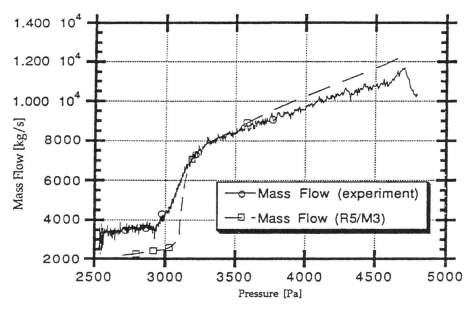

Figure 6.6.2 Comparison of experimental mass flow and RELAP5/MOD3 calculation in the Marviken experiment. (From Ref. 6.6.4.)

In Chapter 5, full-scale counter flooding tests in a reactor downcomer were used to illustrate the shortcomings of multidimensional system computer codes in predicting those results. Here, critical flow through piping cracks and through small breaks in a horizontal pipe are employed to reinforce the conclusion that geometry plays a very important role which can be prescribed initially only from two-phase-flow tests.

6.7.1 Critical Flow through Pipe Cracks

Estimation of leak rate through pipe cracks is important to the application of the leak-before-break (LBB) philosophy to piping integrity safety analyses. Adoption of an LBB approach requires crack-opening-area models and the ability to calculate the critical flow through such openings. In this section, emphasis is placed on flow test data through cracks and slits as summarized and assessed by Paul et al. [6.4.6], who employed the nonthermal equilibrium model of Henry and Fauske [6.4.4] described by Eqs. (6.4.7) to (6.4.10) for their predictions. In defining the various associated pressure losses that appear in Eq. (6.4.11), reference needs to be made first to Figure 6.7.1, which illustrates the geometry of a convergent crack. Three crack areas are depicted along the length L of the crack: an area A_0 at the inlet plane of the crack, an area A_c at the exit or critical plane, and an area A_i corresponding to a distance L_i along the crack where $L_i/D = 12$ and where according to Henry–Fauske, the first vapor is formed. D is the crack hydraulic diameter, and as shown

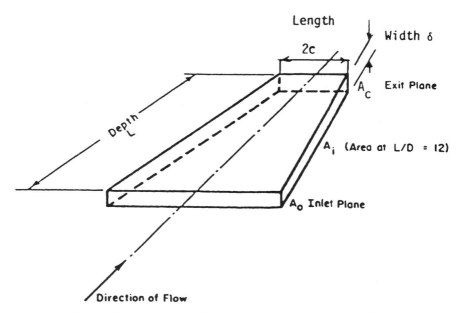

Figure 6.7.1 Geometry of a convergent crack. (From Ref. 6.7.3.)

in Figure 6.7.1, is equal to the rectangular crack opening area $(2\delta c)$ multiplied by 4, and divided by the wetted perimeter $[2(2c + \delta)]$, or

$$D = \frac{4c\delta}{2c + \delta} \qquad (6.7.1)$$

Other crack shapes have been employed, including elliptical and diamond configurations. For the same crack-opening displacement (COD), δ, the area of crack opening (ACO) is 1.0, 0.785, and 0.5 when it is normalized for rectangular, elliptic, and diamond-shaped cracks, respectively.

In order to calculate the leak rate through a crack, it is first important to specify the area of crack opening (ACO). As discussed in Reference 6.4.6, several linear–elastic or pseudoplastic fracture models are available to obtain ACO estimates. In the elastic–plastic regime the ACO evaluations are possible only from numerical analysis techniques, such as the finite element methods (FEM). Kumar et al. [6.7.1, 6.7.2] were able to develop an elastic–plastic fracture mechanics model (EPFM) to calculate the plastic and elastic components of the crack-opening displacement (COD) and to add them to obtain the total COD or δ. This still leaves the problems of specifying the crack length $2c$ and its shape in order to calculate the ACO, and this crack-opening profile is generally left to be assumed by the user.

Another uncertainty is associated with the characteristics of the crack itself (i.e., the surface roughness and number of bends as depicted in Figure 6.7.2). Finally, for very tight cracks, particulates may be able to lodge within the crack and hamper the flow if the size of the particulates is of the same order of magnitude as the

6.7 APPLICATIONS OF CRITICAL FLOW MODELS

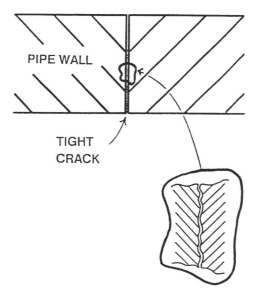

Figure 6.7.2 Two primary factors affecting leakage flow rate are the number of bends and the surface roughness of each crack face. (From Ref. 6.4.6.)

crack opening. In this specific practical application, it is seen that the uncertainties in crack size and crack shape, surface roughness of the crack faces, number of bends, and the presence of particulates plugging can be quite large. All such uncertainties are propagated to the calculation of the various pressure losses appearing in Eq. (6.4.1), and they can overshadow the degree of accuracy desired or expected from the fundamental critical flow model.

The mass flow rates at the inlet, at the location where vapor is first formed, and at the exit of the crack shown in Figure 6.7.1 are defined as G_0, G_i, and G_c and they occur, respectively, at the areas A_0, A_i, and A_c such that

$$G_0 = G_c \frac{A_c}{A_0} \qquad G_i = G_c \frac{A_c}{A_i} \qquad \overline{G}^2 = \frac{G_i^2 + G_c^2}{2} = \frac{G_c^2}{2}\left[1 + \left(\frac{A_c}{A_i}\right)^2\right] \quad (6.7.2)$$

For simplification purposes, Abdollahian and Chexal [6.7.3] assumed no mass or heat transfer between the entrance and $L_i/D = 12$ and neglected the corresponding frictional pressure drop in this region. They defined the pressure losses in Eq. (6.4.11) as follows:

Entrance loss:

$$\Delta p_e = \frac{G_0^2}{2C_D^2 \rho_{L0}} = \frac{G_c^2}{2C_D^2 \rho_{L0}}\left(\frac{A_c}{A_0}\right)^2 \quad (6.7.3)$$

Acceleration loss due to evaporation:

$$\Delta p_{ae} = \overline{G}^2 x_c \left(\frac{1}{\rho_{GC}} - \frac{1}{\rho_{L0}}\right) = \frac{G_c^2}{2}\left[1 + \left(\frac{A_c}{A_i}\right)^2\right] x_c \left(\frac{1}{\rho_{GC}} - \frac{1}{\rho_{L0}}\right) \quad (6.7.4)$$

Acceleration loss due to area change:

$$\Delta p_{aa} = \frac{G_c^2}{2}\left[\frac{1}{\rho_{L0}}\left(1 - \frac{A_c^2}{A_i^2}\right) + \bar{x}\left(\frac{1}{\overline{\rho}_G} - \frac{1}{\rho_{L0}}\right)\left(1 - \frac{A_c^2}{A_i^2}\right)\right] \quad (6.7.5)$$

Frictional pressure drop:

$$\Delta p_f = f\left(\frac{L}{D} - 12\right)\frac{G_c^2}{4}\left(1 + \frac{A_c^2}{A_i^2}\right)\left[(1 - \bar{x})\frac{1}{\rho_{L0}} + \bar{x}\frac{1}{\overline{\rho}_G}\right] \quad (6.7.6)$$

In Eqs. (6.7.5) and (6.7.6) $\bar{x}$ represents the average quality between $L_i/D = 12$ and the exit and $\overline{\rho}_G$ the corresponding average gas density over the same crack span. C_D is the entrance loss coefficient and f the friction factor beyond $L_i/D = 12$.

Paul et al. utilized the same expression for Δp_e, with the exception that they used a value of $C_D = 0.91$ for very tight cracks, (i.e., cracks with crack opening displacements of less than 0.15 mm). For larger cracks, a discharge coefficient between 0.62 and 0.95 was proposed, depending on how round the entrance edges were. Abdollahian and Chexal assumed that $C_D = 0.61$ for all cracks. There are other differences between References 6.4.6 and 6.7.3, including recognition by Paul et al. of the liquid friction pressure drop as well as changes in water density and therefore of corresponding acceleration losses between locations 0 and i. The most significant change is the addition by Paul et al. of a pressure drop term Δp_k to account for losses due to bends and protrusions within the crack, such that

$$\Delta p_k = k_B \frac{\overline{G}_r^2}{2}\left[\bar{x}\frac{1}{\overline{\rho}_G} + (1 - \bar{x})\frac{1}{\overline{\rho}_L}\right] \quad (6.7.7)$$

where k_B is the bend loss coefficient, $\overline{\rho}_L$ the average liquid density, and $\overline{G}_r$ the average mass flux over the entire crack length, or $\overline{G}_r \approx (G_c/2)(1 + A_c^2/A_0^2)$. Another change made by Paul et al. is to recognize that the friction factor f may be higher for flow through cracks. Based on the work of John et al. [6.7.4], they proposed

$$f = \left(C_1 \log\frac{D}{E} + C_2\right)^{-2} \quad (6.7.8)$$

$$C_1 = \begin{cases} 2.00 \\ 3.39 \end{cases} \quad \text{and} \quad C_2 = \begin{cases} 1.14 & \text{for } D/E > 100 \\ -0.866 & \text{for } D/E < 100 \end{cases}$$

where E is the crack face roughness. Equation (6.7.7) and the friction factor of Eq. (6.7.8) for $D/E < 100$ will significantly increase the pressure drop through the crack and reduce the critical flow rate.

The critical flow, crack-opening-area, and pressure drop models have been incorporated into a single computer code called SQUIRT (seepage quantification of upsets in reactor tubes). SQUIRT has been used to predict critical flow in artificially produced slits where the flow geometry is known, and in actual cracks. SQUIRT is compared with the experimental slit data of Amos and Schrock [6.4.7] in Figure 6.7.3, which shows that the predicted flow rate is about half the measured value. This result does not agree with the Amos and Schrock finding that Henry's model overpredicted these results by an order of magnitude. It is believed that this anomaly comes from the different treatment used for the pressure losses in Eq. (6.4.11), which again show their relative importance. Figure 6.7.4 shows the comparison of SQUIRT with the data of Collier et al. [6.7.5] for intergranular stress corrosion cracks with a COD of 50 μm. While, as expected, data for real cracks exhibit increased scatter, SQUIRT does a better job but still tends to underpredict the leakage flow rates. For cracks with a COD of 20 μm, the calculated flow is much larger than the measured flow rate. Collier et al. have attributed this deviation to the

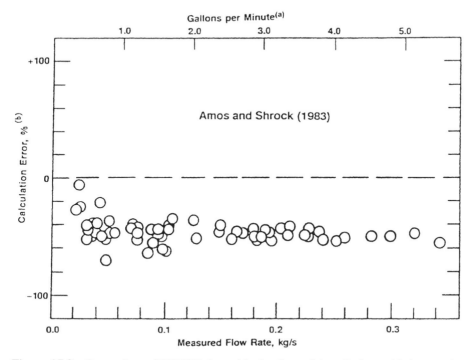

Figure 6.7.3 Comparison of SQUIRT thermal hydraulic model predictions with the experimental data of Amos and Schrock [6.4.7] for flows through artificially produced slits: (*a*) gallons per minute for water at 1 atm and 20°C; (*b*) calculation error, % = predicted minus measured flow rate divided by measured flow rate times 100. (From Ref. 6.4.6.)

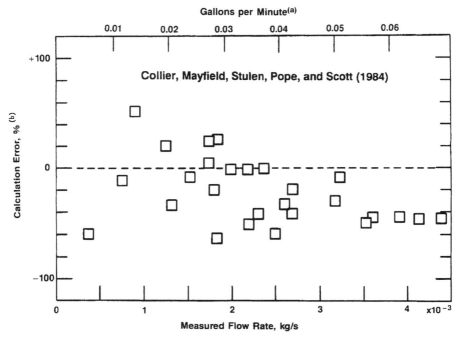

Figure 6.7.4 Comparison of SQUIRT thermal hydraulic model predictions with the experimental data of Collier et al. [6.7.5] for cracks with a COD of 50 μm: (*a*) gallons per minute for water at 1 atm and 20°C; (*b*) calculation error, % = predicted minus measured flow rate divided by measured flow rate times 100. (From Ref. 6.4.6.)

possibility that cracks with small COD could have become partially plugged by particles. This possibility tends to be supported in part by improved agreement between tests and SQUIRT predictions as the COD was increased.

Finally, Paul et al. compared the computer code SQUIRT to leak rates obtained for a circumferential fatigue crack in a carbon steel pipe girth weld. The comparison is reproduced in Figure 6.7.5 as a function of the applied level of bending moment and water pressure. Caution is in order about the favorable comparison exhibited in Figure 6.7.5 because SQUIRT agreed with the data only after a crack pathway loss coefficient k_B of six velocity heads per millimeter of crack wall thickness was introduced into the model. If this loss coefficient had been set to zero, the predicted flow rates would have been high by a factor of 2 to 3, as was the case for the pretest conditions.

In summary, this specific application of critical flow rate through cracks reinforces the previous conclusion that empirical adjustments are necessary where the flow geometry is changed radically. They also reemphasize the importance of entrance, acceleration, exit, and other pressure losses upon the critical flow rates. Finally, they show that geometric characteristics of cracks are quite important in determining leakage flow through actual cracks, and they are not readily predictable.

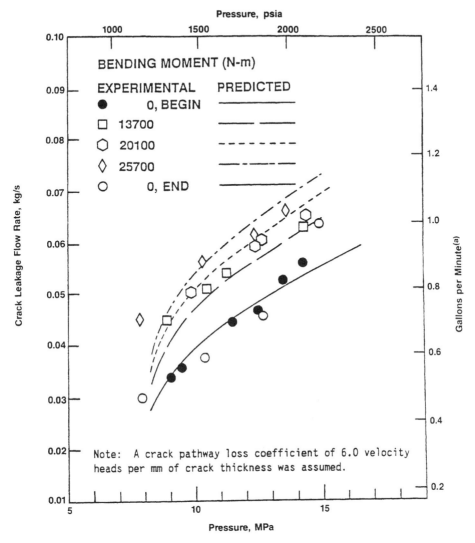

Figure 6.7.5 Comparison of SQUIRT predicted leakage flow rate calculations with the experimental data obtained for the fatigue-generated crack in a girth weld. (From Ref. 6.4.6.)

6.7.2 Critical Flow through a Small Break in a Horizontal Pipe

This second application example deals with a loss-of-coolant accident produced by a small break in a horizontal coolant pipe. As shown in Figure 6.7.6, the coolant escape mechanisms will depend on the location of the break once the flow becomes stratified. If the break is located at the top of the pipe, vapor or gas will be the dominant fluid leaving through the break, as shown in Figure 6.7.6a. Some liquid will be entrained with the vapor or gas due to the vapor or gas acceleration (Bernoulli

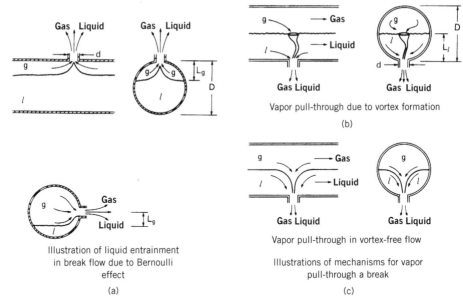

Figure 6.7.6 Mechanism for liquid entrainment and vapor pull-through in break flow. (From Ref. 6.7.6.)

effect) close to the break. If the break is located at the midplane of the pipe, as shown in Figure 6.7.6a, the same basic mechanism will prevail once the stratified liquid level falls below the break. In the case of a break located below the interface, vapor or gas reaches the break because of vortex formation or because it is pulled through a vortex free flow, as illustrated in Figure 6.7.6b. Reimann and Khan [6.7.6] have studied bottom-located breaks and have reported on the onset of gas pull-through and the total break mass flux and quality of air–water mixtures at the break entrance. They have observed two principal flow configurations: (1) the case that is vortex free and where only a portion of the liquid flows through the break and the rest continues to flow in the main pipe; and (2) the case that is subject to vortex flow and where there is no liquid flow perpendicular to the break axis. In the latter case, the flow into the break is symmetrical or favors the upstream side of the pipe, and it results in vortex flow as the stratified liquid level is reduced. The onset of gas pull-through for these two geometries as obtained by Reimann and Khan is shown in Figure 6.7.7, with the vortex-susceptible geometry plotted in Figure 6.7.7b and the vortex-free geometry illustrated in Figure 6.7.7c.

The vortex-free onset of gas pull-through in Figure 6.7.7a is correlated by the relation proposed by Lubin and Hurwitz [6.7.7]:

$$\mathrm{Fr}\left(\frac{\rho_L}{\rho_L - \rho_G}\right)^{0.5} = 3.25\left(\frac{h_b}{d}\right)^{2.5} \tag{6.7.9}$$

6.7 APPLICATIONS OF CRITICAL FLOW MODELS

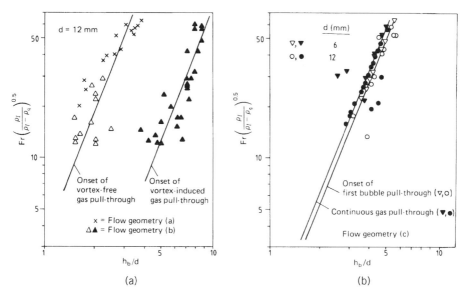

Figure 6.7.7 Onset of gas pull-through for different flow geometries. (From Ref. 6.7.6.)

For the case of vortex-induced gas pull-through in Figure 6.7.7a, the data are best fit by

$$\mathrm{Fr}\left(\frac{\rho_L}{\rho_L - \rho_G}\right)^{0.5} = 0.2\left(\frac{h_b}{d}\right)^{2.5} \quad (6.7.10)$$

while the beginning of first vortex-free gas pull-through in Figure 6.7.7b is fitted by

$$\mathrm{Fr}\left(\frac{\rho_L}{\rho_L - \rho_G}\right)^{0.5} = 0.94\left(\frac{h_b}{d}\right)^{2.5} \quad (6.7.11)$$

and the onset of continuous pull-through is described by increasing the constant 0.94 in Eq. (6.7.11) to 1.1. In Eqs. (6.7.9) to (6.7.11), d is the break diameter, h_b the interface water level where pull-through starts, and Fr is the Froudenumber:

$$\mathrm{Fr} = \frac{\bar{u}_{LB}}{\sqrt{gd}} \quad (6.7.12)$$

with $\bar{u}_{LB}$ being the mean liquid velocity into the break and g the gravitational constant.

In the case of two-phase flow, the break liquid mass flux, G_{LB}, will approach a constant value with the onset of gas pull-through, which means that the liquid flow is

350 CRITICAL FLOW

choked. Figure 6.7.8 shows the air quality $\bar{x}_B$ and the total mass fluxes G_B at the entrance of the break as a function of the dimensionless interface level h/D, where D is the pipe diameter for a test pressure of 0.5 MPa and a constant pressure drop of $\Delta p_{1-3} = 0.315$ MPa. The figure also plots the onset of gas pull-through given by Lubin and Hurwitz and the predictions obtained from models proposed in RELAP4 [6.6.2] and RELAP5 [3.5.14]. In the case of RELAP4, the simplified assumption is made that the gas volume fraction $\bar{\alpha}_B$ is equal to the stratified gas volume fraction $\bar{\alpha}$. For horizontal stratified flow, the RELAP5 break gas volume fraction is given by

$$\bar{\alpha}_B = \bar{\alpha}\left(\frac{\bar{u}_G}{\bar{u}_{GL}}\right)^{0.5} \tag{6.7.13}$$

$$\bar{u}_{GL} = \frac{1}{4\sqrt{2}}(1 - \cos\theta)\left(D\frac{\rho_L - \rho_G}{\rho_G}\frac{2\theta - \sin 2\theta}{\sin\theta}\right)^{0.5}$$

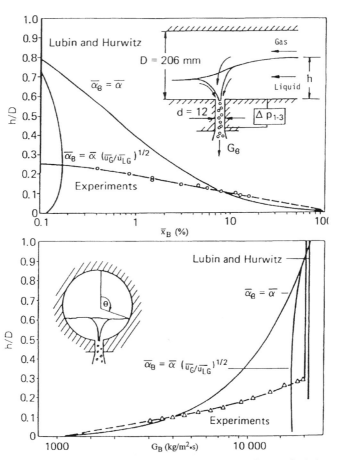

Figure 6.7.8 Total mass flux and quality in the small break for vertical downward flow (air–water flow: $p_1 = 0.5$ MPa and $\Delta p_{1-3} = 0.315$ MPa). (From Ref. 6.7.6.)

where $\bar{u}_G$ is the gas velocity in the stratified flow pattern taken at 1 m/s and θ is the angle as defined in Figure 6.7.8. To calculate the quality at the break, $\bar{x}_B$, the flow was assumed to be homogeneous.

Figure 6.7.8 shows large differences between the test results and the two RELAP models for a bottom break in a horizontal pipe with a stratified flow pattern. Reimann and Khan were able to get all their air water to fall on a single curve:

$$\frac{G_B}{G_{BB}} = \left(\frac{h}{h_B}\right)^{1.62} \tag{6.7.14}$$

where G_{BB} is the flow through the break at the onset of gas pull-through and h_B is the liquid height at that same time, which is obtained from Eq. (6.7.9). However, they cautioned about extrapolating their air–water tests to other fluids and much higher pressures.

In terms of this second specific application, we find again that testing is the only means currently available to get an accurate answer and that the answer will be sensitive to the break location, as illustrated in Figure 6.7.6. If testing is to be carried out, consideration must be given to appropriate scaling as discussed by Zuber [2.6.1]. It is interesting to note that in an uncertainty analysis of the reactor water vessel inventory during a small break LOCA with RELAP5/MOD3 [6.6.4], no special emphasis was put on break location and whether the liquid was being entrained or the vapor was being pulled through. Equation (6.6.1) was employed to calculate the critical flow, and the uncertainty of that correlation was deduced from the Marviken tests, which are geometrically very different from the midplane 2.125-in. break presumed in the 28-in. horizontal pipe. Even though the loss of inventory is most likely greater in the analysis than for the presumed break configuration, the impact of that break simplification is not too important practically because it was found that the reactor vessel water level is much more sensitive to operator action than such key parameters as critical flow rate.

6.8 SUMMARY

1. Critical flow occurs when no information can propagate backward against the flow upstream from the critical location. It is often referred to as choked flow because it reaches a maximum value at the location of minimum area along the flow path.

2. While there exists a basic understanding of the mechanism of choked two-phase flow, its accurate prediction is difficult because of the complex changes taking place at the gas–liquid interfaces. Another reason is the inability to describe the mass, momentum, and energy exchanges at those interfaces with changes in flow geometry, particularly under conditions of nonthermal equilibrium.

3. A large number of simplified models have been developed to calculate critical flow rates without considering the liquid–gas interfaces. They vary the prevailing thermodynamic flow process (e.g., isentropic) and the relative gas velocity with

respect to the liquid. In the case of thermal nonequilibrium conditions, the simplified models require additional empirical adjustments to recognize the actual underpressure or overtemperature necessary to initiate liquid evaporation and its relaxation downstream. The differences among the simplified models are not overwhelming at high pressures where the ratio of liquid to gas density is reduced.

4. The available critical flow database is very large and exhibits considerable scatter (about ±100%). Some of this scatter can be traced to variations in flow geometry. For example, critical flow rate decreases with increased length/diameter ratio of the test section. Similarly, its nonthermal equilibrium behavior is influenced by the location of initial vapor flashing and its dependence on depressurization rate and flow velocity.

5. The simplified models fare reasonably against critical flow data recognizing its inherent scatter; they fare even better if selectivity is applied to the available test results and if some of the pressure loss coefficients (entry, acceleration, friction, and other) are inappropriately adjusted to fit the experiments. The homogeneous (equal gas and liquid velocity) simplified model is used most frequently because of its simplicity, and it is especially favored for nonthermal equilibrium conditions. As stated by Wallis, "the homogeneous equilibrium model is not a bad way of predicting the critical mass flux in long pipes." With empirical adjustments, the homogeneous model can predict critical flow rates to cover other geometries and nonthermal equilibrium conditions.

6. It is important that the various pressure loss coefficients (entrance, acceleration, friction bends, etc.) employed in calculating the critical flow be based on physical grounds, preferably on measurements with single-phase gas and liquid flow. They should not be used as adjustable constants to match the critical flow data. Furthermore, if the simplified model relies on gas slip, care should be exercised to avoid the inconsistency of calculating the pressure losses from homogeneous conditions unless they can be justified experimentally. Finally, the simplified models that utilize all three conservation equations (mass, momentum, and energy) are the most rigorous and should be given preference.

7. The impact of gas–liquid interfaces upon critical flow has been considered and predictions have been made for flow pattern models and for two-fluid models by employing the method of characteristics. To date, the calculated results from such techniques are not superior to the results derived from simplified models because the more complex models also need empirical adjustments to cover changes in flow geometry and nonequilibrium thermal conditions. It is suspected that the closure laws incorporated in two-fluid-system computer codes may not be applicable or may need refinement to describe the special circumstances at the gas–liquid interfaces during critical flow. Also, the numerical solutions of the characteristic equations become more difficult and take increased computing time for most system computer codes. For all those reasons, many system computer codes include simplified model options and even fits to those models to calculate critical flow.

8. Another important factor in predicting critical flow is the uncertainty associated with the break flow geometry and its characteristics. The cases of leakage through

cracks or through a small break during stratified flow in a horizontal pipe were used to illustrate this point and to show that in such practical applications, those geometric uncertainties may overshadow the accuracy of the critical flow model. Prototypical two-phase-flow tests still appear to be the best way to get understanding and accurate critical flow predictions for new practical circumstances.

9. A complete and accurate theory of two-phase critical flow is not yet available. It may come about from improved experimental measurements and treatments of the gas–liquid interfaces. It will take considerable work and time to achieve that goal. In the meantime, simplified models with empirically determined adjustments can provide adequate estimates of critical mass fluxes.

REFERENCES

6.1.1 Saha, P., *A Review of Two-Phase Steam-Water Critical Flow Models with Emphasis on Thermal Non-equilibrium*, BNL-NUREG-50907, September 1978.

6.1.2 Wallis, G. B., "Critical Two-Phase Flow," *Int. J. Multiphase Flow*, **6**, 97–112, 1980.

6.1.3 Abdollahian, D., et al., *Critical Flow Data and Analysis*, EPRI NP2192, 1982.

6.1.4 Brittain, I., et al., *Critical Flow Modeling in Nuclear Safety*, Nuclear Energy Agency, OECD, June 1982.

6.1.5 Hall, D. G., *An Inventory of the Two-Phase Critical Flow Experimental Data Base*, EGG-CAAP-5140, June 1980.

6.1.6 Elias, E., and Lellouche, G. S., "Two-Phase Critical Flow," *Int. J. Multiphase Flow*, **20**, 91–168, 1994.

6.2.1 Shapiro, A. H., *The Dynamics and Thermodynamics of Compressible Fluid Flow*, Ronald Press, New York, 1953.

6.2.2 Binder, R. C., *Advanced Fluid Mechanics*, Prentice Hall, Upper Saddle River, N.J., 1958.

6.2.3 Sallet, D. W., "Thermal Hydraulics of Valves for Nuclear Applications," *Nucl. Sci. Eng.*, **88**, 220–244, 1984.

6.2.4 Ducoffe, et al., *Trans. ASME*, 1349–1357, 1958.

6.2.5 Schiller, W., "Überkritische Entspannung kompressibler Flüssigkeiten," *Forsch. Geb. Ingenieurwes.*, **4**, 128, 1933.

6.2.6 Perry, J. A., *Trans. ASME*, **71**, 757, October 1949.

6.2.7 Bean, H. S. et al., "Discharge Coefficients of Orifice," *Natl. Bur. Stand., J. Res.*, **2**, 561, 1929.

6.2.8 Karditsas, P., and Sallet, W., University of Maryland experimental results (unpublished).

6.3.1 Moody, F. J., "Maximum Discharge Rate of Liquid Vapor Mixtures from Vessels," *Proceedings of the ASME Symposium oin Non-equilibrium Two-Phase Flows*, pp. 27–36, 1975.

6.3.2 Zaloudek, F. R., *The Low Pressure Critical Discharge of Steam–Water Mixtures from Pipes*, Hanford Atomic Products Operation Report HW-68934 Rev., 1961.

6.3.3 Sajben, M., *Trans. ASME*, Ser. D, **83**(4), 619–631, 1961.

6.3.4 Moody, F. J., "Maximum Two-Phase Vessel Blowdown from Pipes," *J. Heat Transfer*, **87**, 285–295, 1966.

6.3.5 McFadden, J. H., et al., *RETRAN-02: A Program for Transient Thermal Hydraulic Analysis of Complex Fluid Flow Systems*, EPRI NP-1850-CCMA, Vol. 1, Rev. 2, 1984.

6.3.6 Fauske, H. K., *Contribution to the Theory of Two-Phase, One Component Critical Flow*, Argonne National Laboratory Report ANL-6633, 1962.

6.3.7 Levy, S., "Prediction of Two-Phase Critical Flow Rate," *J. Heat Transfer*, **82**, 113–124, 1960.

6.3.8 Cruver, J. E., "Metastable Critical Flow of Steam–Water Mixtures," Ph.D. thesis, Department of Chemical Engineering, University of Washington, 1963.

6.3.9 Kroeger, P. G., "Application of a Non-equilibrium Drift Flux Model to Two-Phase Blowdown Experiments," paper presented at the CSNI Specialists' Meeting on Transient Two-Phase Flow, Toronto, Ontario, Canada, 1978.

6.3.10 Wallis, G. B., and Richter, H. J., "An Isentropic Streamtube Model for Flashing Two-Phase Vapor–Liquid Flow," *J. Heat Transfer*, **100**, 595–600, 1978.

6.4.1 Starkman, E. S., et al., "Expansion of a Very Low Quality Two-Phase Fluid through a Convergent–Divergent Nozzle," *J. Basic Eng.*, **86**, 247–256, 1964.

6.4.2 Henry, R. E., and Fauske, H. K., "The Two-Phase Critical Heat Flow of One-Component Mixtures in Nozzles, Orifices and Short Tubes," *J. Heat Transfer*, **93**, 179–187, 1971.

6.4.3 Henry, R. E., "The Two-Phase Critical Discharge of Initially Saturated or Subcooled Liquid," *Nucl. Sci. Eng.*, **41**, 336–342, 1970.

6.4.4 Henry, R. E., and Fauske, H. K., "Two-Phase Flow at Low Qualities, I: Experimental; II. Analysis," *Nucl. Sci. Eng.*, **41**, 79–98, 1970.

6.4.5 *The Marviken Full-Scale Critical Flow Tests; Results of Tests 1-27*, MXC 201-227.

6.4.6 Paul, D. D., et al., *Evaluation and Refinement of Leak-Rate Estimation Models*, NUREG/CR-5128, Rev. 1, June 1994.

6.4.7 Amos, C. N., and Schrock, V. E., "Two-Phase Critical Flow in Slots," *Nucl. Sci. Eng.*, **88**, 261–274, 1984.

6.4.8 Reocreux, M., "Contribution l'étude der debits critiques éconlement diphasique canvapeur," Ph.D. thesis, L'Université Scientifique et médicale de Grenoble, 1974.

6.4.9 Levy, S., and Abdollahian, D., "Homogeneous Non-equilibrium Critical Flow Model," *Int. J. Heat Mass Transfer*, **25**(6), 759–770, 1982.

6.4.10 Alamgir, M. D., and Lienhard, J. H., "Correlation of Pressure Undershoot during Hot Water Depressurization," *J. Heat Transfer*, **103**, 522–525, 1981.

6.4.11 Edwards, A. R., *Conduction Controlled Flashing of a Fluid and the Prediction of Critical Flow Retention in a One-Dimensional System*, UKAEA, Health and Safety Branch, R147, 1968.

6.4.12 Ardron, K. H., and Ackerman, M. C., "Studies of the Critical Flow of Subcooled Water in a Pipe," paper presented at the CSNI Specialists' Meeting on Transient Two-Phase Flow, Paris, 1978.

6.5.1 Richter, H. J., "Separated Two-Phase Flow Model: Application to Critical Two-Phase Flow," *Int. J. Multiphase Flow*, **9**(5), 511–530, 1983.

6.5.2 Sozzi, G. L., and Sutherland, W. A., *Critical Flow for Saturated and Subcooled Water at High Pressure*, GE Report NEDO-13418, 1975.

6.6.1 Micaelli, J. C., et al., "Best Estimate Thermal Hydraulic Code for Safety Reactor Studies, Last Development," paper presented at the International ENS/ANS Conference on Thermal Reactor Safety, Avignon, France, October 1988.

6.6.2 *RELAP4/MOD5: A Computer Program for Transient Thermal-Hydraulic Analysis of Nuclear Reactors and Related Systems, User's Manual,* CR-NUREG 1335, September 1976.

6.6.3 Jones, O. C., *Flashing Inception in Flowing Liquids*, BNL-NUREG-51221, April 1980.

6.6.4 Ortiz, M. G., and Ghan, L. S., *Uncertainty Analysis of Minimum Vessel Liquid Inventory during a Small-Break LOCA in a B&W Plant: An Application of the CSAU Methodology Using the RELAP5/MOD3 Computer Code*, NUREG/CR-5818, 1991.

6.7.1 Kumar, V., et al., *An Engineering Approach to Elastic–Plastic Fracture Analysis*, EPRI Report NP-1931, 1981.

6.7.2 Kumar, V., et al., *Advances in Elastic–Plastic Fracture Analysis*, EPRI Report NP-3067, 1984.

6.7.3 Abdollahian, D., and Chexal, B., *Calculation of Leak Rates through Cracks in Pipes and Tubes*, EPRI Report NP-3395, 1983.

6.7.4 John, E., et al., "Critical Two-Phase Flow through Slits," *J. Multiphase Flow*, 1987.

6.7.5 Collier, R. P., et al., *Two-Phase Flow through Intergranular Stress Corrosion Cracks and Resulting Acoustic Emission*, EPRI NP-3840-LD, 1984.

6.7.6 Reimann, J., and Khan, M., "Flow Through a Small Break at the Bottom of a Large Pipe with Stratified Flow," *Nucl. Sci. Eng.*, **88**, 297–310, 1984.

6.7.7 Lubin, B., and Hurwitz, M., "Vapor Pull-through at a Tank Drain with and without Dielectrophoretic Buffing," *Proceedings of the Conference on Long Term Cryopropellant Storage in Space*, NASA Marshall Space Center, Huntsville, Ala., p. 173, 1966.

CHAPTER 7

THE FUTURE OF TWO-PHASE FLOW IN COMPLEX SYSTEMS

7.1 THE PLAYERS

Before discussing the future of two-phase flow in complex systems, we need first to recognize the variety of players involved in molding that future. They include:

- The *practitioners*, who design, construct, and operate complex systems. They have the lead role because they determine the characteristics and features of the complex system based on their experience with previous models and their efforts to satisfy newly required improvements in performance, reliability, and safety. They are not too particular about what test information and computer codes are utilized as long as they do satisfy the needs of the customers and regulators and as long as the results can be obtained within the prescribed costs and schedules. When information is lacking and cannot be developed within their targets, the practitioners reluctantly will provide margins to compensate for that lack of knowledge rather than yield on their schedule or cost objectives.
- The *fundamentalists* who insist on a good understanding of the phenomena involved and on good test data and adequate models to fit them. They are driven by wanting to capture all the "physics" of two-phase flow even at the expense of costs and schedules if necessary. There are two basic types of fundamentalists: the experimenters, who perform the tests and develop specialized instrumentation to get the necessary data; and the analysts, who generate the models from fundamental principles and show that they can fit the data. In some fortunate cases, the fundamentalists can be adept at both testing and analysis, and those fundamentalists should be nurtured as much as possible. The fundamentalists are not always willing to recognize how difficult some

aspects of two-phase flow can be to predict or test for. In that sense, they are optimistic if not idealistic, as to what can be accomplished. There is clearly some tension between the practitioners and the fundamentalists, but in fact, it is a necessary and healthy tension if it is managed properly.

- The *pragmatists*, who play the go-between role between the practitioners and the fundamentalists. They will take the best information available and develop empirical or semiempirical models partly supported by analysis. They work on identifying the most important phenomena and how they interact in the complex system. In that sense, they try to optimize the application of available resources and work to get the fundamentalists and the practitioners to buy into or accept their approach as a necessary compromise.

- The *customers/regulators*, who need to be convinced that enough has been done to support the bases for the two-phase-flow complex system. The regulators' primary focus is on safety, while the customers' interests include safety as well as performance and costs of the complex system. They both require good reports describing the tests and analyses performed and a convincing story that they are adequate. Currently, the customers/regulators have a preference for realistic system computer codes that can be used to train the operators and to evaluate the uncertainties associated with the performance predictions and safety analyses.

- The *system computer code developers*, who produce a computer simulation that numerically can perform satisfactorily even if they have to compromise some physics to meet that objective. Documentation of the computer code and its validation and verification against available test and complex systems performance data can be a very large task if carried under appropriate control and quality assurance requirements. This task is not a favorite of the computer code developers.

All the foregoing players have a common objective—to assure good performance of the complex system—but there are inherent differences and biases in their approach and desires. Teamwork among all the players would produce the best of both worlds, and more and more such efforts are being sought, particularly in the field of nuclear technology. For example, in 1989, the Organization for Economic Co-operation and Developments (OECD) Nuclear Energy Agency (NEA) Committee on the Safety of Nuclear Installations (CSNI) carried out a review of the current state of knowledge of the thermohydraulics of emergency core cooling in light-water reactors [7.1.1]. A group of experts identified the relevant phenomena, the tests available to assess the phenomenology, and the system computer codes to predict them. Future research activities were discussed as well as the question of how good the system computer codes need to be for each particular application. More recently, on November 5–8, 1996, an OECD/CSNI thermal hydraulic workshop [7.1.2] was conducted. The meeting dealt with (1) current and prospective plans for thermal-hydraulic codes development, (2) current and anticipated uses of thermal-hydraulic codes, (3) advances in modeling of thermal-hydraulic phenomena and associated additional experimental needs, (4) numerical methods in multiphase flows, and (5) programming language, code architecture, and user interface. An important feature of this gathering of international two-phase flow experts was the breakout into specialized sessions to identify and prioritize future needs. The perspectives presented by Städtke and Ishii on the future of two-phase flow modeling in nuclear complex systems are especially noteworthy [7.1.3, 7.1.4].

From February 27 to March 1, 1997, the Institute for Multifluid Science and Technology held its founding meeting at Santa Barbara, California. Their proceedings [7.1.5] deal with the broad subject of fostering scientific understanding of multifluid systems. This initial meeting provided a window to the multiphase flow complex systems found in the chemical, oil, electronic, and nuclear industries as well as to the generation, handling, and use of a variety of particulate materials. All three of the meetings described above represent genuine efforts to deal with multiphase flow in complex systems and how to bring all the various players together to optimize progress.

Recommendation: Meetings to discuss the problems of two-phase flow in complex systems and to improve their understanding and predictions are most valuable, especially if they are conducted on an international scale. It is desirable, if not essential, to augment the participation of designers, constructors, operators, and owners of the complex systems at such meetings.

7.2 THE COMPLEX SYSTEMS

As discussed in Chapters 1 and 2, a complex system is made up of several subsystems, each containing a number of components. Many different phenomena occur in the complex system and their type and importance can vary over time and with the specific event or scenario of interest. As mentioned in Chapter 1, the development of an integrated test and analysis program for the complex system needs to be carried out early in the investigation process to focus on the correct issues and how to resolve them. The key elements of an integrated test and analysis program are:

1. To create a phenomena identification and ranking table (PIRT). By its nature, the PIRT identifies the available and additional separate effects and components tests needed. The scaling uncertainties associated with any new test tend to be reduced by employing the fluids and operating conditions utilized in the prototype complex system. If the tests are performed at reduced scale, the scaling uncertainties can be lowered further by using test facilities at two different scales. When the separate effects phenomenon is very important to the complex system performance, there is merit to carrying out full-scale tests because they yield data and correlations with the best overall accuracy. For example, this is the case for the phenomenon of dryout by steam blanketing of the heat-producing surface in light-water nuclear reactors. The allowable nuclear reactor power level increases with the use of full-

scale tests to predict this phenomenon, which is often referred to as *critical heat flux* (CHF). The prediction uncertainties also decrease.

2. To determine whether there are new steady-state or transient interactions between phenomena, components, and subsystems in the complex system under consideration. If, in fact, they exist, new integral test facilities must be built to investigate them. Such integral test facilities are usually at reduced scale because the costs are too high for full-scale experiments. Also, because of our current inability to scale several of the governing processes from first principles, a report describing the selected scale and scaling approach can be very valuable if subjected to peer review before proceeding with the program. Here again, the use in the integral test facilities of the fluids and operating conditions found in the prototype complex system will help reduce the scaling uncertainties, particularly if the test facility and prototype operate with the same time scale. However, even under such ideal conditions, there remains a scaling uncertainly with reduced scale facilities that needs to be considered and estimated.

3. To select and employ a mature complex system computer code that is capable of predicting a wide spectrum of scenarios in complex systems of the type being assessed. The desired characteristics of the chosen computer system code include good documentation of the code models; favorable prediction of the previous separate effects, components, and integral tests; stable numerical methodology; full reporting of code verification and validation efforts; and acceptable predictions of the new separate effects, components, and integral system test data. It is important to justify the presumption that the system computer code provides an adequate extrapolation from the test facilities to the full-size complex system or to define the scaling and other uncertainties to be taken into account to justify the extrapolation.

Presently, preferred system computer codes are very similar in their formulation of the conservation equations, their spatial- and time-averaging procedures, their numerical techniques, and their use of the two-fluid model and empirical flow regime–dependent closure relations to describe the interfacial transfer of mass, momentum, and energy. For that reason, most system computer codes suffer from the same shortcomings: inaccurate flow pattern maps, idealized representations of the interfacial transfer areas, use of rather large nodes to simulate the prototype system geometry, and departure from test results for nonthermal equilibrium conditions, high pressure gradients, and local geometric variations. The available system computer codes also tend to produce similar and acceptable predictions of the performance of the prototype complex system because they have all been forced to match an enormous amount of results from the same variety and large number of available tests.

In recent years, system computer codes have been utilized to produce best estimate predictions and to estimate their uncertainty. The system computer codes are being praised by the practitioners because they are the only and best available tools to simulate the performance of complex systems. By contrast, the fundamentalists can find many reasons to criticize them and to believe that because they have been fine-tuned to tests, extrapolation beyond those tests is suspicious.

4. Quantitative assessments of two-phase flow in complex systems can be an extremely large and difficult task. The assessments reached a satisfactory level for nuclear power plants only after 30 years of intensive testing and improving the system computer codes. This work was carried out in many different countries. However, there still are shortcomings in our understanding, modeling, and evaluation of performance uncertainties during light-water reactor transients and accidents. Future improvements will require significant resources and may be difficult to justify in terms of degree of improvement or benefits, time to develop them, and their costs.

Recommendations:
- The early formulation of an integral test and analysis program is essential to implementing significant changes in the performance and design of two-phase flow complex systems.
- Similarly, a scaling study and report is needed before proceeding with the construction of new reduced-scale facilities and, particularly, new reduced-scale integral system facilities.
- The system computer code with the best chance of adequately simulating the complex system of interest needs to be selected and controlled (i.e., frozen) before being applied to the newly acquired test data.
- It should be recognized that the tests and the computer code results can only approximate the prototype behavior, and provisions need to be made in the design for uncertainties in the code predictions of the prototype complex system.

7.3 KEY ISSUES IN TWO-PHASE FLOW

The major difficulty in describing two-phase flow is the existence of many different gas–liquid interfaces, which vary in geometry across and along the flow directions. The interfacial geometry not only changes with the flow pattern but also within a specified flow pattern. For example, in the case of bubble flow, the bubbles have different shapes and sizes. They can coalesce or break up. Such behavior is dependent on the liquid flow and its degree of turbulence. In most system codes, bubble flow is idealized by assuming that all the bubbles are spherical and of the same size. Also, the bubbles are all lumped together in order to represent them by a spatial- and time-average cross-section volume fraction for the gas flow. This approach does not allow any representation of the behavior of individual bubbles and of their size, shape variation, and distribution. Similar arguments can be made for dispersed liquid drops during mist flow. The situation gets worse for an annular flow pattern where waves of different size and frequency occur at the liquid-film and gas-core interfaces. The gas core is also known to contain liquid drops of different size and shape. The problem is worst when one has to deal with unsteady and intermittent flows such as slug or plug flow.

At the present time, it is customary to employ idealized interfacial geometries in order to express the interfacial area in terms of the cross-sectional average gas volume fraction. Once the interfacial area is known, the interface drag or friction loss (as well as heat transfer) between the two phases can be calculated if the interfacial drag shear or drag coefficient or heat transfer coefficient are known. They are obtained from empirical correlations of test data with significant scatter and are referred to as closure or constitutive equations. It is important to note that these closure relations are derived from steady-state tests in the fully developed flow conditions and are being applied to unsteady and undeveloped flow conditions and to different geometries than those tested.

Another key issue is that the average gas volume fraction is dependent on the flow pattern and the flow patterns must be specified. Flow pattern maps being employed are derived again from steady-state and fully developed flow conditions. The data do not cover all the flow directions and geometry variations existing in a complex system. Further, most system computer codes employ interpolation procedures to describe transitions from one flow pattern to another. These interpolations are dictated by numerical stability rather than being correct physically.

There are several other important issues in two-phase flow which are not fully understood or accounted for. For example, turbulent momentum and heat diffusion or viscosity and conduction processes across the flow direction are excluded or are poorly defined, yet they determine the velocity and temperature profiles in the fluid cross-sectional area. In the case of two-phase flow, there is an interface-induced turbulence over and above the local shear and heat transfer–produced turbulence. This explains why to date there are no well-accepted or established two-phase flow turbulence models. Another example is the occurrence of nonthermal equilibrium conditions; its representation up to now has been dependent only on empirically based models. One final instance is the case where flow circumstances are three-dimensional in nature, as for countercurrent flow in nuclear reactor downcomer (see Chapter 5) or small-break critical flow in horizontal pipes (see Chapter 6). While there are three-dimensional modules in some system computer codes, they require empirical adjustments because of their inadequate and gross mixing description among adjacent nodes.

The preceding comments are quite damning and may raise considerable doubts about the accuracy of system computer codes. In fact, this is not the case because the preceding issues relate primarily to *microscopic* two-phase-flow details, while the system computer codes focus on *macroscopic* representations of the flow behavior. Rather coarse nodalizations are employed to describe the complex system, and the microscopic details of the flow cannot be considered. A reduction in the nodalization size does not necessarily improve the predictions because the numerical diffusion increases as the number of computational cells rises. Furthermore, the costs of producing the results can jump sharply. It is not surprising, therefore, that decreasing the number of flow patterns or the use of a simplified drift flux model can often yield as good results as the two-fluid model in predicting the performance of complex systems. Städtke [7.1.3] pointed out that the "drift flux model prejudges the resulting slip velocities using highly empirical drift flux correlations which are valid only in case of steady state and fully developed flow which, strictly speaking, never exist in reactor transient situations. However, this does not mean that codes based on the two-fluid model give a priori better results." In other words, the many criticisms advanced by the fundamentalists about current system computer codes may not be relevant as long as low numerical diffusion methods are not available and cannot be incorporated successfully into complex system computer codes. Even then, it is not clear that the predictions would improve unless most of the key microscopic phenomena are included and enough microscopic test data are obtained to represent them accurately. In fact, as indicated in Chapter 3, there may exist good reasons to simplify the present system computer codes to match the scale over which they can describe two-phase-flow phenomena.

Recommendations:

- Our present representation of the physical gas–liquid interfaces, the transfer of mass, momentum, and energy at those interfaces, and the prevailing flow patterns leave much to be desired.
- These shortcomings are mostly of a microscopic nature, and for that reason their impact is minimized by the large nodalization employed in present system computer codes.
- The development of instrumentation to measure the microscopic details and the interfacial mechanisms at those interfaces is necessary to obtain data and develop correlations applicable at the microscopic level. This program will be useful only if it can be coupled with low-numerical-diffusion methodology in future system computer codes and if its development costs are not excessive. A pilot program with one flow pattern to assess costs and benefits may be worthwhile.
- There always will be a need for some empiricism in two-phase flow, and it is important from a cost-benefit viewpoint that the search for two-phase-flow microscopic understanding remains consistent with the remaining necessary empirical treatment (i.e., for nonthermal equilibrium conditions and local geometric changes).

7.4 KEY ISSUES IN COMPLEX SYSTEM COMPUTER CODES

The predominant use of two-fluid models in system computer codes and their large nodalization to overcome numerical diffusion have been touched upon already. It should be noted here that a system computer code provides only a computational means of calculating the performance of a complex system. It relies on geometric idealizations of the system and it employs a prescribed nodalization to subdivide it. The system computer code employs from three to six conservation equations. Three equations are used when the two-phase treatment neglects interfaces and presumes homogeneous flow, with the two phases having equal average velocity and being in thermal equilibrium. Six equations are used to account for different gas and liquid velocities and for all potential mass, momentum, and energy transfer

7.4 KEY ISSUES IN COMPLEX SYSTEM COMPUTER CODES

at the interfaces. In addition, the system computer codes contain a large number of other equations or inputs, which in some cases are derived empirically and are applied beyond their validity range. The numerical integration of all these equations can introduce errors due to numerical diffusion and truncations. Finally, the system code user can have a significant impact by making changes to the code, by tuning the code predictions to match the data, and by employing nodalization changes to get better agreement.

Let us first consider the number of conservation equations and the need to include interface transfer mechanisms. There are many pragmatists who believe that detailed representation of the interface is not necessary. An excellent case for eliminating interfaces is made by Chexal et al. in a nomogram for predicting the average gas volume fraction in several practical applications [7.4.1]. They point out that "the discrete nature of regime based models presents a basic structural problem in the hydraulic modeling of two-phase systems. Crossing a regime boundary may produce discontinuities in steady state thermal hydraulic models, and unmanageable computational instabilities in transient models. Smoothing techniques have been applied to regime based models to force reasonable behavior in both steady-state and transient modeling environment." Given those difficulties, Chexal and Lellouche [7.4.2] developed a gas volume fraction drift flux "model which eliminates the need for a collection of constitutive models (one model for each flow structure) for the interphase area and drag coefficients. It also eliminates the need to define the flow structure in order to select the various constitutive models for interphase friction. In addition, a single correlation eliminates the discontinuities and numerical disharmony that tend to occur when multiple constitutive models are connected." The Chexal–Lellouche model has been introduced in several system computer codes and yields results comparable to those of the two-fluid model for slow transients with thermal equilibrium conditions. Its successful and possibly superior application to steam-water critical flow data is covered in Chapter 6 and Reference 6.1.6. An inherent advantage of such global models is that they are fitted to a lot more test data with different flow directions and test geometries. By contrast, the constitutive equations in two-fluid system computer codes are developed for discrete flow regimes and are based on a much more limited data set and test conditions. The global models therefore appear to be superior when applied to all the available data, but they may not do as well for some fast transients or when they are extrapolated beyond the range of their validity.

To summarize, in this author's opinion, there are two-phase-flow problems where detailed treatment of the interfaces is necessary, but there are also problems where it is not required and the solutions can be simplified considerably by avoiding the treatment of interfaces. The difficulty is being astute enough to decide which way to proceed. As pointed out in Chapter 2, the simplest two-phase homogeneous flow is often used for scaling purposes because of its adaptiveness in producing one-dimensional grouping; similarly, in Chapter 6, it was shown that the homogeneous model dominates current predictions of two-phase critical flows. However, these circumstances are more default positions because our understanding of two-phase flow is inadequate. With better understanding, other more accurate models could emerge besides the homogeneous and drift flux approaches. In the meantime, use of the homogeneous or drift flux models is justifiable where it can be shown that they yield adequate answers by comparison to tests that scale the complex system with acceptable fidelity. Even then, they should be utilized when the range of parameters of the complex system are covered by the tests.

As discussed briefly in Chapter 3, the system computer codes use finite difference schemes to represent the derivation of the key variables. They employ staggered grid and donor cell techniques. Such methods introduce additional mixing due to numerical diffusion and are known, for example, to have difficulties in representing temperature, density, and water steam stratifications. Also, additional terms are added in some cases to the conservation equations to solve for the complex characteristics under critical flow conditions. Most complex system computer codes use the implicit continuous Eulerian (ICE) method, which requires time steps below the Courant limit. Such time-step limitations are not necessary when a fully implicit time integration technique is employed, as is the case for the system code CATHARE [6.6.1]. However, a significant numerical effort is necessary to perform the matrix inversion required by the implicit method, and this has forced CATHARE to return to semi-implicit solutions for many two-phase-flow problems.

The users have a strong role in determining the validity of the results derived from system computer codes. The influence of the users is readily demonstrated by comparing the results different users get on the same two-phase flow problem with the same code. As reported in Reference 7.4.3, the differences can be significant. Another issue is the choice of nodalization, which is often left to the users' discretion. Similarly, users have the right to pick among various options provided for constitutive equations or for empirical correlations. A final and possibly the most serious issue is the users' failure to realize that their mission is to compare the test data to the code results, not to adjust the code and its predicted values to fit the data.

7.5 THE FUTURE OF TWO-PHASE FLOW IN COMPLEX SYSTEMS

There are many opportunities to improve two-phase-flow understanding, behavior, and prediction in complex systems. Some of the changes are of immediate benefit and do not require research and development. For example, they include:

- Standards for establishing plant models and for selecting nodalization
- Preferred choices among empirical correlation options specified in terms of their applicability and/or range of validity and verification
- Benchmark problems to test only the numerical method and to allow users to compare the overall results against solutions provided
- Simplifying the users' workload by automating the introduction of inputs and postprocessing and plotting of the calculated results
- Automatic grid adaptation by employing additional grid points and different numerical schemes in some phases of the flow behavior and by abandoning them when they are no longer required

7.5 THE FUTURE OF TWO-PHASE FLOW IN COMPLEX SYSTEMS

There are other changes of potential short-term value to the present system computer codes, but they require additional investigations. The development of improved thermal nonequilibrium models falls into this category. Subcooled boiling, heat transfer beyond critical heat flux or surface dryout, flashing during critical flow depressurization, and direct contact condensation with and without noncondensable gases are all not well understood and could benefit greatly from additional studies. Correlations that take into account the stream velocity, the degree of nonthermal equilibrium, the presence of noncondensables, and the geometries of interest are needed in the short term. Present system computer codes depend on empirical models validated over a very narrow range of parameters and geometries and approximate relaxation exponential expressions to deal with nonequilibrium conditions. This area deserves high priority.

Another important area that deserves additional testing is the influence on two-phase flow of entry conditions, contractions, expansions, valves, and other sudden changes in geometry. In Chapter 5, different inlet and exit conditions were shown to have a significant impact on countercurrent flow measurements (see Figure 5.2.4). Inlet conditions were found to have a similar role in the case of water injection in a three-dimensional annular downcomer (see Figure 5.2.11). The dominance of flow geometry was particularly obvious in the case of critical flow through pipe cracks and horizontal pipes (see Section 6.7.1). A systematic experimental investigation of single-phase (gas–liquid) and two-phase flow through such undeveloped configurations and good empirical correlations of the test data would be of great help in reducing the system computer code uncertainties in predicting such frequently found conditions in practical complex systems.

There is also a need to improve the simulation of three-dimensional two-phase flow. Such circumstances arise during the injection of cold water in the downcomer of the nuclear reactor pressure vessel after a blowdown of the primary water, or in the upper/lower heads of the pressure vessel and in the inlet/outlet plena of the steam generators. The present system computer codes rely on poorly validated mixing coefficients between cells, and their predictions tend to be questionable unless backed up by full-scale tests. In most cases the tests have produced results with increased margins, and the priority for having accurate three-dimensional models must take into account the very high costs associated with their development and validation through tests.

Finally, there are changes that are considered essential by the fundamentalists. On top of their list is the interfacial area modeling. They would prefer to model the interfaces with an interfacial area transport equation. At the present time [7.1.3], simplified mechanistic models are employed at the interfaces and have the form

$$\sigma^{\text{int}} = c^{\text{tr}} f(\overline{\alpha}) \Delta X \tag{7.5.1}$$

where σ^{int} represents the mass, momentum, and energy source term at the interface (int); $f(\overline{\alpha})$ is a function of the average gas volumetric fraction, $\overline{\alpha}$, which defines the available interface area between the two phases; ΔX corresponds to the driving force for the transport process and c^{tr} is an empirically determined coefficient to fit the data. Correlations of the form (7.5.1) must be developed for each flow pattern.

The proposed interfacial transport model instead would rely on a transient balance equation for the interfacial area concentration a^{int} such that

$$\frac{\partial}{\partial t}(a^{int}) + \nabla(a^{int} \mathbf{V}_{int}) = \sigma^A_{int} \tag{7.5.2}$$

where $\mathbf{V}_{int}$ is the interface velocity vector and σ^A_{int} is the source term for interfacial area concentration. For example, in the case of bubble flow [7.1.4], σ^A_{int} would consist of four terms:

$$\sigma^A_{int} = \phi_{dis} - \phi_{co} + \phi_{ph} + \phi_w \tag{7.5.3}$$

where ϕ_{dis} is the bubble disintegration term due to turbulent fluctuations, ϕ_{co} the coalescence term due to bubble collision and wake entrainment, ϕ_{ph} the phase-change term due to bubble formation within the flow field, and ϕ_w is the bubble source term due to wall nucleation.

This approach clearly will improve the ability to determine when the bubble flow will transition to slug flow under steady-state and transient conditions; however, it will require a turbulent fluctuation model and new source and sink terms, which must rely on local measurements of the flow structure and behavior and which most likely will depend on the flow rate and geometry and the fluid properties. Some early work in this field was noted in Chapter 4, where it was shown that approximate predictions of gas volume fractions and turbulence distributions across the flow direction could be obtained for bubble flow but not without empiricism and simplifications in the bubble geometry (see Section 4.3.8). This concept demands considerable additional development work and time and its final benefits are presently subject to debate. Still, continued fundamental investigations of this topic by academic institutions may be worthwhile if for no other reason than improving our basic understanding of two-phase flow.

As the interest in two-phase flow shifts to local detailed flow structure, it will require low numerical-diffusion methods. Städtke points out that such techniques have been used for gas dynamics calculations and that they include the split coefficient matrix or the flux splitting method. Their application to two-phase flow will be successful only if the governing conservation equations are hyperbolic in nature. Some progress has been made in that direction [7.5.1], but computer cost constraints will limit their application to a few complex system scenarios. However, the continued development of low-numerical-diffusive schemes for two-phase flow is desirable because it will yield a yardstick against which present system computer code predictions can be measured.

There are other suggested improvements. For example, further subdivision of the flow pattern has been proposed by considering the presence of liquid droplets in the gas core of annular flow and of gas bubbles in liquid slugs. This is not a new idea, and as discussed in Section 4.6.7, a droplet field was included in the gas core as well as a wavy interface in the liquid film. This additional detail has merit in evaluating such an important phenomenon as critical heat flux, but it is expected to have a small impact on overall present system computer code predictions.

There are many more recommended improvements, such as dependable flow pattern maps, valid over the entire field of simulation and based on experiments carried out with different geometric configurations and under undeveloped and transient flow conditions; improved closure or constitutive equations which are not derived from steady-state and established flows; closure relations for three-dimensional models; and so on. One of the most frustrating aspects of two-phase flow in complex systems is the inability to assess the costs and benefits of each of the improvements being suggested. It is particularly important to assess whether some changes are being pushed to an accuracy well beyond that achieved in other portions of the system computer code. It is also essential that the intended use of the predicted information in terms of safety and performance determines the accuracy required from the computer system code predictions and not the other way around.

Fortunately, in recent years, new methods have become available to predict the uncertainties in computer system code predictions. The first approach was the result of the code scaling, applicability, and uncertainty evaluation [1.1.5]. Subsequently, a different method was developed by GRS in which the amount of computer effort depends only on the desired confidence level [7.5.2]. Such methods not only determine the overall uncertainty of the predictions but also can also be used together with sensitivity analysis to establish the major sources of uncertainty in the computer system code and to help prioritize future tests and changes of the system computer code most cost-effectively. In all such assessments, one starts by identifying the most important phenomena and the most uncertain associated parameters used in the analysis. Probability distributions are next utilized to define the state of knowledge of the most important parameters. This is often done subjectively based on expert judgment, particularly as it relates to the probability distribution functions. Three principal types of uncertainties usually are recognized: input, modeling, and scaling uncertainties. The input uncertainties are established by defining the range of conditions over which the complex system is expected to perform. The modeling uncertainties are obtained by comparing separate effects test data to the predictions from the system computer code models. The scaling uncertainties are estimated by application of the system computer code to results from integral test facilities and/or by providing additional uncertainties or biases to compensate for lack of full-scale tests. All the uncertainties are combined to obtain the overall uncertainty. The GRS methodology has the advantage of being independent of the number of uncertain parameters and is based primarily on tools from statistics. The values of all uncertain parameters are varied simultaneously for each system computer run, and statistics require that 59 code calculations are necessary if one is to be 95% confident that 95% of the combined influence of all quantified uncertainties are below the limit. For 95% confidence that 90% of all uncertainties are below the limit, the total of computer runs is 45. The GRS method thus requires more runs than the CSAU approach. Another valuable feature of the GRS methodology is that it can determine the degree of influence of uncertain input parameters and rank them in order of importance. Irrespective of the methodology employed, the accuracy of the predicted uncertainties remains subjective and depends heavily upon how good the identification of the dominant sources of uncertainties is and upon the quality of their quantification. A common pitfall is to still use bounding values of parameters or limit type correlations, which tend to distort the calculations from truly being best estimate analyses.

In summary, two-phase flow in complex systems remains an area of significant challenges and great opportunities. To those practitioners, fundamentalists, pragmatists, and system computer code developers who are willing to try to tame it, it can offer great fun for many years, if not generations, to come.

REFERENCES

7.1.1 *Thermohydraulics of Emergency Core Cooling in Light Water Reactors*, CSNI Report 161, OECD Nuclear Energy Agency, October 1989.

7.1.2 *Proceedings of the OECD/CSNI Workshop on Transient Thermal-Hydraulic and Neutronic Codes Requirements*, NUREG/CP-0159, November 1996.

7.1.3 Städtke, H., *Thermal Hydraulic Codes of LWR Safety Analysis: Present Status and Future Perspectives*, NUREG/CP-0159, pp. 732–750, November 1996.

7.1.4 Ishii, M., *Views on the Future of Thermal Hydraulic Modeling*, NUREG/CP-0159, pp. 751–754, November 1996.

7.1.5 *Institute for Multifluid Science and Technology (IMuST) Proceedings*, February 28–March 1, 1997, Santa Barbara, Calif.

7.4.1 Chexal, B., et al., *Void Fraction Technology for Design and Analysis*, EPRI Report TR-106326, March 1997.

7.4.2 Chexal, B., and Lellouche, G., *Void Fraction Model*, EPRI Report TR-106326, pp. 4-1 to 4-23, March 1997.

7.4.3 Aksan, S. N., et al., "User Effects on the Thermal-Hydraulic Transient Code Calculations," *Nucl. Eng. Des.*, **145**, 159–174, 1993.

7.5.1 Städtke, H., et al., "Numerical Simulation of Multi-dimensional Two-Phase Flow Based on Flux Vector Splitting," paper presented at the 7th International Meeting on Nuclear Thermal Hydraulics, Saratoga Springs, N.Y., September 1995.

7.5.2 Glaeser, H., and Pochard, R., "Review on Uncertainty Methods for Thermal Hydraulic Computer Codes," paper presented at the International Conference on New Trends in Nuclear System Thermo-hydraulics, Pisa, Italy, 1994.

CHAPTER 8

OTHER TWO-PHASE-FLOW COMPLEX SYSTEMS

8.1 INTRODUCTION

In Chapters 1 to 7 we have emphasized two-phase flow in complex systems of water-cooled nuclear power plants. In this chapter we deal with nonnuclear systems, and the topics selected for coverage are global climate change and global warming. These are subjects of great current interest and hot debate because they are vital to the future of planet Earth and because they involve a very complex system, if not the most complex system of all. The purpose of this write-up is to identify the similarities between the global climate and nuclear complex systems and how they might benefit from each other's experience, methodology, modeling, and progress in knowledge. Before embarking on that objective, it is important to realize that the plan is only to compare the two different complex systems; there is no claim to resolving the problem of global warming, for the same reason that a climatologist would not try to have answers to nuclear power plant safety issues.

Following the pattern of Chapter 1, the treatment in this chapter is primarily descriptive and avoids the use of numerous or complicated mathematical equations. A definition of global climate change and global warming is offered first; it is followed by a discussion of the global climate system and of its three principal subsystems: sun, atmosphere, and earth. The relevant components and phenomena for each subsystem are covered next as well as the significant interactions and feedbacks. Most of the information presented herein comes from Houghton's book on global warming [8.1.1], the excellent work of the Intergovernmental Panel on Climate Change (IPCC), (e.g., see Reference 8.1.2), and the extensive comments generated by IPCC publications (typically, References 8.1.3 to 8.1.6). As the write-up proceeds, it will become apparent that there are many features, strategies, models, and lessons transferable between the global climate and the nuclear power plant complex systems. Those findings, summarized in the final section of the chapter, support previous statements that Chapters 1 to 7 apply to other nonnuclear complex systems.

8.2 GLOBAL CLIMATE CHANGE AND GLOBAL WARMING

Local changes in temperature and pressure occur in the atmosphere all the time, from minute to minute, hour to hour, day to day, and season to season; they determine the local weather and its changes with time. *Climate* is defined as the average of the weather over a long period of time and over a geographical region; *global climate* corresponds to the same averaging over time but over the entire earth. Because of the long period of time involved (i.e. decades), the global climate system must include "the totality of the atmosphere, hydrosphere, biosphere, and geosphere and their interactions" as proposed by the United Nations Framework Convention on Climate Change. The convention defined *global climate change* as a change "attributed directly or indirectly to human activity that alters the composition of the global atmosphere and which is in addition to natural climate variability observed over comparable time periods."

Global warming is a special scenario of global climate change. It results from the continuous addition of greenhouse effect gases (GEGs) to the environment due to human industry and other activities, such as deforestation. Greenhouse gases absorb some of the thermal radiation leaving the surface of the earth and they produce the equivalent effect of a blanket. As the concentration of GEGs rises, so does the blanketing impact, and the chances for global warming increase. Global warming has been assessed primarily by looking at global annual average temperatures as a function of time [8.1.1, 8.1.2]. The average temperatures are estimated from records of temperature over land and within the sea. Land temperatures are derived from weather stations, while sea temperatures are estimated from observations from ships. The available data are first averaged over squares of about 1° latitude by 1° longitude; next, the average temperatures within the squares are combined to get the global average temperatures. Both average earth surface, sea, and combined temperatures and average temperatures within the lower-atmosphere region (referred to as the *troposphere*) have been employed. The troposphere information is obtained from weather balloons or from microwave sounding units on orbiting satellites. The satellites have the advantage of providing global data directly; however, they have become available only for the last 20 years.

8.3 GLOBAL AVERAGE TEMPERATURES

Global average temperatures are often described as anomalies because they are expressed relative to average values measured over a prior specific period of time.

Figure 8.3.1a shows the annual global average surface air temperature anomalies for land areas, 1861 to 1994, relative to the 1961–1990 averages [8.1.2]. Corresponding sea surface and global temperatures are plotted in Figure 8.3.1b and c. The solid curves represent smoothing of the temperatures to suppress yearly time variations [8.3.1 to 8.3.3].

There is a significant variability in Figure 8.3.1 not only from year to year but also from decade to decade. Yet there appears to be an increasing trend in temperature from about 1910 to 1940 and again from 1975 to 1995. The increasing trends are not uniform and undergo short periods of cooling and leveling. The variations could be caused by external causes (e.g., volcanic eruptions, El Niño) or most probably by the large temperature variations existing between different parts of the globe and their variations over time.

8.4 GLOBAL CLIMATE SYSTEM

The global climate system consists of the *sun*, which emits radiation energy toward the *earth*. Between the earth and sun there is the *atmosphere* across which terrestrial and solar radiation travel in opposite directions. If that exchange of energy is not equal, the earth temperatures would change to equalize it. The three principal subsystems of the global climate system, together with their components and interactions, are illustrated in Figure 8.4.1. The sun provides the energy to keep the earth warm, and together with the atmosphere and earth, it determines the amount of incoming solar radiation, its variation over time, and its distribution over the globe surface. In Figure 8.4.1 the lower atmosphere is subdivided into two components: the troposphere, where the temperature declines with altitude, and the stratosphere, where it rises.

As shown in Figure 8.4.1, the atmosphere has strong interactions with most of the components of the earth. The most relevant components are the oceans, the land surface, portions of the globe covered with ice and snow (cryosphere), and the vegetation and other living systems on the land and in the ocean (biosphere). For completeness purposes, the roles of human activity and such natural forcing functions as volcanic eruptions are included in Figure 8.4.1. In the sections that follow, the three global climate subsystems are discussed in more detail.

8.5 SUN

The sun is at a temperature of about 6000°C and emits radiation in the visible shortwave region. A surface of 1 m^2 directly facing the sun receives 1370 W. Figure 8.5.1 shows that the incoming solar radiation reaches part of the earth and only during daylight hours. On the other hand, the outgoing radiation takes place at all hours over the entire surface of the earth. The temperatures at the earth surface and within the lower atmosphere vary locally. They are low (about 200 to 300 K) and

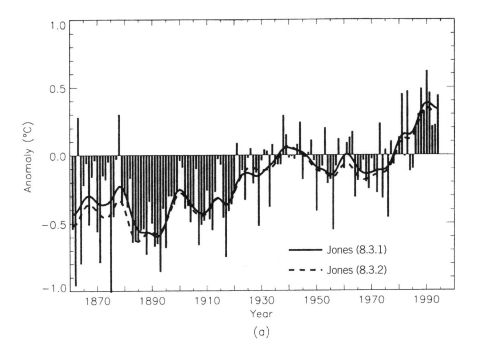

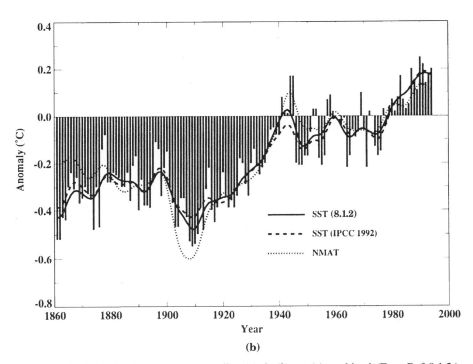

Figure 8.3.1 Global surface average anomalies: (*a*) air, (*b*) sea, (*c*) combined. (From Ref. 8.1.2.)

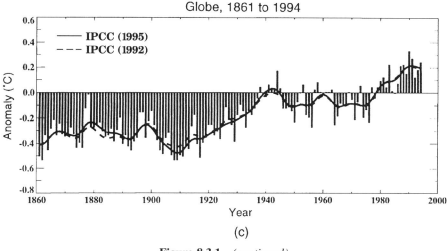

Figure 8.3.1 *(continued)*

the radiation they emit lies in the infrared invisible region (sometimes referred to as longwave radiation, to distinguish it from the sun's shortwave radiation).

8.5.1 Sun Influence over Climate

The *net radiation* (absorbed solar radiation minus outgoing terrestrial radiation) averaged over latitude and over a complete year is shown on the left side of Figure

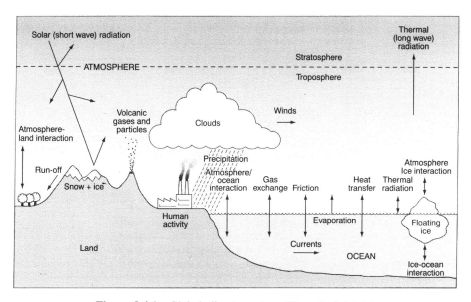

Figure 8.4.1 Global climate system. (From Ref. 8.1.1.)

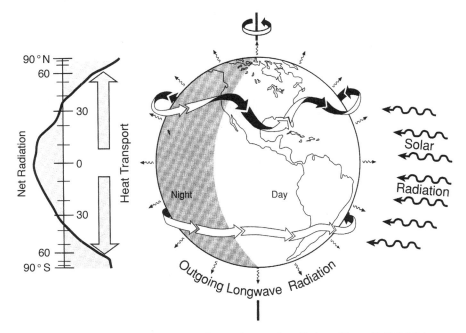

Figure 8.5.1 Incoming solar and outgoing earth radiation. (From Ref. 8.1.2.)

8.5.1. It is seen to vary with latitude and to be different for the northern and southern hemispheres. At the equator there is an excess net radiation of 68 W/m² and there are deficits of -100 W/m² at the South Pole and -125 W/m² at the North Pole. The result is that high temperatures prevail at the tropics and cold temperatures at the high latitudes. Also, the excess of net radiation energy reaching the equatorial regions is transported through atmospheric motions and ocean currents to the poles to satisfy the conservation of energy.

This heat transport mechanism combined with the earth's rotation, results in a broad band of midlatitude westerlies in the troposphere in which there are embedded large weather systems and a jet stream that is represented in both hemispheres by broken ribbons with arrows in Figure 8.5.1. Furthermore, the westerlies and jet stream give a geographic spatial structure to the climate because their flow over underlying surfaces produces waves in the atmosphere that recognize the land–ocean contrasts and the presence of such obstacles as mountains. The waves shift with time as heating patterns change in the atmosphere. So does the weather over typically large geographic regions.

Figure 8.5.1 demonstrates that there is strong coupling at the subsystem stage among the sun, atmosphere, and earth. The sun's impact on the global climate is far from uniform, and it can be expected to become even more variable as longitude and all components of the earth and atmosphere are considered. Some have suggested that natural changes in the sun's output radiation may also have a noticeable effect on the global climate, and this topic is examined next.

8.5.2 Variations in Sun Output Energy

There are two reasons for changes in the sun's radiation: (1) long-term oscillations in the earth's orbit, and (2) short-term variations in the sun's activity or number of sunspots and solar flares. The changes in earth orbit take place over tens of thousand years and are generally well understood. The orbit is an ellipse, and the eccentricity of the ellipse varies with a period of 100,000 years. The earth also spins on its own axis, and the axis of spin is tilted with respect to the axis of the earth's orbit. The tilt ranges between 21.6 and 24.5° with a period of 41,000 years. Finally, the time of year when the earth is closest to the sun (earth's perihelion) moves through the months of the year with a period of 23,000 years.

All three orbital periods are evident in an 800,000-year reconstructed record of global ice volume and in analyses of polar ice cores over the past 220,000 years. Ice ages followed by interglacial periods have occurred close to every 100,000 years. During the ice ages, the mean surface temperature decreases by about 4 to 5°C over the globe and by twice that amount in the polar regions. The increasing interglacial temperatures last not as long a time (a few to about 10,000 years) before the cycle is repeated. Superimposed on this sawtooth temperature pattern are other variations of 0.5 to 1°C produced by the other two orbit oscillations [8.5.1].

Despite the good correlation between ice ages and the earth's orbital oscillation, the climate changes are much larger or different than might be expected from only variations in the solar energy reaching the earth [8.5.1]. For example, the 100,000-year orbital oscillation predominates the climactic history, but its incident solar radiation changes by only 0.1%. Another possibility is that the climate changes over those long periods of time are due to significant differences in the distribution of the solar radiation with latitude and season. For instance, between 22,000 and 11,000 years ago [8.5.1], there was an increase in summer sunshine of around 10% at 60°N at the end of the last ice age. However, there was a decrease in summer radiation of 10% at 60°S at the same time, even though polar ice and glaciers were retreating in the southern hemisphere; feedback mechanisms besides solar energy input change and redistribution are therefore needed to explain climate cycles of the long past.

Over shorter periods of time, the sun activity and brightness vary with a period of about 11 years. In Figure 8.5.2 the sun magnetic cycle lengths are plotted for the period 1750–1990. The shorter the magnetic cycle length, the more active and brighter is the sun and the warmer is the earth. Figure 8.5.2 also shows the moving 11-year average terrestrial northern hemisphere temperature as a deviation from the 1951–1970 corresponding mean temperature. Robinson et al. [8.1.6] concluded from the information in Figure 8.5.2 that there is a strong correlation between the fluctuations of sun activity and global temperature and that it might be relevant to global warming studies. On the other hand, the IPCC view [8.1.2] is that the sun solar output has been measured accurately with satellites since 1978 and that the sun solar output changed at most by about 0.1%, which is not enough to explain the surface temperature rise during the most recent cycles of solar magnetic activity.

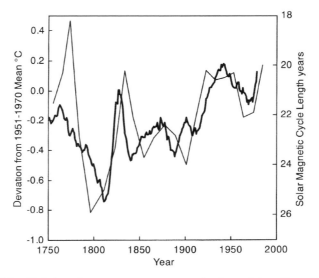

Figure 8.5.2 Variations in mean temperatures and solar activity. (From Ref. 8.1.6.)

Similarly, very few sunspots were recorded during the cold period known as the Maunder Minimum in the seventeenth century. Yet, according to Houghton [8.1.1], the solar energy output decreased by only 0.4% or about 1 W/m². This reduction in solar energy did contribute to the period of reported cold temperatures, but again it had to be reinforced by other feedback processes. In summary, changes in solar energy alone are insufficient to match the short-term or long cold climate variations of the past.

8.5.3 Uniform Net Radiation: Single-Temperature Model

In this oversimplified model, a completely absorbing surface at a constant effective temperature T_e (in K) is assumed to exist within the atmosphere and to receive a uniform net radiation input by equating the radiation energy emitted to the solar radiation absorbed; there results

$$4\pi a^2 \sigma T_{ee}^4 = \pi a^2 (1 - A) F \quad (8.5.1)$$

where a is the radius of the radiating surface and σ is the Stefan–Boltzmann constant ($\sigma = 5.67 \times 10^{-8}$ J/m² · K⁴ · s), A the *albedo* or ratio of reflected (scattered) radiation to incident radiation, and F the solar flux ($F = 1370$ W/m²). If the albedo is set at 0.31 (see Section 8.6), the effective temperature is calculated to be 254 K ($-19°C$), which compares favorably with the corresponding measured temperature T_m of about 250 K ($-23°C$). Temperatures T_e and T_m are both below values prevailing near the earth surface, where the annual mean measured temperature is about 288 K (15°C).

The difference between 15 and $-19°C$, as explained later, comes about from temperatures falling off with height from the earth surface. A temperature of $-19°C$ is reached at an altitude between 5 and 6 km above the earth surface.

Equation (8.5.1) neglects transient and feedback mechanisms and the wide variation of temperature conditions over the earth surface. Yet it provides a very simplified means of predicting a single, steady-state average global temperature very dependent on the albedo value chosen. For example, if the albedo decreases from 0.31 to 0.3, the radiation energy increases from 236 W/m² to 240 W/m², and the effective temperature rises by 1°C, to 255 K ($-18°C$). As discussed later and as one might infer from Figure 8.3.1, that is a significant change in global average temperature, with potentially serious consequences.

In many ways, the single-temperature model of Eq. (8.5.1) resembles the top-down spatially uniform system discussed in Section 2.4. Furthermore, one could consider the atmosphere to be a slab of thickness H with a uniform density ρ and specific heat c_p, radiating energy like a blackbody at a temperature T to surrounding layers both above and below the slab. If this atmospheric slab is subjected to a small change of temperature ΔT, its rate of temperature change ΔT with time t is

$$\rho c_p H \frac{d\,\Delta T}{dt} = 8\sigma T^3\,\Delta T \qquad (8.5.2)$$

and the top-down time constant for radiative processes in the lower atmosphere is $\rho c_p H/8\sigma T^3$. With T set at 270 K (close to halfway between 250 and 288 K), $H = 6$ km, and the density ρ calculated at a pressure of 50 kPa, the time constant for the lower atmosphere is about 4.5 days. This time constant is large enough to neglect radiative processes for short-term minute, hourly, even daily weather changes, but to have them play a dominant role over long-time climate changes.

While the use of average temperatures over the globe or large segments of it makes it easy to explain the problem of global warming to political leaders and nonscientists, it also reduces the temperature variations involved, and it might smear the contribution from different sources to the rise or decline of the average temperatures. It might be preferable to amplify such impact by dealing with the extremes of the distribution (e.g., regions of highest and coldest temperature) and limiting their surface areas of influence to where they are relatively constant. It is worthwhile to note that such an approach was used successfully in Chapter 1, where a loss-of-coolant accident (LOCA) is postulated to occur in nuclear power plants. The peak fuel clad temperature rather than the average temperature is employed not only for safety reasons but also because it enhances the comparison of the analytical predictions to tests.

The rates of temperature change vary significantly over the globe and can be related to population centers. One particular region exhibiting much higher tempera-

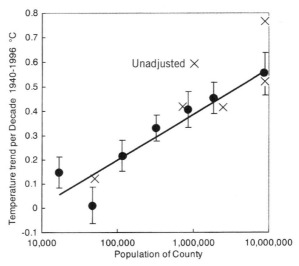

Figure 8.5.3 Temperature trend from six measuring stations in Los Angeles county from unadjusted and mean values. (From Ref. 8.1.6.)

ture change than the global average is Los Angeles County as illustrated in Figure 8.5.3. The surface temperature data for the period 1940–1996 from 107 measuring stations in 49 California counties were utilized to calculate the means of the trends in each county and to combine counties of similar population and plot them in Figure 8.5.3 as closed circles along with their standard deviation. In the case of Los Angeles County, with a population of 8.9 million, there are six measuring stations and their data show that the mean temperature has been increasing by 0.55°C per decade. This is much higher than the global average. It is well known that Los Angeles County has one of the largest numbers of operating motor vehicles, which generate significant amounts of greenhouse effects gases (GEGs) [8.5.2]. It is also suggested that other surface energy releases in the Los Angeles County environment are among the highest. There could be considerable merit to setting up a one-dimensional radiative–convective model for Los Angeles County with and without a hydrological cycle, as suggested by Renno et al. [8.5.3].

The data available from the six county measuring stations as well as energy usage information could help guide the one-dimensional model formulation until it reproduces the results of Figure 8.5.3. This exercise could provide a strong confirmation of the impact of GEGs because their concentrations must be amplified in Los Angeles County. It is also customary to neglect other energy release into the atmosphere when compared to the solar radiation input. Although this is justified on a global basis, it may deserve checking for highly populated areas where the surface energy release is magnified significantly.

8.6 ATMOSPHERE

The atmosphere is made up of two principal components: nitrogen (78% by volume) and oxygen (21%). Other components are water vapor (a few percent), argon (1%), carbon dioxide (CO_2, about 0.035%), and other important trace gases, such as ozone (O_3), methane (CH_4), nitrous oxide (N_2O), and chlorofluorocarbons (CFCs). The atmosphere also contains small suspended particles (aerosols), which arise from such sources as forest fires, biomass burnings, and volcanic eruptions, and sulfate particles, generated by burning coal and oil in power plants and other industries.

8.6.1 Atmosphere Components

The atmosphere is subdivided into regions stacked vertically above the surface of the earth. The regions are associated with different modes of heat transfer, as illustrated by the temperature structure sketched in Figure 8.6.1. Within the lowest region of the atmosphere (10 to 15 km) referred to as the *troposphere*, convection is the dominant process. The incoming radiation heats the ground and the air close to it. The heated air rises because of its lower density and it is replaced by descending cooler air; the air is continually being turned over and the temperature in the troposphere falls with height at a rate determined by such convective and gravitational processes.

The rate at which the temperature falls with increasing altitude is known as the environmental lapse rate (ELR). On average, it is about 6.4°C/km and this pattern continues up to the top of the troposphere or *tropopause*. The ELR is affected by

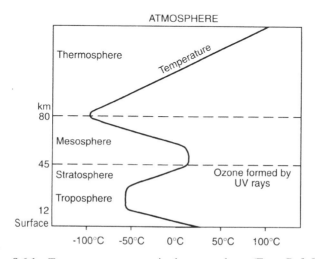

Figure 8.6.1 Temperature structure in the atmosphere. (From Ref. 8.6.1.)

atmospheric and surface conditions. For example, when strong winds are present or the surface heating is high, the ELR increases (i.e., air temperatures drop faster with height). Above the tropopause, the characteristics of the temperature structure are determined by radiative processes. As the radiation is scattered or absorbed, there is a warming of the atmosphere, as shown for the *stratophere* in Figure 8.6.1. The temperature peak shown at about 50 km (known as the *stratopause*) is caused by absorption of solar radiation by ozone. Above the temperature peak is the *mesosphere*, where the temperature falls again with height but at a reduced rate by comparison to the troposphere. Above the lowest temperature at the *mesopause* (about 80 km), solar ultraviolet radiation is absorbed by molecular and atomic oxygen and the temperature rises sharply to between 500 and 2000 K because of the very low gas density in that region, known as the *thermosphere*. Within the thermosphere, one reaches a height (120 km), where molecular diffusion becomes important. That region is referred to as the *turbosphere* or *homosphere*. Within the turbosphere and above it separation of the air contituents takes place. For example, oxygen becomes the dominant contituent at about 170 km and helium at about 500 km above the surface of the earth.

As the altitude increases, both the atmosphere pressure and density decrease. Above 80 km there is only 10^{-5} of the atmosphere mass, and above 30 km there is only 1%. The atmosphere behavior is thus controlled by the regions closest to the earth. Similarly, close to the earth surface, turbulent mixing dominates the convective transport of heat. The coefficient of eddy viscosity is typically between 1 and 100 m^2/s. The corresponding diffusion coefficient at atmospheric pressure is about 10^{-4}m^2/s, and diffusion can be neglected. However, the diffusion coefficient reaches about 100 m^2/s at a pressure of 10^{-6} atm or an altitude of 96 km where its role becomes important.

The variation of temperature with latitude in the atmosphere is shown in Figure 8.6.2. The isotherms of Figure 8.6.2 tend to confirm the schematic temperature structure of Figure 8.6.1. They also show not only differences with respect to season but also with latitude. The temperatures at the equators and the reduced temperatures at the poles near the earth surface are related to the variation of solar radiation input with latitude (see Section 5.5.1). However, the reversal of this pattern at increased distances from the earth can be explained only by atmospheric motions, which are discussed in Section 8.6.6.

8.6.2 Simplified Treatment of Dry Air Vertical Temperature

Vertical movement of the atmosphere by natural circulation close to the earth surface was noted previously. If that vertical motion is presumed to be small, one can write the following simplified one-dimensional steady-state momentum or hydrostatic equation:

$$dp = -g\rho \, dz \quad (8.6.1)$$

where p is the pressure at altitude z, ρ the corresponding atmosphere density, and g the gravitational constant. If the perfect gas equation of state is assumed to apply:

$$\rho = \frac{\text{Mr}_a p}{RT} \tag{8.6.2}$$

where Mr_a is the molecular weight of the atmosphere, R the gas constant per mole, and T its temperature. By combining Eqs. (8.6.1) and (8.6.2), there results

$$\frac{dp}{p} = -\frac{\text{Mr}_a g}{RT} dz \qquad p = p_0 \exp\left(-\int_0^z \frac{\text{Mr}_a g}{RT} dz\right) \tag{8.6.3}$$

where p_0 is the pressure at $z = 0$ and $l_{NC} = RT/\text{Mr}_a g$ is a *scaling length for natural circulation* in the lower atmosphere. At $T = 210$ K, $l_{NC} = 6$ km and $l_{NC} = 8.5$ km at $T = 290$ K.

The total energy E_t per unit mass of atmosphere dry air is

$$E_t = c_p T + gz + \frac{u^2}{2} \tag{8.6.4}$$

where $c_p T$ is the sensible heat content, c_p the specific heat at constant pressure, gz the potential energy, $u^2/2$ the kinetic energy, and u the velocity. If the airmass moves adiabatically (i.e., does not exchange heat with neighboring air and is not affected by solar or terrestrial radiation), Eq. (8.6.4) yields, when the kinetic energy is neglected,

$$-\frac{dT}{dZ} = \frac{g}{c_p} = \text{DALR} \tag{8.6.5}$$

where DALR is the *dry adiabatic lapse rate*. The DALR is constant at about 9.8°C/km and describes the rate at which air cools with height as it expands adiabatically. As discussed previously, the surrounding air cools at the environmental lapse rate (ELR) due to convective processes, and this is shown schematically in Figure 8.6.3. The ELR can change with altitude due to radiative and convective heat transfer and the temperature decrease can be reduced and even increased with altitude, producing a temperature inversion as illustrated in Figure 8.6.3. The airmass rising adiabatically will have the same temperature and density as that of the surrounding air when the DALR and ELR intersect; the airmass will then stop rising and reach a stable state. In more general terms, stable conditions will occur when ELR > DALR, and they will be unstable when ELR < DALR.

This simplified model has many shortcomings, including the exclusive transport of heat by natural convection and the adiabatic process of the rising atmosphere. Yet it bears a good relation to the circumstances in the real atmosphere, its stability, and its division into the troposphere and the stratosphere.

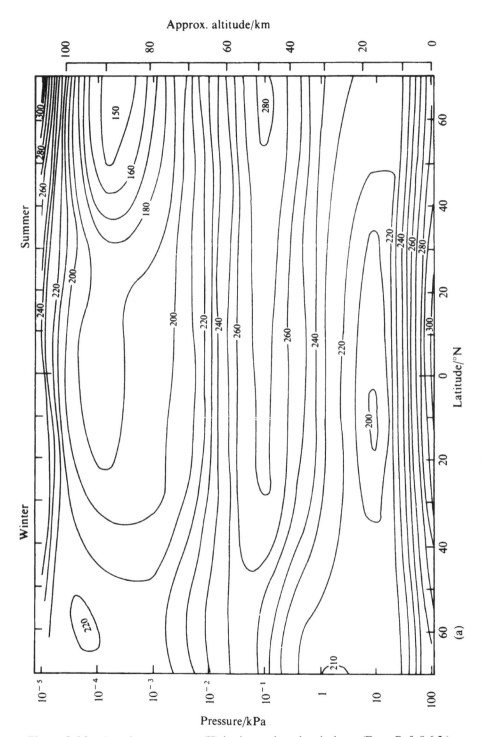

Figure 8.6.2 Actual temperatures (K) in the northern hemisphere. (From Ref. 8.6.2.)

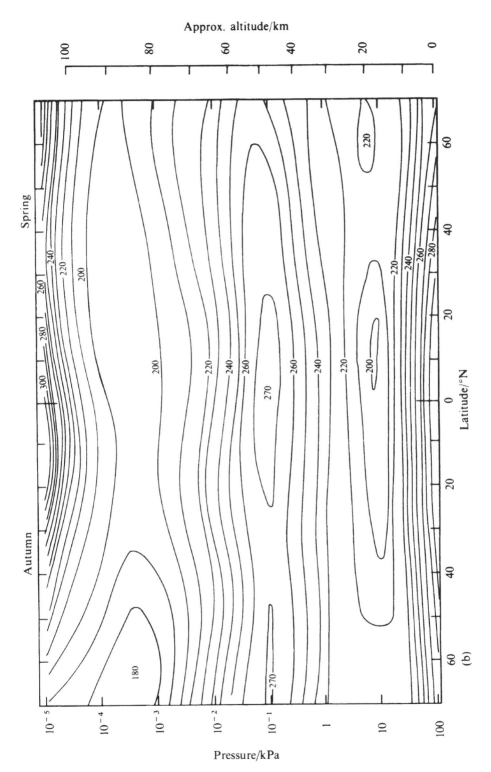

Figure 8.6.2 *(continued)*

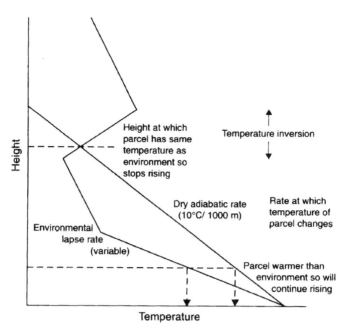

Figure 8.6.3 Thermal rise of dry air parcel. (From Ref. 8.6.1.)

8.6.3 Simplified Treatment of Vertical Movement of Air with Moisture

Contrary to the dry air assumption of Section 8.6.2, practically all air contains moisture. The moisture comes from water evaporated from the oceans, from the land, or transpired by vegetation. If the air is cooled to the temperature where it cannot hold water vapor, condensation starts and water droplets are formed. If the air is heated, the liquid droplets evaporate subsequently and water vapor is reformed. The vertical movement of air containing moisture is illustrated in Figure 8.6.4. As long as condensation does not occur, the air rises along the dry air lapse rate derived previously (about 10°C/km) because the amount of moisture is usually small enough not to change the gas properties of the atmosphere. When condensation takes place, the air not only has to carry the extra weight of the liquid droplets but also has to deal with the latent heat produced by converting water vapor to its liquid state. This latent heat warms the air and therefore reduces the rate of temperature decline with height. This new lapse rate, called the *saturated adiabatic lapse rate* (SALR), is shown in Figure 8.6.4. The corresponding ELR is plotted above the condensation level and is labeled SELR (saturated environmental lapse rate). Unlike the ELR and DALR below the condensation height, the SLAR and the SELR above it are no longer constant because they are influenced by the amount of latent heat released by condensation, and therefore by the moisture content and temperature of the air as a function of altitude.

When the air temperature is high, its ability to hold moisture increases. Warm air will produce more precipitation and release more latent heat after reaching the

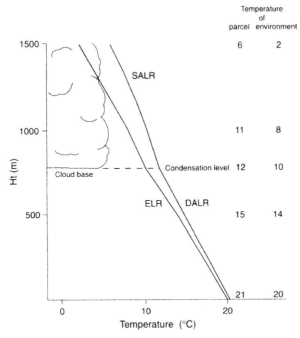

Figure 8.6.4 Thermal rise of moist air parcel. (From Ref. 8.6.1.)

condensation level than cool air, which explains the increased amount of rainfall prevailing in the tropics. Reference 8.6.2 provides an expression for the SALR:

$$\text{SALR} = \text{DALR}\left(1 + \frac{Le \cdot \text{Mr}_v}{pRT}\right)\left(1 + \frac{Le \cdot \text{Mr}_v}{pRT}\frac{\varepsilon L}{c_p T}\right)^{-1} \quad (8.6.6)$$

where L is the heat of vaporization, e the saturation water vapor pressure, Mr_v the water vapor molecular weight, and $\varepsilon = \text{Mr}_v/\text{Mr}_a$, with Mr_a being the air/atmosphere molecular weight. According to Eq. (8.6.6), SALR < DALR. It should also be noted that L and e in Eq. (8.6.6) depend on the temperature T.

As for dry air, stable conditions result when SELR < SALR, and unstable conditions prevail when SELR > SALR. In the case of Figure 8.6.4, since SELR > SALR, the cloud will continue to grow with altitude until instability sets in. The situation is different in Figure 8.6.5, where due to temperature inversion, the SELR and SALR intersect and growth of the cloud is arrested. This is again a simplified model of cloud formation because it presumes that the air remains saturated and that evaporation and condensation occur at thermal equilibrium conditions. Another shortcoming is the assumption that formed water drops will always move with the air. Still, the model provides a good uncomplicated picture of cloud behavior, particularly of their formation and instability.

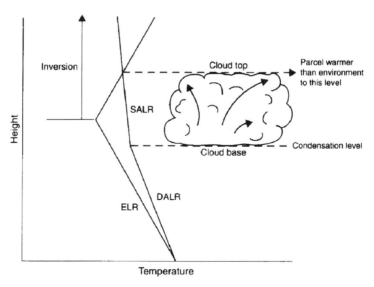

Figure 8.6.5 Thermal rise of moist air parcel with temperature inversion. (From Ref. 8.6.1.)

8.6.4 Clouds

Ascent of moist air produces clouds in the atmosphere. The clouds, in turn, have a great influence on the global climate because they reflect, transmit, and absorb solar and terrestrial radiation. Depending on their characteristics, clouds can cool or warm the earth. The net radiative impact of clouds is influenced by many factors: degree of cloud sky coverage (global coverage is about 50%) and such cloud properties as type, height, thickness, water and ice content, and average size of cloud particles. It is therefore not surprising that the IPCC panel on climate change stated that "one of the major areas of uncertainty in climate models concerns clouds and their radiative effects."

Generally, low clouds cool the earth by reflecting more solar radiation back to space and by reemitting absorbed terrestrial radiation at a higher temperature. High clouds warm the earth because they reflect less radiation back to space and reemit absorbed terrestrial radiation at a reduced temperature. Overall, clouds are believed to have a cooling effect. The earth radiation budget experiment (ERBE) and other satellite measurements [8.6.3] show that the greenhouse effect of clouds amounts to about 31 W/m^2, versus a reflective ability of 48 W/m^2. However, there are large variations in those numbers, depending on the cloud types and their physical characteristics.

To date, clouds have been classified by their appearance: layered clouds fall into the *stratiform* class, while clouds where the vertical extent is important fall into the *cumuliform* class. The stratiform clouds are classified further according to their height: *stratus* clouds at low levels (<2000 m) which are composed primarily of water droplets; and *cirrus* clouds at high levels (>3000 m), which are composed of a mixture of water droplets and ice crystals between 2000 and 6000 m and of ice crystals above 6000 m. Overlapping the stratus and cirrus clouds are the *alto* clouds. The stratiform clouds are characterized by widespread and slow upward motion. The cumiliform clouds are the result of convection and instability and they occur over a wide range of heights; when they are associated with flow over mountains they become elongated and assume a lenticular shape. Cloud heights vary with the local climatic conditions: They tend to be the highest in the tropics, lowest in the arctic, and in between for temperate regions.

In the early assessments of global climate, great simplification was used to describe the clouds and their impact on radiation. They were assigned heights for their lower and upper surfaces, and relationships were specified between cloud amount and relative humidity. Alternatively, cloud amounts were derived from climatological records in terms of height, latitude, and season. Condensed water was created or removed in all such models in amounts dependent only on the water vapor saturation temperature. All clouds were given an albedo of 0.6, except for high-level clouds, which were assigned an albedo of 0.2. All clouds were presumed to absorb 10% and to transmit 30% of incident radiation. The corresponding numbers for high clouds were 5% and 75%. The clouds were presumed to act as blackbodies except for high clouds, which were assigned an emissivity of 0.75 [8.6.2]. Furthermore, if there were several layers of clouds, multiple reflections were taken into account.

This empirical approach produces significant variations in the prediction of global climate when key parameters are changed. For example, in Reference 8.1.1, it is reported that increasing the low cloud coverage by 3% reduces the global average temperature by 1°C, which just about cancels the temperature increase resulting from doubling the concentration of CO_2. Similarly, reference 8.1.4 states that the various feedback mechanisms present while increasing greenhouse gases concentrations can generate a gain between 0.4 and 0.78 in the global average temperature. Most of this gain range is attributed to uncertainty in the role of clouds. The total effect of doubling the CO_2 concentration which alone raises global average temperature (8.1.4) by about 1°C is therefore estimated to be between $1°C/(1 - 0.4) = 1.6°C$ and $1°C/(1 - 0.78) = 4.5°C$ due to cloud uncertainties.

To reduce the preceding cloud uncertainties, physically based models are being introduced to determine the amount of water in clouds and its subdivision between liquid and ice water. This partitioning between liquid and ice particles is necessary because they undergo different thermodynamical processes. For example, when the temperature falls below 0°C, the water droplets, because of their small size, may remain unfrozen in a supercooled state. When the temperature reaches about $-10°C$, ice crystals start to develop, and a mixture of water and ice may coexist. Because the saturation vapor pressure over water drops is about 10% higher than that over ice crystals, the air can be saturated for ice when it is not saturated for water. At $-10°C$, air saturated with respect to water is supersaturated relative to ice, and the ice crystals in the clouds will grow at the expense of water droplets. This transition of water to ice actually is carried out under nonthermal equilibrium conditions and is often dealt with empirically. Comparable nonthermal equilibrium conditions were discussed in Chapter 5 and treated empirically for water vapor formation under critical two-phase-flow conditions.

When ice crystals descend into the lower portions of the cloud, they become exposed to temperatures below freezing, but they tend to stick together to form snowflakes. The supercooled droplets of water present in the cloud act as an adhesive and help this process. If the snowflakes continue to drop to lower layers, they may melt and form drops that can grow by collision with other cloud droplets. This mechanism of melting ice crystals to form raindrops is referred to as the Bergeron–Findeisen process. It occurs when cloud temperatures are well below freezing. Again, it involves the coexistence of ice and water under nonthermal equilibrium conditions, and it can influence the amount of water formed and the optical properties of the cloud.

Cloud drops are first formed by condensation on submicron aerosol particles. The droplets or ice crystals have a typical initial diameter of about 10 μm. The number of cloud particles formed depends on the concentration of cloud condensing nuclei (CCN) or aerosols, which in turn depend on the degree of prevailing air pollution and the convective upward flows in the atmosphere. After the cloud droplets are formed, they initially grow by diffusion of water vapor to their surface. When they reach about 30 μm in diameter, they grow primarily by collision and coalescence. The collision efficiency increases with the size of the droplets, the rising velocity of the air, and the depth of the cloud. When the drops are large enough to overcome the lift of the rising air, they fall as raindrops through the clouds into the nonsaturated dry air below. The raindrops are subject to minimal evaporation before they reach the ground.

Precipitation occurs in the form of fog, rain, snowflakes, and hail. They all involve condensation, solidification, evaporation, and countercurrent flow of air and water droplets and ice crystals. The conditions are similar to the droplet flow pattern discussed in Section 4.5.6 except that they are now more complex because they involve all three phases: gaseous, liquid, and solid. They are in some other ways less complicated because there is no solid surface to perturb the flow velocity field and because the updraft velocities are reduced (10 to 50 cm/s). Those circumstances, however, are overshadowed by the difficulties of predicting the global distribution of cloud water and ice concentration (i.e., the cloud net radiative impact).

As pointed out in Reference 8.1.2, the physically based models have led to improvements in the global temperature predictions, but they still do not agree and do not always compare satisfactorily with the limited amount of available measurements. In Reference 8.1.1 it is reported that the empirical models predicted a 5°C in global average temperature rise with doubled CO_2 concentration. The physically based models reduce that rise to between 2 and 3°C [8.6.4].

8.6.5 Aerosols

The presence of aerosols in the atmosphere is important because they act as nucleation sites for cloud droplets and ice crystals. Aerosols also directly scatter and absorb radiation as well as modify the optical properties and lifetime of clouds. Naturally

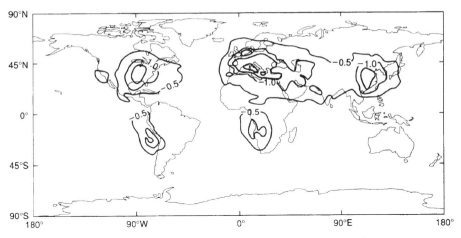

Figure 8.6.6 Cooling effects of sulfate aerosols: radiative forcing (W/m^2). (Reprinted from Ref. 8.5.1 with kind permission from Kluwer Academic Publishers.)

formed aerosols account for 90% of the atmosphere aerosols, and their dominant sources are soil dust and sea salt. Of the natural and humanmade aerosols, sulfate particles are the largest contributor to global cooling by promoting cloud formation and changing the size and distribution of cloud droplets and hence the radiative properties of clouds. These effects are illustrated in Figure 8.6.6, where they are observed to be the highest in the northern hemisphere. Reductions in solar radiation can reach up to 2 W/m^2 in Figure 8.6.6. However, there is no adequate correlation yet between the available quantity of sulfate aerosols and the cooling they might produce. The reasons are that the aerosols are short lived (4 to 5 days), their distribution is uncertain, and their chemical reactions cannot be segregated from other atmospheric reactions. The second most important humanmade aerosol is combustion related (e.g., soot, which tends to have the opposite effect of sulfate). The spatial and temporal distribution of different aerosols is therefore important, to establish their overall impact. It is interesting to note that glacial age ice samples contain 30 times more dust particles than current values. This suggests the possibility of increased desert land areas during glacial periods, which might generate additional dust particles and accelerate their cooling effect. At the present time, the consensus is that aerosols have a net cooling effect on the globe, but the degree of that cooling is uncertain.

8.6.6 Atmosphere Circulation

In Sections 8.6.2 and 8.6.3, simplified models were developed for upward movement of the atmosphere with and without moisture. Most of the atmospheric motion, however, is quasi-horizontal, and such circulation occurs to respond to pressure and temperature gradients produced by unequal solar radiation input to different regions of the globe. This process generates winds and storms, and it is responsible for the large variety of weather conditions reported over the earth. By contrast to the vertical air motion, where the velocities are small (typically, a few centimeters per second), the horizontal velocities are high (on the order of 10 m/s and above in the upper atmosphere). Also, a model of the entire globe is required to predict the weather because, for example, the circulation in the northern hemisphere affects the southern hemisphere weather a few days later. This means that the length scale in atmospheric circulation simulation must be quite large, about 100 to 1000 km. For a thunderstorm with a space mesoscale of 100 km and a horizontal velocity of 10 m/s, the corresponding time scale is a few hours; for a low-pressure system with a synoptic space scale of 1000 km, the time scale is several days. Thus the atmosphere adjusts itself to disturbances rapidly, in a few days, as illustrated in Figure 8.6.7. That figure shows the length and time scales of other key components contributing to the global climate: i.e. soil moisture, snow and sea ice, upper and deep oceans and ice sheets, as well as the residence times of aerosols, water vapor, and greenhouses gases. Figure 8.6.7 reveals that the deep oceans and ice sheets have long time scales (10 to 1000 years) because their heat

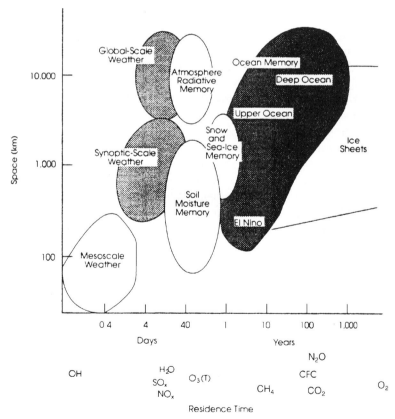

Figure 8.6.7 Time and space scales of climate. (Reprinted from Ref. 8.6.1. with kind permission from Kluwer Academic Publishers.)

capacity is much larger than the atmosphere. They will therefore determine global temperature changes over decades and centuries (i.e., global warming).

Most of the airflow in the atmosphere is driven by pressure gradients. The air moves from areas of high pressure to areas of low pressure. Such movement can be split into two circulation patterns: the airflow within the main wind belts and the Rossby waves in the upper atmosphere. Both forms of circulation are transformed into more complex patterns by friction near the earth surface and the Coriolis force produced by the earth rotation.

The friction force reduces the atmosphere velocity near the earth surface; it varies with the type of earth surface (i.e., ocean versus land) and with the surface nonuniformity (i.e., plains versus mountains). The Coriolis force changes with latitude and air velocity and is equal to $-2\Omega V \sin \phi$, where Ω is the rate of rotation of the earth, V the air velocity, and ϕ the angle of latitude. The ratio of the acceleration to Coriolis forces is called the Rossby number, Ro,

$$\text{Ro} = \frac{V^2/L}{2\Omega V \sin \phi} = \frac{V}{2L\Omega \sin \phi} \qquad (8.6.7)$$

For $\phi = 30°$, $\Omega = 7.29 \times 10^{-5}$ s^{-1}, $V = 10$ m/s, $L = 1000$ km, and Ro $= 0.14$. The Coriolis force tends to produce spiraling clouds by deflecting air movement toward the right in the northern hemisphere and toward the left in the southern hemisphere.

Some of the modes of energy transfer within the atmosphere are illustrated in Figure 8.6.8 developed by Palmer and Newton [8.6.5]. Near the equator, the mechanism of energy transfer is through the Hadley cell. The solar input is the highest at

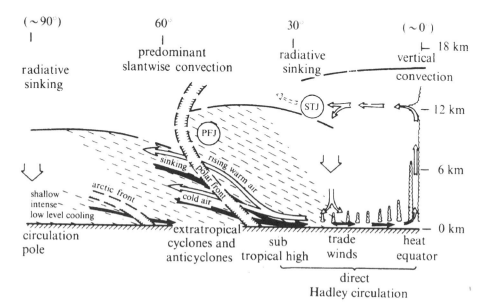

Figure 8.6.8 Schematic features of atmospheric transport. (From Ref. 8.6.2.)

the equator, and a low-pressure trough exists at that location. As shown in Figure 8.6.8, hot air rises to high altitudes and attempts to flow toward the pole. However, its temperature is reduced with latitude due to radiational cooling, and the air sinks at latitudes between 20 and 30°. Some of the descending air is returned to the equator to complete the Hadley cell. In the temperate midlatitudes, the transfer of potential to kinetic energy is by sloping convection, where large-scale eddies and rotating storms are dominant.

Three main wind belts are shown in Figure 8.6.8. Between the subtropical high and the equatorial trough, a belt of easterly trade winds prevails. Beyond the subtropical high lies a belt of westerlies (labeled extratropical cyclones and anticyclones). Beyond the latitude of the westerlies and close to the pole, a belt of shallow polar easterlies occurs. Within the tropical easterlies and the westerlies, jet streams exist at high altitudes. The jet streams are narrow currents of high velocity (up to 130 m/s), and Figure 8.6.8 shows the subtropical jet (STJ) associated with the break in the troposphere at a latitude of 30° and the polar front jet (PFJ) associated with a polar front.

In the upper atmosphere, particularly between the temperate midlatitudes and the pole, the presence of upper winds is noticeable because their flow direction is different from the surface winds. These upper winds become stronger with height, reaching a mean velocity of about 35 m/s up to an altitude of about 12 km. They flow in a series of waves called the Rossby waves which change their amplitude and wavelength over a period of three to eight weeks, as shown in Figure 8.6.9. The Rossby waves are important because they account for the energy exchange between the polar and the midlatittude temperate regions.

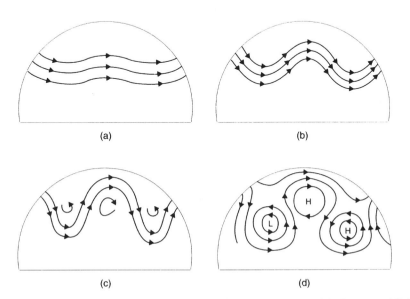

Figure 8.6.9 Rossby waves as their amplitudes increase from (*a*) to (*d*) before reestablishing type (*a*). (From Ref. 8.6.1.)

All the preceding atmospheric movements are important to predicting local surface weather conditions and their changes with time and seasons. Fortunately, dynamic models, called general circulation models (GCMs), have been developed to simulate such atmospheric conditions.

8.6.7 Atmosphere General Circulation Models

General circulation models (GCMs) for the atmosphere are based on the fundamental conservation equations of mass, momentum, and energy. They are written in three dimensions usually directed toward the east, north, and vertical or perpendicular to the earth surface. In most models, the hydrostatic equation (8.6.1), which neglects vertical acceleration, is employed. The differential equations predict the key parameters of interest (pressure, temperature, wind velocity, humidity) as a function of time and location. Because the models need to cover the entire globe, they require large grid dimensions or nodes. A typical spacing between nodes would be about 100 to 250 km and about 1 km in the vertical direction. The solutions are started off with a set of initial conditions representing the prevailing global weather. The amount of data provided is enormous and the treatment of the data to fit the grid size selected requires great care.

The GCM equations are very similar to those discussed in Chapter 3 for nuclear system computer codes, with the exception that the GCMs include a Coriolis force in the momentum equation and employ a much larger node size, dictated by the available computer power. The GCMs equations are also nonlinear and, as discussed in Chapter 3, can be subject to numerical instability and diffusion when they are solved by finite difference methods. The potential chaotic behavior of GCMs was first recognized by Lorenz in 1961 [8.6.6], when his computer weather predictions grew farther and farther apart, even though they utilized nearly the same starting point. In other words, weather predictions are very sensitive to initial starting conditions, and as pointed out by Houghton [8.1.1], even with virtually perfect initial data, the predictability of weather with GCMs can be extended up to only about 20 days because of its chaotic behavior.

Both the nuclear system codes and the atmosphere GCMs rely on closure equations (or parametrizations, as they are referred to by the IPCC). Some of the parametrizations needed to predict the atmospheric weather have been touched upon earlier. They include:

- Two-phase processes, such as evaporation, condensation, ice formation, and melting
- Cloud formation, coverage, and dispersal and their characteristics, such as water and ice content and their optical properties
- Momentum and heat transfer exchanges at the interfaces (ocean–atmosphere and land–atmosphere)
- Absorption, emission, and reflection of solar and terrestrial radiation

- Convective processes (responsible for showers and other precipitation)
- Aerosol generation and distribution

Another important difference between GCMs and the nuclear system computer codes of Chapter 3 is the high degree of coupling of the atmosphere with the earth components (i.e., oceans, land surface, cryosphere, and biosphere). The coupling in nuclear systems is rather limited (e.g., containment and primary system during a loss-of-coolant accident, nuclear and thermal-hydraulics in boiling water reactors during transients) and much easier to account for. The difficulties in resolving GCMs coupling come about for two primary reasons:

- A major complication in predicting global climate is how to link the rapid atmospheric circulation with the considerably slower reactions of the oceans and ice sheets (see Figure 8.6.7). This was first handled by *periodic flux adjustments* at the atmosphere–ocean interface. Today, of the 16 atmospheric GCMs, half no longer require such flux adjustments [8.1.2].
- Perturbation of the GCMs, for example by rising CO_2 concentrations, increases other global parameters, such as precipitation, but it also changes its distribution over the earth [8.1.3]. This, in turn, increases the uncertainties about water availability, soil moisture, and even deep-ocean thermohaline circulation due to variations in saltwater concentration.

Reference 8.1.2 provides an excellent summary of the performance of available GCMs. It states that GCMs simulate well the main circulation features of the atmosphere. Furthermore, atmospheric GCMs can predict the weather satisfactorily for short periods of time (about 10 days); with adequate boundary conditions and careful placement of the boundaries of a region, the models can adequately describe regional climates for a season. Seasonal results based on the international Atmospheric Model Intercomparison Project (AMIP) [8.6.7] are shown in Table 8.6.1,

Table 8.6.1 Hemispheric Mean Seasonal Root-Mean-Square Differences between Observations and the Mean of the AMIP Models

Variable	DJF		JJA	
	NH	SH	NH	SH
Mean sea level pressure (hPa)	1.4	1.4	1.3	2.4
Surface air temperature (°C) (over land)	2.4	1.6	1.3	2.0
Zonal wind (m/s) (at 200 hPa)	2.4	1.8	1.8	2.4
Precipitation (mm/day)	0.80	0.71	0.62	0.77
Cloudiness (%)	10	21	14	16
Outgoing longwave radiation (W/m^{-2})	2.8	3.2	2.9	5.5
Cloud radiative forcing (W/m^{-2})	9.1	20.5	16.2	6.5
Surface heat flux (W/m^{-2}) (over ocean)	22.5	27.3	30.5	17.2

Source: Ref. 8.1.2.

where the area root-mean-square (RMS) differences between simulated and measured variables are listed. The RMS values are given for the northern hemisphere (NH) and the southern hemisphere (SH) over periods of three months: December to February (DJF) and June to August (JJA).

The large deviations for the cloud radiative forcing and the surface heat flux at the ocean–atmosphere are worthy of note. The high RMS values for the surface air temperature during the winter are also important because they equal and may even exceed the effect of doubling the amount of CO_2.

For very long periods of time (decades or centuries), the success of GCMs becomes heavily dependent on their coupling with earth components and the accuracy of the coupled models, particularly of the oceans. These are discussed in the next section.

8.7 EARTH

In terms of climate, the most relevant components of the earth subsystem are the oceans, the cryosphere, the land surface, and the biosphere (see Figure 8.4.1). The oceans have the greatest influence because their large heat capacity moderates any potential global warming and because they are the source of most precipitation. The cryosphere includes both sea ice and snow-covered lands. They affect the climate because ice and snow are good reflectors of radiation. When they melt as a result of increased global temperature, additional solar radiation would be absorbed and global warming enhanced. Land surfaces are important because of their role in determining the biosphere level (crops, vegetation, and forestation) and the ground moisture, which affect the land radiative properties.

8.7.1 Oceans

The oceans and seas cover 70.8% of the earth's surface. They contain 97% of all the terrestrial water (1.4 billion km^3). If their water was equally distributed over the entire globe, it would have a depth of around 2750 m. For comparison purposes, the polar ice caps and glaciers account for 2.1% (60 m) of the world's water, the groundwater for 0.7% (20 m), the rivers and lakes for 0.01% (26 cm), soil moisture for 0.002% (4.8 cm), and atmospheric water for only 0.001% (2.8 cm) [8.5.1]. The average daily rain and snowfall amounts to an equivalent depth of 0.3 cm, which means that the atmosphere water is turned over in 10 days, whereas the ocean water would require close to 2500 years. Such turnover statistics provide another strong contrast between the atmosphere's rapid and the ocean's very slow response to perturbations.

Ocean water has a density of 1.03 g/cm^3 and an alkaline pH of 7.8 to 8.4. These characteristics are due to the presence of a variety of elements dissolved or suspended in the ocean water. Eleven elements are responsible for 90% of the solutes: Cl, Na, Mg, K, Ca, Si, Cu, Zn, Co, Mn, and Fe in order of mass. Sodium chloride (NaCl) accounts for over 90% of the solution by mass, and it is primarily responsible for

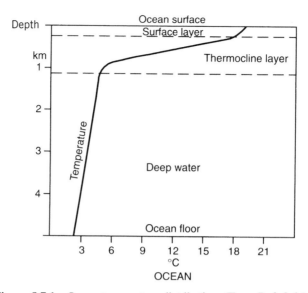

Figure 8.7.1 Ocean temperature distribution. (From Ref. 8.6.1.)

the salinity (percent of solutes by weight in parts per thousand) of ocean water. Salinity values range from about 1 part per thousand in fresh water to between 32 and 37.5 parts per thousand in oceans. Salinity of ocean water increases with depth due to temperature and density stratification away from the surface. However, surface salinity can rise sharply due to increased solar evaporation or low rainfall. A typical temperature distribution within the oceans is shown in Figure 8.7.1. There is a surface layer of 100 to 200 m, where due to solar radiation the temperature is highest: 0 to 30°C, depending on the globe location. Temperatures within the surface layer undergo mixing due to wind and ocean surface currents. The steepest temperature gradients fall between 0.4 and 0.7°C per 10 m below about 100 m and up to 500 to 1200 m. Below this thermocline region, the temperature decreases much more slowly and thermal mixing between the deep water and the thermocline layer is limited by the warmer temperature of the upper hot layer. Also, not shown in Figure 8.7.1, is the photic zone, which allows enough light to penetrate for photosynthesis by phytoplankton. This zone varies between 10 and 200 m, depending on the turbidity of the water. Typical ocean surface temperatures are shown in Figure 8.7.2 and their average is about 17°C. Annual changes in that temperature range from 2°C in the tropics to 4 to 8°C in midlatitude and polar waters.

Oceans are subject to surface circulation due to wind forces and deep-ocean thermohaline circulation due to convective temperature buoyancy and salinity forces. The ocean surface circulation is illustrated in Figure 8.7.3. It is produced by frictional shear stress applied to the ocean surface by winds. The forces are proportional to the square of the wind velocity and depend on the size and form of waves atop the ocean surface. Below the waves there exists a current moving at about 3 to 5% of the wind speed, and it extends 50 to 100 m within the surface layer. Once they are

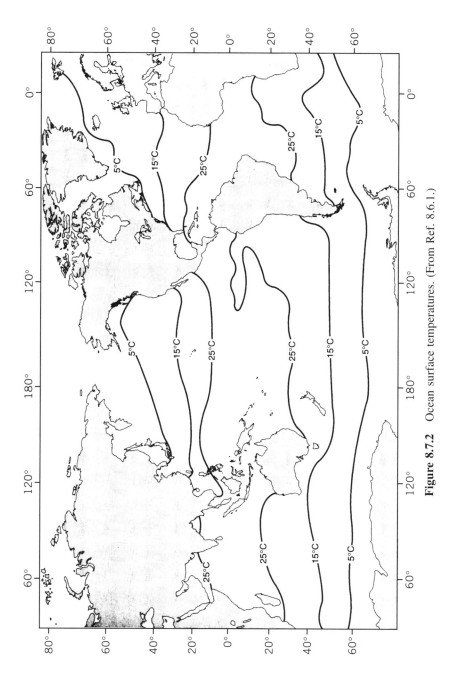

Figure 8.7.2 Ocean surface temperatures. (From Ref. 8.6.1.)

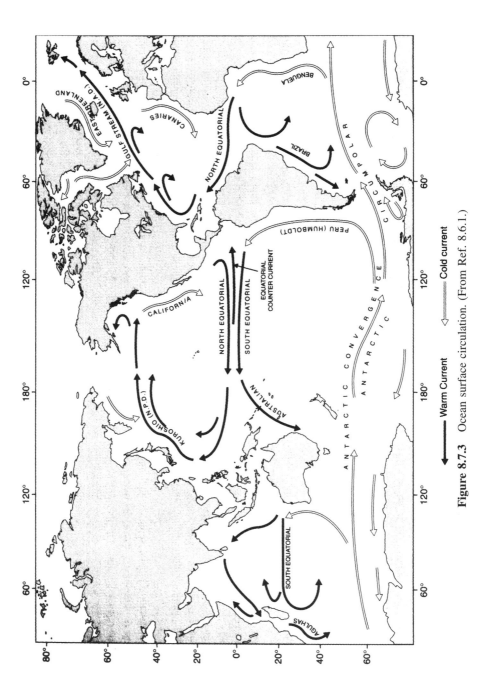

Figure 8.7.3 Ocean surface circulation. (From Ref. 8.6.1.)

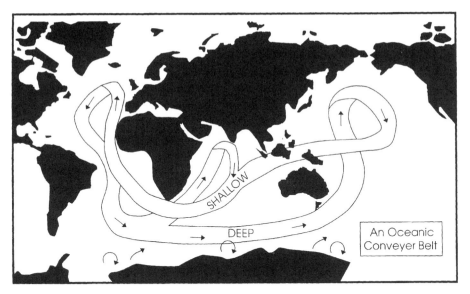

Figure 8.7.4 Ocean thermohaline circulation. (Reprinted from Ref. 8.5.1. with kind permission from Kluwer Academic Publishers.)

out of the friction zone, the layers are deflected further by the Coriolis force: toward the right in the northern hemisphere and toward the left in the southern hemisphere. Warm equatorial waters are pushed west by the trade winds and are redirected toward the poles by the shorelines. Returning cold currents from the poles move east to complete the circulation *gyres* (ocean term comparable to *cells* in the atmosphere). Figure 8.7.3 shows two gyres in both the Atlantic and Pacific oceans and two other major northern major latitude currents: the Gulf Stream or North Atlantic Drift (NAD) and the Kuroshio or Northern Pacific Drift (NPD). The Indian Ocean has a single gyre, and the polar oceans supply cold water through the Antarctic Convergence and Circumpolar Currents and the East Greenland Currents. Although these currents move slowly and are shallow, they transport significant amounts of water and energy (150×10^6 m^3/s at a velocity of about 1.5 m/s for the Gulf Stream). The upper layers of the oceans therefore have a significant interaction with the atmosphere in determining short-term (about 10 years) climate changes and their regional variations.

For longer time durations, the deep-ocean waters become important. Figure 8.7.4 shows the thermohaline circulation or global ocean conveyor belt, which is driven by differences in salinity and temperature of the ocean water. This circulation has a cycle time of 500 to 2000 years and moves at speeds of about 10 to 50 km/yr. It involves all the oceans. The moving surface ocean water becomes cooler and saltier as it travels north, and it sinks and travels south as a deep current. The salty deep currents eventually move northward, mix, and move upward to resurface far away from their original sink locations. The ocean global conveyor belt has a substantial role in long-term heat transport, but it is subject to variations from precipitation, evaporation, and melting or freezing of ice at high latitudes. (The differences in the thermohaline circulations depicted in References 8.1.1, 8.5.1, and 8.6.1 are a testimony to such variations.) Predictions of the North Atlantic maximum thermohaline circulation show considerable scatter and range from 2 to 26 $\times$ 10^6 m^3/s versus an accepted range of 13 to 18 $\times$ 10^6 m^3/s [8.1.2].

For completeness purposes, it should be noted that tides and waves also transmit mass and energy from one part of an ocean to another location near the coastline. Another important biological process in the oceans is worth mentioning. It is referred to as the *biological pump* in Reference 8.1.1 because of its influence on the carbon cycle to be discussed in Section 8.8. In the spring there is a sharp increase of living material, particularly of the plankton plant population, in the ocean surface water. As insects and others eat the plankton, the debris they produce sink partly to the ocean floor and the amount of carbon in the surface water is reduced and more carbon dioxide is drawn down from the atmosphere. Furthermore, when they die or are eaten, some species of phytoplankton produce the gas dimethyl sulfide (DMS). In June 1991, field measurements indicated that algal blooms producing DMS released about 215 metric tons of sulfur per day over an area of 250,000 km^2. Such a release with its potential to exert a cooling feedback is another example of how the biosphere can impact global climate.

In this author's opinion, there is no question that the topic of global oceans is the neglected half of the coupled atmosphere–ocean GCMs. The ocean GCMs adequately portray the large-scale behavior of gyres and the gross features of the ocean conveyor belt, but as pointed out by Gates et al [8.6.7], they have difficulties simulating ocean mixing and mesoscale eddies, the strength of the polar and meriodinal heat transport systems, the temperature and salinity of deep waters, and the surface heat fluxes at the ocean–atmosphere interface. Reference 8.1.2 notes that the available observational ocean data "have significant spatial and temporal shortcomings, with many regions not sampled." The same comments could be made about the El Niño–Southern Oscillation (ENSO), which causes warm top waters to appear in the tropical Pacific Ocean about every 3 to 5 years. The high temperatures spread with time are illustrated in Figure 8.7.5 and they can reach temperatures about 7°C above normal. The floods and droughts associated with the 1982–1983 El Niño and illustrated in Figure 8.7.5 demonstrate how fast and far an ocean temperature disturbance can be propagated by atmospheric circulation. As pointed out in Reference 8.1.2, several efforts have been made to predict ENSO, but they reproduce only "some aspects of the observed variability." Because ENSO produces westward surface-wave motions which are reflected eastward when they reach the ocean western boundaries, the El Niño cycle takes about 4 years (i.e., the time required by the surface waves to travel back and forth). Simulation of waves demands high resolution (i.e., small nodes), which presently limits El Niño modeling to limited ocean regions. As noted in References 8.1.1 and 8.1.2 such models can reproduce *some* of the characteristics of the El Niño disturbance.

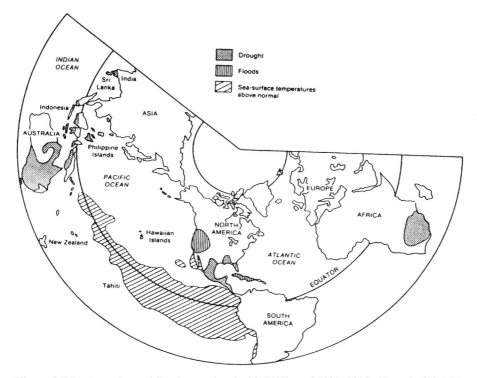

Figure 8.7.5 Droughts and floods associated with El Niño of 1982–1983. (From Ref. 8.1.1.)

8.7.2 Snow and Ice

Snow and ice can accumulate in different forms. Ice sheets and glaciers cover 11% of the land surface and account for most of the 35 km^3 of the earth ice, which is split into 32 km^3 in the Antarctic and 3 km^3 in Greenland. Such continental ice sheets are important to climate changes over thousands of years. Sea ice covers 7% of the ocean global surface area and has a mean thickness of about 2.5 m. Sea ice and snow affect the climate over years to decades. Both snow and ice have a role in climatology because they are good reflectors of solar radiation. When they melt, less solar radiation is reflected. They also have an effect on soil moisture, sea level rise, ocean salinity, and the earth temperatures. For global warming, sea ice and soil moisture are especially relevant. Soil moisture is discussed in the next section.

Dynamic/thermodynamic sea ice models have been able to describe the seasonal variation of the ice extent and its edge location [8.1.2]. The simplest available model employs a stagnant ice layer whose thickness changes due to the heat conduction through the ice. Several more sophisticated snow and ice models have been developed, as discussed in Reference 8.1.2, and their incorporation in the coupled GCMs tends to reduce the climate sensitivity to sea ice.

8.7.3 Land Surface

Land occupies 29.2% of the earth's surface. Its spatial structure is not uniform, which, as noted previously, influences the atmosphere air currents, formation of clouds, and rainfall distribution. Land has many other important impacts on the climate, due to changes in the vegetation, forests, and the fresh water it provides to support life and other needs over the globe. Presently, close to 10% of the land surface is cultivated and another 30% is occupied by forests. By changing the vegetation and/or removing some of the forests, humans affect the prevailing regional climates. For example, deforestation and its replacement by bare soil or grass reduces the local rainfall because tree leaves increase water transpiration and moisture in the atmosphere. Trees also absorb more radiation than is absorbed by bare soil and offer more drag resistance to winds, both of which tend to increase atmosphere convection and rainfall.

Plants and trees play another significant role by taking in carbon dioxide in the presence of light and releasing oxygen back into the atmosphere. This process of photosynthesis helps plant and trees grow while reducing the CO_2 concentration in the air. Some have even argued [8.1.6] that such a CO_2 fertilization benefit might overcome global warming by stimulating even more photosynthesis and permitting the plants to fix more carbon. Reference 8.1.2 by the IPCC reports field studies and analyses of such CO_2 fertilization and concludes that the magnitude of this effect is highly uncertain. A plausible range of its benefit was estimated to be between 0.5 and 2.0 Gt [gigatonnes (thousand million metric tons)] of carbon (C) per year, which falls well below the current yearly rate of carbon addition (see Section 8.8.1). Furthermore, the continual increase in world population is encouraging conversion of forests to other purposes (e.g., agriculture land or bare soil to satisfy the additional needs for food and hardwood). During the 1980s, deforestation accounted for an average loss of about 1% of forest surface area per year, and unless this trend is reversed in the future, the fixation of carbon by forest trees will go down.

Even though the majority of atmospheric water vapor comes from the ocean and returns to that location, there are other relevant water recirculations over land. For example, large amounts of water are extracted from the ground by plants, which release it by evapotranspiration through their leaves to the atmosphere. About 32,000 km^3 per year of water is involved, which is slightly below the 40,000 km^3 of water returning to the ocean from river runoff and water underground percolation. This overall global water cycle is shown in Figure 8.7.6. It illustrates the strong ties provided by water between land and sea and between the earth and atmosphere. The amount of rainfall over land (111,000 km^3) is particularly important because it exceeds the water vapor leaving the land by only 40,000 km^3. That excess amount of water is used (but not readily available) to satisfy the rising demand for fresh water for drinking, food production, and industrial applications. Fresh water is obtained direct from rivers, or by pumping from the underground, or from natural or human-built storage reservoirs.

Fresh water is most important; in fact, the availability of fresh water has much more to do than the earth temperature with the existing forms of life on earth and

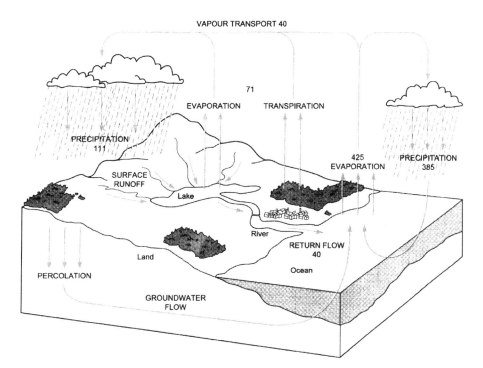

Figure 8.7.6 Global water cycle (thousands of km^3/yr). (From Ref. 8.1.1.)

their survival over time. Worldwide, the total use of water is increasing at an alarming rate; in the last 100 years, it has jumped from 500 km^3/yr to about 5000 km^3/yr, and it is far from leveling. In 20 years it could reach 10,000 km^3/yr and start to stretch or even locally exceed the amount that can be obtained from the available 40,000 km^3. Because the climate determines the precipitation level, its distribution over the land surface, and its variation with time, the climate controls the hydrological processes depicted in Figure 8.7.6. Rainfalls are known to be highly variable from region to region and from year to year. In fact, in recent times, several parts of the globe have been subjected to droughts and unexpected shortages of fresh water.

This heterogeneity in precipitation is compounded by the nonuniformity of soil properties. Land has a limited ability to store water and it varies with the soil permeability, which determines its capacity to absorb and retain water underground. At high latitudes, the amount of frozen soils becomes important. River flows and runoffs also depend on the distribution of rainfall, soil moisture, type of soil, and land topography. Unfortunately, there is no good scaling methodology for the variety of involved processes. Also, the heterogeneity cannot be represented well by the present large horizontal resolution (250 to 1000 km) employed in ocean–atmosphere GCMs. The current GCMs do not predict the global water cycle adequately. Furthermore, they exhibit significant regional differences when their results are compared during global warming.

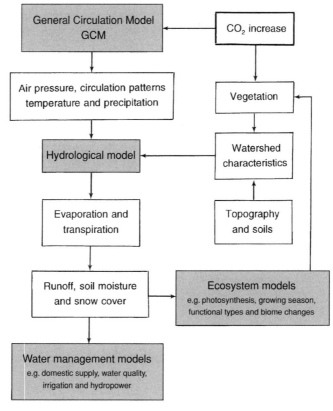

Figure 8.7.7 Linkages between climate, hydrological, and ecosystem models. (Reprinted from Ref. 8.5.1. with kind permission from Kluwer Academic Publishers.)

Global warming will modify the precipitation distribution over land. Also, changes in plant evapotranspiration, snowmelt, and river flows are likely. Soil moisture will therefore be affected. Reference 8.1.3 points out that GCMs use different parametrizations for evapotranspiration and that their projections of future water availability disagree. A total of five formulations are discussed, and they are found to be quite different in their physics assumption, their complexity, and their comparison to the few available measurements. In Reference 8.1.2 the IPCC generally agrees with that conclusion and emphasizes the "need for improved treatment of soil moisture and runoff."

Improved soil–vegetation–atmosphere models are being developed and their features are shown in Figure 8.7.7. Linkages are provided for GCMs to hydrological, ecosystem, and water management models. This extra complexity in simulation will be useful only if the parametrization (closure) equations of the GCMs are improved and if the models can deal with the heterogeneity of the land, hydrological, ecosystem, and climate processes.

By contrast to the oceans, land primarily conducts heat and its thermal conductivity is low. Also, as already noted, the soil has a limited ability to hold water and to store heat, which explains the diurnal changes in land surface temperatures. It also means that the exchanges of heat and moisture between land and atmosphere will take place on a short time scale near the land surface and that they will add further to the difficulties of developing a realistic integrated model of the type illustrated in Figure 8.7.7.

From time to time, the land can impose its own radiative forcing on the world climate. A major perturbation of global temperatures occurs whenever a volcano becomes active. For example, in June 1991, the eruption of Mount Pinatubo in the Philippines injected close to 20 million metric tons of sulfur dioxide and considerable amounts of dust aerosols into the atmosphere. The sulfur dioxide undergoes photochemical reactions using the sun energy to form sulfuric acid and sulfate particles. The particles ejected or formed in the lower atmosphere do not stay there for more than a few weeks before they fall down or are rained out (see Figure 8.6.7). The particles reaching the stratosphere can stay there for several years. Initially, the impact of the material ejected is to cool the atmosphere because it lowers the solar radiation reaching the earth. At its peak, the net global radiative forcing was about -3.5 W/m^2 a few months after the eruption of Mount Pinatubo, and the temperature anomaly was $-0.5°C$. After the amount of volcanic aerosols decreased, the global temperature rose and returned to its original value about 3 years later. Figure 8.7.8 shows the observed surface and night marine air temperatures anamoly and the corresponding GCMs predictions. Figure 8.7.8 shows that the models are able realistically to predict short-time perturbations in radiation forcing and to reproduce the observed phase and magnitude of the global average temperature anomaly.

This completes the summary presentation of the global climate system, its subsystems, and their principal components. It is recognized that it is far from complete, but sufficient details have been provided to draw conclusions and comparisons with nuclear complex systems. These are covered in the final section of this chapter after discussing the topics of greenhouse gases and their potential to cause global warming.

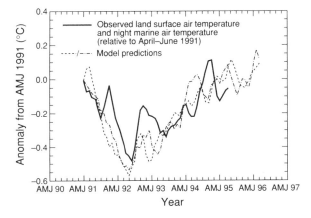

Figure 8.7.8 Predicted and measured global land and ocean temperatures after Mount Pinatubo eruption. (From Ref. 8.1.2.)

8.8 GREENHOUSE GASES

The ability of the atmosphere to act as a barrier to the loss of heat from the earth surface has been called the *greenhouse effect* and the atmosphere gases responsible for it are referred to as *greenhouse gases*. Greenhouse gases have been present in the atmosphere for a very long time. Without their presence, the average earth surface temperature would be close to an untenable temperature of $-23°C$ as determined by the single-temperature model of Section 8.5.3. The principal concern at this time is the continued increase in the concentration of greenhouse gases due to human intervention. Water vapor is one of the most relevant greenhouse gases. It has a life of a few days (see Figure 8.6.7) and its concentration does not appear to be rising due to direct human activities. Water vapor role is discussed in Sections 8.6.3 to 8.6.6, and as mentioned there, humans can have an indirect impact on atmospheric water vapor (e.g., by changing vegetation or encouraging global warming).

The other greenhouse gases that are influenced by human activities are carbon dioxide (CO_2), methane (CH_4), nitrous oxide (N_2O), chlorofluorocarbons (CFCs), hydrochlorofluorocarbons (HCFCs), and hydrofluorocarbons (HFCs). CO_2 is covered first because it is the most dominant of greenhouse gases. The other greenhouse gases come next. This is followed by a discussion of the concept of radiative forcing or of the change produced by greenhouse gases in average net radiation (either solar or terrestrial in origin) at the top of the troposphere. Projections of global warming and the models and uncertainties associated with them are presented last.

8.8.1 Carbon Dioxide

Of all the greenhouse gases, CO_2 contributes most to increasing the greenhouse effect. Its residence time in the atmosphere ranges from 50 to 200 years. Atmospheric CO_2 concentration has increased from about 280 parts per million by volume (ppmv) in 1800 to 358 ppmv in 1994, as illustrated in Figure 8.8.1. This increase comes from combustion of fossil fuel (oil and coal) and to a lesser extent from cement production. The annual cycles shown from 1960 to 1994 result from seasonal variations in plant use of CO_2. The amount of CO_2 currently released by human activities is 5.5 ± 0.5 GtC/yr (gigatonnes of carbon per year). The rate of its atmospheric increase is 3.3 ± 0.2 GtC/yr. The difference between those two numbers is the result of exchanges of carbon between the atmosphere, oceans, and land. Those exchanges and the size of the reservoirs involved are shown in the global carbon cycle depicted in Figure 8.8.2. Some of the exchanges were touched upon previously: photosynthesis, ocean biological pump, deforestation, plant respiration, and land use. What was not discussed before are the significant magnitudes of some of the carbon reservoirs and the *large* rates of exchanges between them; the net transfers of CO_2 between sources and sinks are rather *small* and the uncertainties rather large, as listed in Table 8.8.1. The bottom of the table shows an imbalance of 1.3 ± 1.5 GtC/yr, which has been the subject of a quarter-century debate [8.1.4]. Recent measurements of variations of O_2 in the atmosphere and $^{13}C/^{12}C$ ratios in atmosphere CO_2 point to a northern hemisphere terrestrial sink. Most predictions of future CO_2 assume CO_2 fertilization to be the cause of this last sink and extrapolate

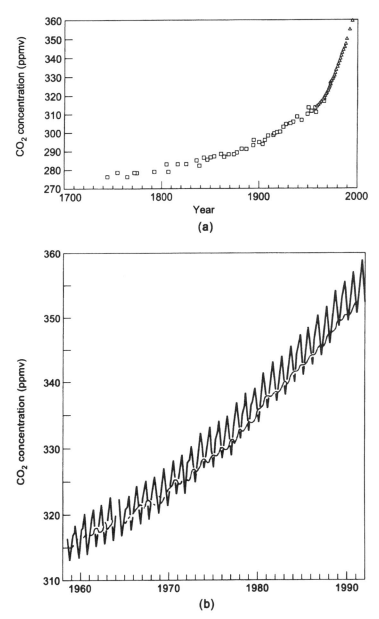

Figure 8.8.1 Atmospheric CO_2 concentrations: (a) since 1700; (b) since 1959. (From Ref. 8.1.1.)

it monotonically with changes in CO_2 concentrations. That strategy may be risky, since the processes and physics presumed for the inferred sink are not fully validated.

There are definite annual variations in the atmosphere CO_2 concentrations, but their timing and size remain controversial. There are much stronger correlations between average global temperature and atmospheric CO_2 concentrations over longer

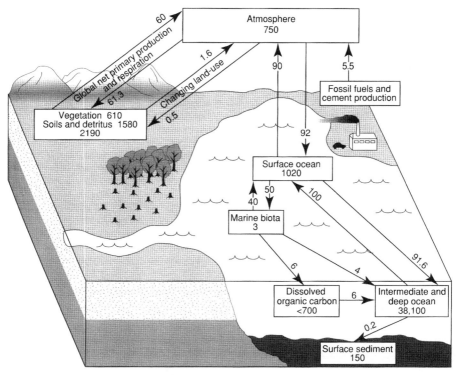

Figure 8.8.2 Global carbon cycle, showing reservoirs (in GtC) and fluxes (in GtC/yr). (From Ref. 8.1.2.)

periods of time. The ice core data show that during warming periods of the ice age cycles, the atmospheric concentration of CO_2 increases and leads or remains in phase with the global temperature rise. During cooling conditions, the atmospheric concentration of CO_2 decreases and lags the temperature change by up to 4500 years. Similar correlations are noticeable over the last several hundred years; CO_2 concentrations dropped during the little ice age of the seventeenth century. Finally, over the most recent decades, there appears to be an adequate correlation between the CO_2 and the temperature anomalies. The reasons for this coupling involve more than just CO_2 and the global temperature. There are numerous ecosystem feedbacks from the terrestrial biosphere that affect the climate change and atmosphere CO_2, as illustrated in Figure 8.8.3. The feedbacks can be positive (+), or negative (−), or doubtful (+/−). Most of the current global warming models do not include ecosystem feedbacks; they include only the long-term (10 years) effects and presume that the exchanges between the different reservoirs depend only on their carbon contents. The combined effects of terrestrial ecosystems, climate, and CO_2 need to be taken into account to get good future simulations of the atmosphere or of the terrestrial ecosystems. The same statement could be made about the marine biogeochemical processes.

Table 8.8.1 Annual Average Anthropogenic Carbon Budget for 1980 to 1989[a]

CO_2 sources	
Emissions from fossil fuel combustion and cement production	5.5 ± 0.5[b]
Net emissions from changes in tropical land use	1.6 ± 1.0[c]
Total anthropogenic emissions = (1) + (2)	7.1 ± 1.1
Partitioning among reservoirs	
Storage in the atmosphere	3.3 ± 0.2
Ocean uptake	2.0 ± 0.8
Uptake by northern hemisphere forest regrowth	0.5 ± 0.5[d]
Inferred sink: 3 − (4 + 5 + 6)	1.3 ± 1.5[e]

Source: Ref. 8.1.2.
[a]CO_2 sources, sinks, and storage in the atmosphere are expressed in GtC/yr.
[b]For comparison, emissions in 1994 were 6.1 GtC/yr.
[c]Consistent with Chapter 24 IPCC WGII (1995).
[d]This number is consistent with the independent estimate, given in IPCC WGII (1995), of 0.7 ± 0.2 GtC/yr for the mid- and high-latitude forest sink.
[e]This inferred sink is consistent with independent estimates, given in Chapter 9, of carbon uptake due to nitrogen fertilization (0.5 ± 1.0 GtC/yr), plus the range of other uptakes (0–2 GtC/yr) due to CO_2 fertilization and climatic effects.

8.8.2 Other Greenhouse Gases

Methane (CH_4) is the second most important greenhouse gas. Its atmospheric concentration was about 0.7 ppmv in 1800 and it is currently above 1.7 ppmv. CH_4 grows at a rate of 0.8% per year or twice that of CO_2. Even though the atmospheric concentration of methane is much lower than CO_2, its enhancing greenhouse effect is important because it is 7.5 times that of CO_2 per molecule.

The principal natural source of methane is wetlands. It is generated from decomposition of organic matter in rice paddies, landfills, and during the digestive process of ruminants. CH_4 is also released during natural gas drilling and biomass burning. Methane is removed from the atmosphere by chemical reaction with hydroxyl radicals formed from oxygen, ozone, water vapor, and sunshine. Hydroxyl radicals have a lifetime of 1 s, and their fast reaction time limits the lifetime of methane to about 12 years in the atmosphere. The same strong coupling, reported for CO_2, exists between CH_4 and global temperature. For example, CH_4 concentration was about 300 to 400 ppb (parts per billion) during glacial periods and 600 to 700 ppb during interglacial periods. Methane production and consumption depend on a much smaller number of feedback systems: soil temperature, soil moisture, and land topography (change in water table).

Nitrous oxide concentrations have increased from about 275 ppb in preindustrial times to about 310 ppb, and their rate of increase is 0.25% per year. N_2O comes from combustion, fertilizers, and biomass burning. It is removed from the atmosphere by photodissociation, which breaks apart molecules of N_2O with energy from solar radiation. Nitrous oxide has a lifetime of about 120 years.

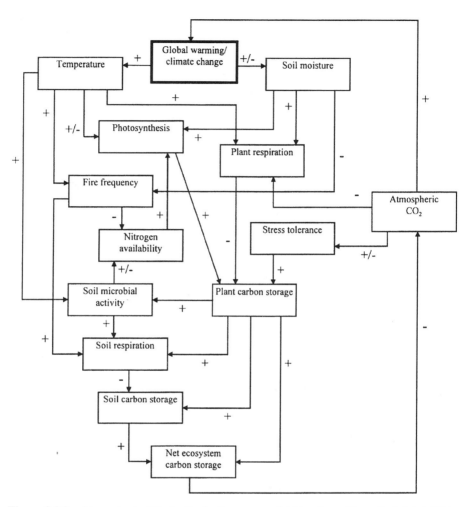

Figure 8.8.3 Atmospheric CO_2 feedbacks from terrestrial biosphere. (From Ref. 8.1.4. With permission, from the *Annual Review of Energy and the Environment*, Volume 22, © 1997, by Annual Reviews.)

The chlorofluorocarbons (CFCs) are used as propellants, refrigerants, and fire retardant. Their current concentration in the atmosphere is low, 500 parts per trillion by volume (pptv), but their concentration was increasing at the yearly high rate of 4%. In addition to enhancing the greenhouse effect, they can lead to the destruction of the ozone layer after they are subjected to photodissociation. For that reason, their use is being eliminated and their replacements are the hydrochlorofluorocarbons (HCFCs) and the hydrofluorocarbons (HFCs), which do not destroy ozone. Their current concentration is at 110 pptv, but their concentration is increasing at the dramatic yearly rate of 7%. While the depletion of the stratospheric ozone layer is serious environmentally, its destruction has been shown to increase cooling of the stratosphere and to partly compensate for the presence of CFCs.

The radiative forcing of the various greenhouse gases is not the same per unit mass. Relative to a unit mass of CO_2, a unit mass of CH_4 is 56 more effective in enhancing the greenhouse effect; NO_2 is 280 times more effective and the CFCs are 2000 to 6000 times more effective. In terms of greenhouse contribution, the gases rank in the order they were introduced in the text. As noted in Figure 8.6.7, their lifetimes are different, but they are large and range from 12 years for methane to 100 and more years for the others. All the greenhouse gases are chemically active with the exception of CO_2, and it is important to work out their interactions among themselves and with other atmosphere constituents to achieve acceptable greenhouse predictions. For example, carbon monoxide (CO) emitted by motor vehicles forms CO_2 due to chemical reactions in the atmosphere and can affect the greenhouse effects.

8.8.3 Radiation Budget

A global average radiation budget can be constructed as shown in Figure 8.8.4. This is a modified picture of the balance developed by Kiehl and Trenberth [8.1.2, 8.8.1]. It starts from the average amount of solar radiation incident on the entire earth surface, or 342 W/m². The incoming solar radiation is reflected by clouds, aerosol, and atmosphere at the rate of 77 W/m². In addition, 30 W/m² is reflected by the earth surface. Of the remaining 235 W/m² incoming radiation, 67 W/m² is absorbed by the atmosphere and 168 W/m² reaches the earth surface. Some of that 168 W/m² radiation is retransmitted unimpeded through the atmosphere from areas where there are no clouds. It is referred to as the *atmospheric window* and amounts to 40 W/m². The rest, 128 W/m² reaches the atmosphere through sensible heating (thermals) accounting for 24 W/m², evapotranspiration (latent heat) responsible for

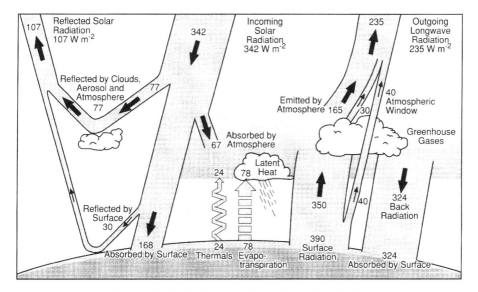

Figure 8.8.4 Radiation budget. (From Ref. 8.1.2.)

78 W/m², and net radiation of about 26 W/m². Of the total energy of 195 W/m² absorbed into or transmitted to the atmosphere, 30 W/m² is emitted to the upper atmosphere by clouds. The balance 165 W/m² is emitted to space by the atmosphere containing greenhouse gases, which act as a blanket for the earth. As the concentration of greenhouse gases increases, the 165 W/m² emitted radiation will rise and so will the blanket effect. It is interesting to note that the latent heat of 78 W/m² shown in Figure 8.8.4 corresponds at an average temperature of 10°C to a total evaporated water amount of 505,000 km³/yr, which is very close to the total value of 496,000 km³/yr appearing in Figure 8.7.6.

8.8.4 Radiative Forcing of Greenhouse Gases

Global mean radiative forcing offers a simplified approach to assess the relative climatic importance of various mechanisms (e.g., different greenhouse gases). When the concentration of a greenhouse gas in the atmosphere is changed, it affects the incoming solar radiation and the outgoing terrestrial radiation at the top of the troposphere. The net change is called *radiative forcing*. It is measured in watts per square meter, and can be positive or negative.[†] Variations in the concentrations of greenhouse gases depend on changes in their sources and sinks. Future projections require a good understanding of the processes producing and removing the greenhouse gases. In Reference 8.1.2 it is noted that 18 different models for the carbon cycle gave future CO_2 concentrations, which varied ±15% about the median value of the models.

Figure 8.8.5 shows the radiative forcing of the greenhouse gases from 1850 to the present. The contribution of individual greenhouse gases is indicated in the first bar and their total is 2.45 W/m² with an uncertainty of ±15%, depicted by an error vertical line at the middle of the first bar. In Figure 8.8.5, the calculation of radiative forcing allows the stratospheric temperatures to readjust to a radiative equilibrium (on the order of a few months) but with the surface and tropospheric temperature and atmospheric moisture kept fixed. The next two bars show the radiative forcings associated with stratospheric and tropospheric ozone. The following three bars cover the direct contributions of sulfate, fossil fuel soot, and biomass burning aerosols. The indirect tropospheric aerosol effect arises from their potential impact on cloud properties. As pointed out in Reference 8.1.2, the "understanding of this process is limited" and only an uncertainty range is shown. The final bar deals with natural changes in solar output over the same period of time, from 1850 to the present. Future projections of radiative forcings have been made [8.1.1]. For example, in the case of CO_2, the radiative forcing grows by a factor of about 4 from 1990 to the year 2100. It goes without saying that the uncertainties of such projections are much larger. Often, it is desirable to compare the relative radiative forcing of different greenhouse gases between now and specified future time. The global warming

---

[†]Even though water vapor has been often referred to as a greenhouse gas, its impact is evaluated from GCMs and it is considered to be a climate feedback rather than a radiative forcing.

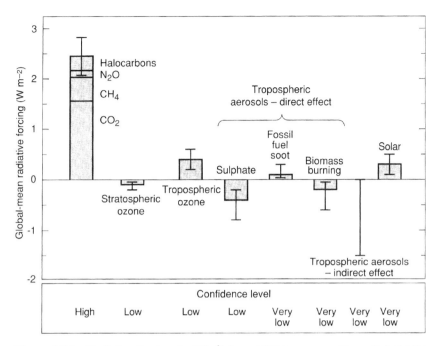

Figure 8.8.5 Radiative forcing (in W/m^2) from 1850 to present. (From Ref. 8.1.2.)

potential (GWP) provides such an index. It is usually expressed relative to CO_2, and typical current values of GWP were provided toward the end of Section 8.8.2.

Both the radiative forcings and GWPs must be used with caution. They both deal with global circumstances and do not recognize spatial or regional patterns. In particular, the addition of positive and negative radiative forcing functions is to be avoided because the gases and aerosols may be distributed differently and unevenly. Such projections are especially unsatisfactory when carried out into the future because their future values and interactions are much less certain. The alternative and preferred approach is to employ more complex models (i.e., GCMs) to guide the projections and to consider coupling terrestrial ecosystems and marine biogeochemical processes.

8.8.5 Global Warming Projections

If human activities continue to add greenhouse gases under a "business as usual" scenario, the vast majority of available studies project an increase in global mean surface air temperature relative to 1990 of about 2°C by 2100 [8.1.2]. This projection includes the effects of future increases in aerosols. If the aerosols remain at their current level, the temperature rise would be about 2.5°C. The models also project [8.1.2]:

- More warming of the land than that of the oceans

- Minimum warming where deep oceanic mixing occurs (e.g., Antartica and the North Atlantic)
- Maximum warming in high northern latitudes due to reduced sea ice cover (it is the largest in late autumn and winter)
- An increase in global mean precipitation
- Little seasonal variation of the warming in low latitudes or over the southern circumpolar ocean
- Increased frequency of extreme events (e.g., intense precipitation)
- Sea level rise of about 50 cm

In developing such projections, the IPCC goes to considerable length to identify the many prevailing uncertainties and the lack of present understanding and sufficient measurements. Many of these shortcomings have been identified in previous items. They include concern about radiative performance of clouds (see Section 8.6.4); the level and distribution of aerosols (see Section 8.6.5); the complex nature of atmosphere circulation (see Section 8.6.5); the number and validity of the parametrizations or closure equations in GCMs (see Section 8.6.6); the difficult exchanges of momentum and energy at the interfaces of the atmosphere with the land and oceans (see Section 8.6.6); the large cells employed by the GCMs and their difficulty to account for local heterogeneity and to agree upon regional trends (see Section 8.6.6); the ability to predict the surface and deep-water thermohaline circulations of the oceans (see Section 8.7.1); the simplified snow and ice representations (see Section 8.7.2); the changes in the global water cycle (see Section 8.7.3); the impact of periodic disturbances of ENSO and volcanic eruptions (see Sections 8.7.1 and 8.7.3); the small net exchanges between large sources and sinks in the global carbon cycle (see Section 8.8.1); and the large number and poor understanding of the land ecosystems and marine biotic feedbacks (see Section 8.8.1).

Some scientists have argued that the IPCC projections overestimate the global warming. Their main argument is that the involved processes are very complex and cannot be simulated accurately as one might infer from the preceding long list of potential concerns. They also point to the fact that over the past century, global warming was only about 0.5°C and that it could be due exclusively to natural variability, which is not considered in the IPCC projections. Some have gone further and have proposed that global warming may have benefits and that efforts to cut back increased emissions of greenhouse gases will particularly cripple developing countries [8.8.2]. The debate is best left to others. Instead, in this section, some of the physics of global warming are discussed because at the end they will determine the actual trend. The debate, in my opinion, will be resolved when the relevant physics are put on a much more solid ground than they are today by additional measurements and improved models.

The concentrations of greenhouse gases in the atmosphere are rising (e.g., see CO_2 behavior in Figure 8.8.1) due to increased emissions from human activities. About 45% of the CO_2 emissions stay in the atmosphere, and at their current level, they would lead to an increase in concentration through at least the next century, due to the long life of greenhouse gases. The increased concentrations produce a positive radiative forcing on the lower atmosphere which is reduced by the direct and indirect effects of aerosols (see Figure 8.8.5). The presence of aerosols in the atmosphere is also the result of human intervention, and the impact of some of the aerosols is anticipated to decrease with time for two reasons: The ozone destruction in the stratosphere will be lowered if the emission of CFCs is reduced as planned; and the sulfate aerosols come from emission of sulfur dioxide from coal-burning power plants and from volcanic eruptions. The sulfate aerosols not only scatter solar radiation but also modify the behavior of clouds to reflect more solar radiation. Sulfate aerosols have a short lifetime (see Figure 8.6.7) and their concentration in the atmosphere could decrease sharply if efforts to reduce acid rain and to employ low-sulfur coal or improved scrubbers are implemented in fossil power plants.

The impact of increased concentrations of greenhouse gases is expected to overcome the effects of aerosols and to lead to global warming. In fact, according to Reference 8.1.2, the average global surface temperatures have increased by about 0.3 to 0.6° from 1850 to the present and by about 0.2 to 0.3°C over the last 40 years (see Figure 8.3.1). A temperature increase of 0.3°C according to the single-temperature model of Section 8.5.3 requires a radiative forcing of 1.1 W/m^2, while an increase of 0.6°C requires a radiative forcing of 2.25 W/m^2. While these radiative forcings fall within the anticipated range of Figure 8.8.5, they fail to account for two important factors: (1) the feedback processes, which can amplify or lower the change in global mean temperature, and (2) the role of the oceans in moderating the global climate change.

Cloud feedback is one of the most important feedbacks because it influences the solar radiation input to the earth and the radiative heat loss to space. The feedback mechanisms are rather complicated and involve cloud coverage, altitude, water and ice content, and the presence and distribution of aerosols. Recent measurements indicate that the current characteristics of clouds cool the globe. If the cloud coverage is reduced, their greenhouse effect would be decreased and the feedback negative. If the clouds move to lower altitudes, the feedback would again be negative; if the cloud water and ice content changes, the resulting feedback is not agreed upon by existing models. The overall effect of clouds is assessed by GCMs to be positive in a warming climate, but the physics to support that result are still lacking.

Another important feedback mechanism is associated with water vapor. When the surface temperature rises, more water is evaporated and the air can hold more water. Since water vapor acts as a greenhouse gas, the feedback is anticipated to be positive as long as that excess water vapor does not influence the cloud properties (e.g., reduce their altitude or their water content). This is an important characteristic of the global climate where one feedback mechanism affects another (e.g., aerosol and water vapor influence on clouds).

The snow–ice feedback is straightforward, it is positive because with increased temperature some snow–ice will melt and less solar radiation would be reflected. Similarly, there is a net overall positive feedback between global temperatures and concentrations of greenhouse gases even though the mechanisms are complex (see Figure 8.8.3). In general, most feedback systems evaluated tend to amplify the increase in global temperature. That temperature rise is, however, moderated greatly by the ocean circulations. The oceans will provide a significant time lag to the rise in temperature and it will depend on the degree of ocean water mixing and the volumetric flow rates involved. Here again the physics are far from established (see Section 8.7.1).

In summary, human activities are increasing the amount of greenhouse gases in the atmosphere, and the resulting trend is anticipated to be an increase in global temperatures. The rate of temperature rise is presently uncertain and will be firmed only after such physics as those involved with aerosols, clouds, water vapor, and ocean circulations are understood and established.

8.9 SUMMARY

8.9.1 General Comments

1. If nuclear systems are complex, the global climate system pushes complexity to the ultimate level. The coupling between subsystems and components is very strong and the number of feedback systems very large.

2. An intensive and coordinated international program (IPPC) is in place with regularly scheduled meetings and published proceedings. That degree of coordination and cooperation was never achieved in nuclear systems, due to commercial competition and different national objectives. A recommendation is made in Section 7.1 to hold international meetings for all complex systems. That recommendation is all the more valid now, particularly the concept of increasing the participation of practitioners and pragmatists.

3. An emotional debate about global warming would be very unfortunate. There are enough uncertainties that one could make the case one way or the other by employing selective information. Emphasis needs to be placed on the physics involved and to improve their understanding.

8.9.2 Comments about International Climate Change Program

1. As stated in Section 7.2, all complex systems need to create a phenomena identification and ranking table (PIRT). This is particularly important for the super complex global climate system. Once the important phenomena are identified, the available and additional necessary separate effects and components tests need to be listed. Because of the large scale of the global climate system and its interacting feedbacks, reduced-scale integral tests are not very likely. Very little scaling analysis of global climate was found. Scaling analysis of past and future climate tests and models is essential.

2. The only way to study global climate changes is with highly sophisticated computer system models (i.e., GCMs). As indicated in Section 7.2, the models need

to be documented as well as their predictions and their changes controlled. There are too many global computer system models, and an international effort to reduce that number may be beneficial as long as the models selected are made available to all the countries agreeing to sponsor and to improve them.

3. The range of time and length scales are enormous in the global climate system. As stated in Chapter 2, there is merit to grouping the phenomena which have the same time and length scales and time phasing their occurrences so that one "can avoid doing with more what can be done with less." Figure 8.6.7 was particularly useful to this author in segregating the fast climate system to predict the weather from the slow climate system to project global warming. It is much more important to simulate the significant issues rather than all the potential issues. This may be particularly advisable in representing the biosphere.

4. The GCMs suffer from the same problem as nuclear system computer codes: They both use relatively large nodes and a large number or variety of approximate closure equations or parametrizations. Because of their coarse current nodalization, reality demands that emphasis be put on macroscopic rather than microscopic details of the simulation.

5. The global climate models could benefit from the experience of nuclear system codes: benchmark problems to test the numerical method, standards for establishing models and for selection of nodalization, doubling and halving the nodalization to assure numerical stability, automating the introduction of inputs, grid adaptation to different nodes and varying numerical schemes when necessary and abandoning them when they are not required, development of low-numerical-diffusion methods, and so on.

6. The global climate models need to evaluate the uncertainties in their projections. It is not clear that the IPCC considers all the uncertainties of inputs, model, parametrization, measurements, and so on.

8.9.3 An Outsider's Comments

1. Too much emphasis is put on the atmosphere by comparison to the oceans, which deserve more attention because they will determine the global temperature rise over the first few decades if not the first century. The benign neglect of oceans is obvious if one looks at the poor comparison of the ocean thermohaline circulations predicted by several models.

2. The next important area is the global water cycle because it has the greatest impact on ecosystems and human activities. Current models differ in their predictions of overall precipitation and even more widely in their predictions of regional precipitations. Changes to uses of water and to the availability of fresh water can lead to catastrophic regional conditions unless they are managed properly. Detailed evaluations of ecosystems will have significance only when the global water cycle projection is accurate.

3. The carbon cycle requires agreement on the various exchanges involved and how they will adjust as the carbon concentrations in the atmosphere increase.

4. Simple climate models are most appropriate to assess global warming. The simplification in number of dimensions and the enlarging of time scales will provide improved answers when they are applied to all subsystems and components (i.e., atmosphere, land, and oceans). The lumped-parameter approach used in Chapter 2 may deserve examination. For example, lumping atmospheric parameters in three latitude groupings as done in Figure 8.6.5 may be desirable.

5. Although paleoclimate comparisons are worthwhile, they are limited by missing information. Increased emphasis should be placed on recent events, particularly such extreme events as volcanic eruptions and ENSO. Additional measurements and improved instruments need to be considered to identify such parameters as the cause and total energy released by ENSO and the spreading temperature distributions with time. Ocean temperature and flow measurements lag considerably behind the atmospheric data.

6. Instead of focusing on average global temperatures, the attention needs to be shifted to extreme regional temperatures (e.g., tropics and arctic), which will provide an earlier and more accurate assessment of global warming. They should not be as susceptible to the natural variability of the climate. For example, use of the geoscience laser altimeter system starting in 2001 to measure temperatures at the north and south poles is welcome [8.9.1].

7. The following phenomenological issues deserve special considerations: cloud processes, ocean circulations, precipitation amounts and patterns, momentum and energy transfers at atmosphere–ocean and atmosphere–land interfaces, CO_2 net interchanges, and soil moisture. The physics involved are not understood well enough.

8. The variety and number of parametrizations or closure equations for the same phenomena should be reviewed and reduced. Some closure equations, such as those dealing with nonequilibrium conditions in the clouds, are best left in an empirical form.

9. The nonlinearity and chaotic nature of the global climate conservation equations should be stressed in order to anticipate unstable thermal-hydraulic behavior such as possible flip-flops in the deep-ocean natural circulation. Recent ice cores from central Greenland give credence to such flip-flops and the concept of sudden climate changes [8.5.1].

10. The current international program needs to be reexamined periodically to simplify and coordinate all the ongoing efforts and to increase their coherence [8.9.2].

If I were a betting person, I would be on the side expecting global warming as the concentration of greenhouse gases increases. After working for over 45 years on the safety of nuclear power plants, I have learned not to bet on results but rather to emphasize conservative decision making. There are *practical* corrective actions that can be taken now while monitoring the trend of global temperatures. That is the conservative and appropriate way to proceed.

REFERENCES

8.1.1 Houghton, J., *Global Warming*, 2nd ed., Cambridge University Press, New York, 1997.

8.1.2 Intergovernmental Panel on Climate Change (IPCC), *Climate Change 1995: The Science of Climate Change*, Cambridge University Press, New York, 1996.

8.1.3 Rind, D., et al., "The Role of Moisture Transport between Ground and Atmosphere in Global Change," *Annu. Rev. Energy Environ.*, **22**, 47–74, 1997.

8.1.4 Lashof, D. A., et al., "Terrestrial Ecosystem Feedbacks to Global Climate Change," *Annu. Rev. Energy Environ.*, **22**, 75–118, 1997.

8.1.5 Parson, E. A., and Fisher-Vanden, K., "Integrated Assessment Models of Global Climate Change," *Annu. Rev. Energy Environ.*, **22**, *589–628, 1997.*

8.1.6 Robinson, A. R., et al., *Environmental Effects of Increased Atmospheric Carbon Dioxide*, George C. Marshall Institute, January 1998, Cuba City, Oregon.

8.3.1 Jones, P. D., "Hemispheric Surface Air Temperature Variations: A Reanalysis and an Update to 1993," *J. Climate*, **7**, 1794–1802, 1994.

8.3.2 Jones, P. D., "Hemispheric Surface Air Temperature Variations: Recent Trends and an Update to 1987," *J. Climate*, **1**, 654–660, 1988.

8.3.3 Houghton, J. T. (ed.), *Climate Change, 1992*, Supplementary Report to the IPCC Scientific Assessment, Cambridge University Press, New York, 1992.

8.5.1 Münn, R. E., et al., *Policy Making in an Era of Global Environmental Change*, Kluwer Academic Publishers, Norwell, Mass., 1996.

8.5.2 Transportation Research Board, *Addressing the Long-Term Effects of Motor Vehicle Transportation on Climate and Ecology*, National Academy Press, Washington, D.C., 1997.

8.5.3 Renno, N. O., et al. "Radiative-Convective Model with an Explicit Hydrologic Cycle, 1: Formulation and Sensitivity to Model Parameters," *J. Geophys. Res.*, **99**(7), 14429–14441, July 1994.

8.6.1 Briggs, D., et al., *Fundamentals of the Physical Environment*, 2nd ed., Routledge, New York, 1997.

8.6.2 Houghton, J. T., *The Physics of Atmospheres*, 2nd ed., Cambridge University Press, New York, 1986.

8.6.3 Ramanathan, V., et al., "Cloud-Radiative Forcing and Climate: Results from the Earth Radiation Budget Experiment," *Science*, **243**, 57–63, 1989.

8.6.4 IPCC, *Climate Change: Radiative Forcing of Climate Change*, Cambridge University Press, New York, 1994.

8.6.5 Palmer, E., and Newton, C. W., *Atmospheric Circulation System*, Academic Press, San Diego, Calif., 1969.

8.6.6 Gleick, J., *Chaos: Making a New Science*, Viking Penguin, April 1988.

8.6.7 Gates, W. G., et al., *An Intercomparison of Selected Features of the Control Climates Simulated by Coupled Ocean–Atmosphere General Circulation Models*, World Climate Research Programme WCRP-82, WMO/TD 574, World Meteorological Organization, Geneva, 1993.

8.8.1 Kiehl, J. T., and Trenbeth, K. E., "Earth's Annual Global Energy Budget," *Bull. Am. Meteorol. Soc.* (to be published).

8.8.2 Moore, T. G., *Climate of Fear: Why We Shouldn't Worry about Global Warming*, Cato Institute, Washington, D.C., 1998.

8.9.1 Carraline, L., "Taking Earth's Temperature," *Mech. Eng.*, September 1998.

8.9.2 *Global Environmental Change: Research Pathways for the Next Decade: Overview*, National Academy Press, Washington, D.C., 1998.

INDEX

Aerosols, 379, 388–389, 394, 400, 413
Annular flow
   closure equations, 248
   continuous liquid layer, 227, 228
   deposition rate, 245, 248
   drift flux model, 120, 238
   entrainment rate, 135, 240, 245–247, 254
   gas core with entrained liquid, 238, 248–251, 255
   heat transfer in annular flow, 252, 253, 255
   ideal annular flow analysis, 222–226
   interfacial area, 133
   interfacial flooding waves, 222, 237, 254
   interfacial friction factor, 135, 137, 229, 232, 234, 235–238
   interfacial ripple waves, 230
   interfacial roll waves, 222, 232
   liquid droplet characteristics, 241–243
   liquid droplet drag, 134, 239, 242, 243
   mixing length for gas core, 250
   modifications to ideal annular flow, 226, 227
   momentum equation for liquid droplets, 244
   transition from annular to mist flow, 128–131, 253
   two-layer liquid film, 227, 230
   types of annular flow, 221
Atmosphere
   adiabatic lapse rate, 381, 384, 385
   circulation, 389–393
   components, 379–380
   Coriolis force, 391
   Hadley cell, 391–392
   Rossby number, 391
   Rossby waves, 392
   sloping convection, 392
   temperature structure, 379–380
   temperature variation, 380, 382, 383
   winds, 374, 392

Bottom-up scaling
   containment phenomena, 68–76
   peak fuel clad temperatures, 60–68
Bubble flow
   batch process, 162, 170
   bubble collision and clusters, 181, 182, 189, 190
   bubble distribution, 177
   bubble drag correlation, 164–169, 190
   churn turbulent flow, 163
   cocurrent upward flow, 170, 184, 185

Bubble flow (*continued*)
  countercurrent flow, 170, 184, 185, 264
  drag coefficient/correlation, 134, 173, 191
  frictionless bubble flow analysis, 169
  ideal bubble flow, 162
  interfacial area/shear, 132, 134, 173, 175, 177, 178, 180
  multidimensional bubble flow, 182–187, 191
  one-dimensional bubble flow analysis, 174–176
  Sauter mean bubble diameter, 177–182
  single bubble rise, 164–167
  transition from bubble to slug flow, 128–131, 175, 187–190
  types of bubble flow, 162–164
  wall friction, 174, 176

Carbon Dioxide (CO_2)
  concentration, 406–407
  cycle, 406–408
  deforestation, 402
  feedback, 408, 410
  fertilization, 402
  photosynthesis, 402
  radiative forcing, 412
Circulation models, general
  atmospheric model intercomparison, 394, 395
  equations, 393, 417
  parameterizations, 393
  performance, 394
  periodic flux adjustment, 394
Closure laws
  definition and role, 29, 57, 99–104, 131–143, 155, 361
  interfacial heat transfer, 134
  interfacial heat transfer to bubbles and liquid drops, 140, 141
  interfacial shear area, 132–134, 361
  interfacial shear stress, 134–136, 361
  subcooled condensation, 141
  wall boiling, 142, 143
  wall heat transfer, 139
  wall shear, 136–138

Clouds
  albedo, 387
  classification, 386, 387
  heights, 386, 387
  ice content, 386–388
  precipitation, 387, 388
  radiative forcing, 394, 395
  sky coverage, 386
  water content, 386–388
Complex systems
  alternate test and analysis program, 11–13
  characteristics, 13, 16, 43, 65, 80
  integrated test and analysis program, 1, 13–16, 29, 83, 360
  phenomena importance and ranking, 17–23, 358
  scaling, 13, 28, 29, 32–34, 358, 360
  scenarios, 13, 16, 17, 83
  system computer codes, 2, 28, 86–154, 359
  test matrix and plan, 9, 13, 23–25, 359
Conservation laws with interface exchange
  two-fluid model, 125–127
Conservation laws with no interface exchange
  average fluid density, 90, 91, 94
  average kinetic energy density, 92–94
  average momentum density, 91, 92, 94
  conservation of energy, 92
  conservation of mass, 90
  conservation of momentum, 91
  nondimensional groupings, 95–98
  steady-state flow, 93–95
Countercurrent flooding conditions
  annular geometry, 283–287
  boiling water reactor upper plenum, 287–289
  definition, 264–265
  experimental setups, 266–268
  flow patterns in downcomer, 283–286
  nonequilibrium thermal conditions, 287
  pressurized water reactor downcomer, 291–293

pressurized water reactor emergency
    cooling ports, 290, 291
pressurized water reactor upper
    plenum, 289, 290
tube geometry, 265, 268–273
Countercurrent flooding predictions
    Kutateladze correlation, 268
    system computer codes, 293–295
    Wallis correlation, 268
Countercurrent mechanisms
    change in wave motion, 275, 276
    droplet entrainment, 281–283
    kinematic waves, 273–275
    upward liquid film flow, 276–281
Critical flow
    single-phase
        gas flow with friction, 305–306
        ideal compressible flow, 301–305
        nonideal compressible flow,
            305–310
        through sharp-edged orifice,
            308–310
        through valve, 306–309
    two-phase/nonthermal equilibrium
        frozen homogeneous model,
            327–329
        Henry–Fauske model, 329–330
        other nonequilibrium models, 333
        pressure undershoot model,
            331–333
    two-phase/thermal equilibrium
        homogeneous equilibrium model,
            311–318
        Moody model with friction,
            321–324
        Moody model with slip, 318–321
        Richter model, 333–337
        separated flow models, 324, 326
Critical flow data and predictions
    Marviken experiment, 340
    nozzles, 337
    pipe cracks, 341–347
    Reocreux experiment, 331, 336
    slits, 345
    small break in horizontal pipe,
        347–351
    system computer codes, 337–340, 351
    vapor pull-through, 348–351

Dispersed flow, 87, 132, 160, 161,
    255–257
Drift flux model
    applications, 56, 118–125, 173, 238
    EPRI correlation, 121–124, 363
    Zuber and Findley model, 119

Earth
    biosphere, 402–404, 410
    cryosphere, 401
    land surface, 402, 405
    oceans, 395–401

Feedbacks
    carbon dioxide, 410
    cloud, 415
    water vapor, 415
Flow patterns
    annular flow, 87, 128–131, 160, 161,
        201, 215, 221–252
    bubble flow, 87, 128–131, 160,
        162–192
    slug flow, horizontal and vertical, 87,
        128–131, 160, 161, 192, 214,
        215
    spray, mist, dispersed flow, 87,
        128–131, 160, 161, 255
    stratified flow, 87, 128–131, 161, 215
Future of complex systems
    future of two-phase flow, 364–368
    key issues in system computer codes,
        362–364
    key issues in two-phase flow, 360–362
    players, 356–358

Global climate
    average temperatures, 370–373
    carbon dioxide cycle, 406, 408
    climate, 370
    climate change, 369, 370, 416
    climate system, 369, 371
    paleoclimate, 389 (glacier age), 418
    time and space scales, 377, 390
    warming, 369, 370
    warming potential, 413
    water cycle, 403, 404
Greenhouse gases
    carbon dioxide, 406–409
    others, 409–410

Greenhouse gases (*continued*)
    ozone layer, 410, 415
    radiative forcing, 412, 413
    water vapor, 386

Homogeneous/uniform property flow
    closure laws, 99–104
    conservation equations, 98–99
    critical flow, 311–318
    natural circulation, 54–56
    thermal conductivity, 106
    viscosity, 104, 105
Horizontal stratified flow
    interfacial friction, 220, 221
    interfacial waves, 218, 219, 221
    smooth stratified flow, 215–218
    transition to slug flow, 128–131

Loss-of-coolant accident
    emergency cooling, 5
    fuel peak clad temperature, 11, 60–68
    large break, 4–11, 238
    phenomena, 9, 10

Natural circulation
    single-phase flow, 51–54
    two-phase homogeneous, 54–56
    two-phase nonhomogeneous, 33, 56, 57
Nuclear power plants
    boiling water reactors, 51, 68
    containment, 40, 47, 48, 60, 68–79
    decay heat, 44, 64, 65, 68, 77
    nuclear reactors, 1, 40
    passive containment, 68–72
    pressurized water reactors, 4, 32, 153

Oceans
    currents, 398, 399
    El Niño Southern Oscillation, 400–401
    gyres, 399
    photic zone, 396
    photoplankton, 396
    salinity, 396
    sea ice, 401
    surface layer, 396
    temperatures, 396, 397
    thermocline layer, 396
    thermohaline circulation, 399

Phenomena
    blowdown, 3, 19, 41, 43–45, 47, 50, 70, 77, 78
    condensation, 5, 9, 30, 31, 71, 77, 79, 87, 138, 141, 252, 287, 294
    countercurrent flooding, 5, 31, 87, 123, 264–296
    critical flow, 2, 5, 43–45, 50, 300–353
    critical heat flux, 5, 138, 142, 143, 255, 359
    deentrainment, 6, 8
    deposition, 245, 248, 255
    entrainment, 6, 10, 72, 73, 75, 79, 135, 136, 245–247, 254, 255, 281
    flow coastdown, 5
    flow patterns, 2, 32, 87, 128–131, 155, 160–192, 214, 215, 221–252, 255
    flow reversal, 5, 265, 290
    gap conductance, 60, 65
    interfacial area modeling, 365, 366
    oscillations, 7, 10, 53, 290, 294
    refill, 6, 20, 143
    reflood, 6, 21, 143
    slip, 95, 318
    stored energy, 9, 48, 60, 67
Pressure drop, 48, 94
    acceleration, 50, 94, 344
    frictional, 50, 94, 344
    hydrostatic, 52, 94
Pressurized water reactors
    accumulator, 5, 34
    core, 4, 32, 34
    downcomer, 31, 32, 34
    emergency core cooling, 4
    fuel assembly, 35
    fuel rod, 32
    large break, 34
    loop, 32, 33
    lower plenum, 34
    pump, 5, 32
    reactor pressure vessel, 4, 33
    steam generator, 4, 33
    upper plenum, 34

INDEX

Radiation
  albedo, 376, 387
  budget, 411
  forcing, 412
  uniform radiation model, 376, 377

Scaling
  analysis, 13
  biases, 34, 80–83
  bottom-up scaling, 29, 59–80, 83
  catastrophe functions, 58, 59
  guidelines for scaling, 29–32
  hierarchical two-tiered scaling, 3, 29, 34–40
  nondimensional groupings, 28, 38, 39, 45, 57, 61–63, 68, 72, 74, 76, 95–98
  scaling interactions, 32
  top-down scaling, 29, 40–59, 60, 83, 107
Separated flow models
  drift flux models, 118–125, 155
  Martinelli models, 109–117
  single-phase type models, 108, 109
Slug flow
  horizontal
    interfacial area and shear, 202–206
    slug flow analysis, 202
    slug formation, 193
    slug frequency, 205, 206
    slug length, 192, 203, 205, 214
    transition to intermittent annular flow, 138–141, 207–209
    wall friction, 202
  vertical
    interfacial area and shear, 132, 137, 198, 199, 202
    single slug velocity, 194
    slug flow analysis, 195
    slug formation, 200, 212
    slug length, 198, 199, 201
    terrain slugging, 211–214
    transition to annular flow, 138–141, 201
    wall friction, 197, 198

Sun
  influence, 371, 373, 374
  radiation, 371, 373, 374, 380, 386, 389, 396, 401
  variations in energy, 375, 376
System computer codes
  accuracy, 14, 367
  nodalization, 146, 147
  numerical solution, 143–146, 362, 364, 366
  subchannel representation, 147–153, 155
  use, 2, 28, 86–155
  validation, 153–156, 364, 367

Testing
  component performance tests, 2
  integral system tests, 2
  prototype operational tests, 2
  separate effects tests, 2
  testing program, 2, 366–367
Top-down scaling
  fluid transfer between volumes, 48–51
  natural circulation systems, 51–57
  reactor coolant blowdown, 43–45
  spatially uniform containment, 47, 48
  spatially uniform system, 40–43
  vapor volume fraction, 45–47
Two-fluid models
  closure laws, 131–143
  conservation laws, 125–127
  flow pattern maps, 2, 128–131
Two-phase density
  average fluid density, 90, 91, 94
  average kinetic energy density, 92–94, 324
  average momentum density, 91, 92, 94, 324

Vertical air movement
  air with moisture, 384–386
  dry air, 380–384